AF291513

ENVIRONANOTECHNOLOGY

Edited by

MAOHONG FAN
University of Wyoming, Laramie

CHIN-PAO HUANG
University of Delaware, Newark

ALAN E. BLAND
Western Research Institute, Laramie

ZHONGLIN WANG
Georgia Institute of Technology

RACHID SLIMANE
Gas Technology Institute

IAN WRIGHT
Oak Ridge National Lab

Amsterdam • Boston • Heidelberg • London • New York • Oxford
Paris • San Diego • San Francisco • Singapore • Sydney • Tokyo

Elsevier
The Boulevard, Langford Lane, Kidlington, Oxford OX5 1GB, UK
Radarweg 29, PO Box 211, 1000 AE Amsterdam, The Netherlands

British Library Cataloguing in Publication Data
A catalogue record for this book is available from the British Library

Library of Congress Cataloging-in-Publication Data
A catalog record for this book is available from the Library of Congress

For information on all **Elsevier** publications
visit our web site at elsevierdirect.com

Typeset by diacriTech, India

Printed and bound in Great Britain
10 10 9 8 7 6 5 4 3 2 1

ISBN: 978-0-08-054820-3

CONTENTS

14. On the Relationship between Social Ethics and Environmental Nanotechnology

Janne Nikkinen

CONTRIBUTORS

Igor E. Agranovski
Griffith School of Engineering, Griffith University, Brisbane, 4111 QLD, Australia.

Sinan Akgöl
Department of Chemistry, Adnan Menderes University, Aydin, Turkey.

Vlamidir I. Anikeev
Boreskov Institute of Catalysis SB RAS, Novosibirsk, 630090 Russia.

Hsunling Bai
Institute of Environmental Engineering, National Chiao Tung University, Hsinchu, Taiwan.

Alan E. Bland
Western Research Institute, 365 N. 9th Street, Laramie, WY 82072, USA.

Lucija Boskovic
Griffith School of Engineering, Griffith University, Brisbane, 4111 QLD, Australia.

Wenfa Chen
Department of Environmental Engineering, National Chung Hsing University, 250 Kuo Kuang Road, Taichung 402, Taiwan.

Hsun-Wen Chou
Department of Civil and Environmental Engineering, University of Delaware, Newark, DE 19713, USA.

Donald W. Collins
Western Research Institute, 365 N. 9th Street, Laramie, WY 82072, USA.

Giovana de Fátima Lima
Departamento de Ciências Exatas, Universidade Federal de Alfenas, Unifal-MG, Alfenas, MG 37130-000, Brazil.

Polyana Maria de Jesus Souza
Departamento de Ciências Exatas, Universidade Federal de Alfenas, Unifal-MG, Alfenas, MG 37130-000, Brazil.

Adil Denizli
Department of Chemistry, Hacettepe University, Ankara, Turkey.

Mladen Eić
Department of Chemical Engineering, University of New Brunswick, P.O. Box 4400, Fredericton, NB, E3B 5A3, Canada.

L. T. Fan
Department of Chemical Engineering, Kansas State University, Manhattan, KS 66502, USA.

Maohong Fan
Department of Chemical and Petroleum Engineering, University of Wyoming, Laramie, WY, USA.

Héctor Guillén-Bonilla
Departamento de Física C.U.C.E.I. Universidad de Guadalajara, Blvd. Marcelino García Barragán 1421, 44410 Guadalajara, Jalisco, México.

Sukai Hu
Department of Environmental Engineering, National Chung Hsing University, 250 Kuo Kuang Road, Taichung 402, Taiwan.

Chin-Pao Huang
Department of Civil and Environmental Engineering, University of Delaware, Newark, DE 19713, USA.

Qinglin Huang
Department of Chemical Engineering, University of New Brunswick, P.O. Box 4400, Fredericton, NB, E3B 5A3, Canada.

Kelvin R. Johnston
Department of Environmental Engineering, National Chung Hsing University, 250 Kuo Kuang Road, Taichung 402, Taiwan.

Yu-Kuan Lin
Department of Environmental Engineering, National Chung Hsing University, 250 Kuo Kuang Road, Taichung 402, Taiwan.

Chungsying Lu
Department of Environmental Engineering, National Chung Hsing University, 250 Kuo Kuang Road, Taichung 402, Taiwan.

Pedro Orival Luccas
Departamento de Ciências Exatas, Universidade Federal de Alfenas, Unifal-MG, Alfenas, MG 37130-000, Brazil.

Alma H. Martínez
Departamento de Física C.U.C.E.I. Universidad de Guadalajara, Blvd. Marcelino García Barragán 1421, 44410 Guadalajara, Jalisco, México.

Carlos R. Michel
Departamento de Física C.U.C.E.I. Universidad de Guadalajara, Blvd. Marcelino García Barragán 1421, 44410 Guadalajara, Jalisco, México.

Janne Nikkinen
Department of Systematic Theology, P.O. Box 33 (Aleksanterinkatu 7), Faculty of Theology, FI-00014, University of Helsinki, Finland.

Nevra Öztürk
Department of Chemistry, Adnan Menderes University, Aydin, Turkey.

Richard A. Pethrick
WestCHEM, Department of Pure and Applied Chemistry, University of Strathclyde, Thomas Graham Building, 295 Cathedral Street, Glasgow G1 1Xl Scotland.

Krystyna Pyrzynska
Department of Chemistry, Warsaw University, Pasteura 1, 02-093 Warsaw, Poland.

David G. Rickerby
Institute for Environment and Sustainability, European Commission Joint Research Centre, 21020 Ispra VA, Italy.

Mariana Gava Segatelli
Instituto de Química, Departamento de Química Inorgânica, Universidade Estadual de Campinas, Unicamp, Campinas, SP, 13083/970, Brazil.

Alessandra M. Serventi
INRS-Énergie, Matériaux et Télécommunications, 1650 boulevard Lionel–Boulet, Varennes, Québec J3X 1S2, Canada.

Fengsheng Su
Department of Environmental Engineering, National Chung Hsing University, 250 Kuo Kuang Road, Taichung 402, Taiwan.

César Ricardo Teixeira Tarley
Departamento de Ciências Exatas, Universidade Federal de Alfenas, Unifal-MG, Alfenas, MG 37130-000, Brazil.

Yao-hsing Tseng
Department of Chemical Engineering, National Taiwan University of Science and Technology, Taipei, Taiwan, ROC.

Deniz Türkmen
Department of Chemistry, Hacettepe University, Ankara, Turkey.

Walter P. Walawender
Department of Chemical Engineering, Kansas State University, Manhattan, KS 66502, USA.

Bilen Wu
Department of Environmental Engineering, National Chung Hsing University, 250 Kuo Kuang Road, Taichung 402, Taiwan.

Anna Yermakova
Boreskov Institute of Catalysis SB RAS, Novosibirsk, 630090 Russia.

Tengyan Zhang
Western Research Institute, 365 N. 9th Street, Laramie, WY 82072 and Department of Chemical Engineering, Kansas State University, Manhattan, KS 66502, USA.

Tianming Zuo
Western Research Institute, 365 N. 9th Street, Laramie, WY 82072, USA.

PREFACE

Understanding and utilizing the interactions between environment and nanoscale materials is a new way to resolve the increasingly challenging environmental issues we are facing and will continue to face. Therefore, the applications of nanotechnology in environmental engineering have been of great interest to many fields, and consequently, a fair amount of research on the use of nanoscale materials for dealing with environmental issues has been conducted.

The aim of this book is to report on the results recently achieved in different countries. We hope that the book can provide some useful technological information for environmental scientists and assist them in creating cost-effective nanotechnologies to solve critical environmental problems, including those associated with energy production.

Maohong Fan*
C P Huang
Alan E. Bland
Zhonglin Wang
Rachid Slimane
Ian Wright

*mfan@uwyo.edu, (307) 766-5633; mfan3@mail.gatech.edu, (404) 385-4577

Responses of *Ceriodaphnia dubia* to Photocatalytic Nano-Titanium dioxide Particles

Chin-Pao Huang[*]**, Hsun-Wen Chou**[*]**, Yao-hsing Tseng**[**] *and* **Maohong Fan**[***]

  [*] Department of Civil and Environmental Engineering, University of Delaware, Newark, Delaware, USA

  [**] Department of Chemical Engineering, National Taiwan University of Science and Technology, Taipei, Taiwan, ROC

  [***] Department of Chemical and Petroleum Engineering, University of Wyoming, Laramie, Wyoming, USA

Contents

1. INTRODUCTION

Nanotechnology is fast growing in the past decade. Due to unique physical and chemical properties, nano-sized materials have found many applications in many fields, including electronics, manufacturing, medicine, and daily goods. Among various common nanomaterials, titanium dioxide (TiO_2) is becoming one of the most commonly deployed due to its photoactive property. Many applications, such as manufactured semiconductor, solar cell, and environmental remediation are made of titanium dioxide [1–3]. Household products, such as self-cleaning surfaces and antifogging mirrors have also been made by coating nano-TiO_2 to improve the super-hydrophilic property that provides the characteristics of water-repellent and low-particle adhesion on material [4]. However, benefits brought by nanotechnology might come with dangers. Extensive applications of nanomaterials would lead to their release into environment eventually. The release of nanomaterials to the environment can have severe ecological and health consequences. This is of particular concern as the nanomaterials benign in their bulk phase can be toxic to the aquatic organisms due to their unique physical, chemical and biological properties. There are a number of studies on the effect of toxicity of nanoparticles on animals, but only few studies are available on ecotoxicology, with the effect of fullerenes (C_{60}) and titanium dioxide being the most extensively investigated. Several authors have reported the impacts of C_{60} on aquatic organisms [5–7]. Kerstin and Markus [8] and Lovern and Klaper [9] studied the toxicity of TiO_2 to daphnia and algae. While results clearly showed significant impacts of nanomaterials on aquatic organisms, little is dealt with the effect of particle size on aquatic organisms.

In this study, experiments were conducted to assess the effect of particle size on the toxicity of nano-TiO_2 to *Ceriodaphnia dubia*. *C. dubia* is very sensitive to environmental changes (listed as one of the United States Environmental Protection Agency [USEPA]-recommended test organisms for toxicity) and has been used in many toxicity studies, i.e. pesticides, herbicides, heavy metals, and many other toxic substances [10–13]. Furthermore, *C. dubia* is present commonly in freshwater pools and lakes around the world and plays an important role in ecosystems. The diets of *C. dubia* contain algae and many consumers of higher trophic levels, such as fish, amphibians, and aquatic insects. It is a primary consumer in a very important ecological position, which links the primary producers and higher animals. Impacts on the survival and reproduction of *C. dubia*, directly or indirectly will affect the stability of ecosystem.

2. MATERIALS AND METHODS

2.1. Test Organism and Culture Maintenance

Test organism, *C. dubia*, was purchased from Aquatic BioSystem Inc. (Fort Collins, Colorado). Cultures maintenance and preparation of dilution water or "synthetic, moderately hard, reconstituted water" followed the USEPA guidelines [14]. In brief, mass and individual cultures were incubated in the growth chamber, inside a climate control room with a light intensity of 70–120 ft-c, followed by a 16-h light/8-h dark photoperiod and a room temperature of 24 °C. Both mass and individual cultures were daily fed with the green algae *Selenastrum capricornutum* (renamed as *Pseduokirchneriella subcapatitata*) and with a combination of yeast, cerophyll and trout chow (YCT), also purchased form Aquatic BioSystem Inc. Individually cultured organisms were raised in 30-mL plastic cups with 15 mL dilution water and were daily fed with 100 μL of green algae and 100 μL of YCT. A mass culture was raised in 1-L beaker with 1 L of dilution water and was daily fed with 4–6 mL of green algae and 4–6 mL of YCT. The water was changed every 2 and 7 days in individual and mass cultures, respectively.

2.2. Effect of Particle Size

To understand the effect of particle size on the survival rate of *C. dubia*, 11 particle sizes, ranging from 4.7 to 1467 nm, in the form of TiO_2 were applied in the 24-h acute toxicity test. Reade5 (5.2 nm) was purchased from Nanostructured & Amorphous Materials Inc (Houston, Texas), UV-100 (4.7 nm) was purchased from Hombikat Inc. (Japan), ST-01 (5.3 nm) and ST-21 (23 nm) were purchased from Ishihara Sangyo Kaisha LTD. (Japan) and P25 (34 nm) was purchased from Degussa Corporation (Frankfurt, Germany). Five different particle sizes of TiO_2, namely, 46, 116, 204, 636 and 1467 nm were made in our laboratory using thermal-treatment of P25 (Y660, Y780, Y840, Y970, and Y1100) [15]. Thermal-treatment was also used to generate 13 nm particles (Y350) by heating UV-100 at 350 °C. Particle size was determined by Brunauer-Emmiet-Teller (BET) measurements. Table 1.1 lists the nanoparticles used in toxicity tests, their crystal composition and their primary and secondary particle sizes.

Nine concentrations (0, 10, 30, 60, 100, 200, 400, 800, and 1000 mg/L) were used to determine the dose–response curves for all particles. TiO_2 can be easily suspended in water solution; therefore special preparation for test suspensions is not required. Test suspensions were freshly prepared by mixing the given amount of TiO_2 particles and the dilution water, right before use. The concentrations from 10 to 800 mg/L were diluted to 15 mL

Table 1.1 Summary of particle information, including particle name, primary particle size, secondary particle size, rutile component (%), and sources of particle

Particle	Primary particle size (nm)	Secondary particle size (nm)	Rutile component (%)	Source[a]
Rease 5	5.2	749	0.5	NAM
ST-01	5.3	700	0	ISK
UV-100	4.7	715	0	HB
Y350	13	1752	0	UD
ST-21	23	905	0	ISK
P25	34	1859	27	DC
Y660	46	672	27	UD
Y780	116	682	28	UD
Y840	204	644	47	UD
Y970	636	900	51	UD
Y1100	1467	773	100	UD

[a] NAM, Nanostuctured & Amorphous Materials Inc; ISK, Ishihara Sangyo Kaisha LTD; HB, Hombikat Inc; DC, Degussa Corporation; UD, University of Delaware.

of total volume in 30–mL plastic cups (test chamber) from 100 mL of stock solution, which was 1000 mg/L in concentration. To design an environmentally relevant protocol, no surfactant was applied to stock solutions for particle dispersion. All stock solutions were treated with ultrasound at a power of 24 W for 1 minute.

Each experimental set included nine test chambers for nine concentrations in each particle. Each test chamber included 15 mL of test suspension and five *C. dubia* neonates that were less than 24 h old. Each experimental set was repeated four times and treated as one replicate. At least, two and up to four replicates were applied to each particle size. As per USEPA guidelines, each concentration required only 20 neonates to give reliable statistic analysis. During experiment, different individual culture boards incubated in different time periods might have different LC50 results for same particle size. To eliminate the errors rising from the four experimental sets per particle size in one individual culture board, only two experimental sets for each particle size were run from one individual culture board at a minimum replicate of two.

All test chambers were placed in the environmental chamber and covered with transparent plastic plate to prevent water loss from evaporations. The conditions of environmental chamber are as follows. Light intensity was 50–100 ft-c and the photoperoid was 16-h light and 8-h darkn and was controlled by automatic timers. The lamps (SoLux MR-16 Display Lamp with 4700 lamp color temperature and 36° beamspread) were purchased from Wiko Ltd that provided the closest spectrum to sunlight. The temperature of environmental chamber was 24 °C in dark and 25.5 °C in light. The temperature varied by light heating.

After 24 h, survival rate from each concentration was recorded. Because particles settled on the bottom of test chamber after 24 h, dropper was used to resuspend particles in the test chamber before pouring out suspensions. After pouring out the suspension and *C. dubia* in Petri dish, 2–4 mL of clean dilution water was added to the test chamber to rinse out residual particles and dead *C. dubia*. Immobilized *C. dubia* was counted as dead.

2.3. Effect of Photoperiod

The mechanism of hydroxyl radicals generated by irradiated TiO_2 is fully understood. Generally, when photon energy is adsorbed by photocatalyts, electrons are excited from the valence band (VB) to the conduction band (CB). There is transfer of electron to oxygen molecule to form superoxide ion radical ($\cdot O_2^-$) and transfer of electron from water molecule to the VB hole to form hydroxyl radical (OH) [4]. Without modification, pure TiO_2 is not able to store photon energy under dark condition. In other words, without light, pure TiO_2 cannot generate hydroxyl radicals. If the death of *C. dubia* was caused by hydroxyl radical attack, then longer photoperiod means longer hydroxyl radicals attack period. Consequently, longer photoperiod should cause higher mortality rate and yield lower LC50 values.

TiO_2 nanoparticle 25 (34 nm), purchased from Degussa Corporation (Frankfurt, Germany) was used in the experiment. Nine concentrations (e.g., 0, 10, 30, 60, 100,200, 400, 800, and 1000 mg/L) were used to obtain dose–response curves for all particle sizes. Test suspensions were freshly prepared by mixing TiO_2 particles and dilution water, right before use. The concentrations from 10 to 800 mg/L were diluted to 15 mL of total volume in 30-mL plastic cups (test chamber) from 100 mL stock solution. All solutions were irradiated with ultrasound at a power of 24 W for 1 minute as to bring the particles to fine suspension.

Each test chamber included five *C. dubia* neonates that were less than 24 h old. All test chambers were placed in environmental chamber and covered by transparent plastic plate to prevent evaporations. Each replicate also included four experimental sets. However, unlike particle size effect experiment, four experimental sets were conducted in the same individual board. Three replicates were run in each photoperiod. Eight different photoperiods, 0-h light/24-h darkness, 1-h light/23-h darkness, 2-h light/22-h darkness, 4-h light/20-h darkness, 6-h light/18-h darkness, 8-h light/16-h darkness, 12-h light/12-h darkness, and 18-h light/6-h darkness, were applied. After 24 °h, survival rate from each concentration was recorded.

2.4. Data Analysis

Mortality data obtained from 24-h acute toxicity tests were used to calculate the endpoints, LC50, No Observed Effect Concentration (NOEC) and Lowest Observed Effect Concentration (LOEC), by point estimation techniques. LOEC and NOEC values were calculated by the Fisher's exact test. Toxicity Relationship Analysis Program (TRAP), version 1.00 from the USEPA's National Health and Environmental Effects Research Laboratory (NHEERL) was used to plot dose–response curves of survival rate in log concentration scale. LC50, and 95% confidence interval were calculated using piecewise regression tailed in TRAP. The equation for the dose–response model is as follows:

$$Y = \frac{Y_o}{1 + e^{4S(X - X50)}} \tag{1}$$

where Y is the response, Y_o is the response of the control, S is the slope of the curve, X is the dose concentration, and $X50$ is the dose which has a 50% effect on the organism.

2.5. Effect of Secondary Particle Size

The presence of ionic strength can bring about particle aggregation, which will modify the particle size. However, the relationship between secondary particle size and toxicity has not been addressed. Thus, experiments were conduced to assess the effect of secondary particles on the response of aquatic organisms in 24-h acute toxicity tests. Stock solutions (100 mg/L) of nanoparticles were freshly prepared in 250-mL flasks with *C. dubia* culture media and mixed vigorously by magnetic bar to suspend TiO_2 particles. A volume of 3 mL of stock solution was transferred to 150-mL flask and diluted to 30 mg/L test suspensions whose total volume was 100 mL. All test suspensions were ultrasound irradiated with 24W of power for 1 min. A volume of 3.5 mL of finely suspended test solution was placed in aplastic cuvette and the aggregated particle size was measured with time by dynamic light scattering (DLS; Malvern Zetasizer, Southborough, MA). Due to different particle characteristics, particles of different primary sizes might aggregate, settle, or stabilize after sonication. Therefore the stable secondary particle size was defined when the particle size (Zave) reached stable value without continuously increasing or decreasing over 1 h.

2.6. Sedimentation of Nanoparticles

Sedimentation experiments were conducted for nine particles with size less than 300 nm in acute toxicity test. Particle Y970is ideal for comparison because of its high LC50 value and large primary particle size. The procedure for the preparation of test suspension was similar as that of aggregation experiments. Concentration of particles used was 50 mg/L. A volume of 4 mL of the test suspension was placed in a plastic cuvette for absorbance measurements with UV–Vis spectrophotometry (HP Hewlett Packard 8452A diode array spectrophotometer) at a wavelength of 600 nm. In the first 12 h, absorption was measured every hour and every 3 h during the remaining time beyond 12 h. The settling curves were analyzed for the kinetics of aggregation according to the following first–order equation:

$$\ln \frac{\tau}{\tau_0} = -kt \tag{2}$$

where τ and τ_0 are the turbidity at time t and 0, respectively. By plotting graph, $\ln \frac{\tau}{\tau_0}$ versus time, t, the slope of regression line is equal to $-k$. Only data from the first 600 min were used to calculate the k value for each particle, because after 600 min most particles already settled to the bottom of the cuvette. The rate constant, k for each particle size is related to its settling characteristics. Larger k values mean faster settling rate and shorter suspension time.

2.7. SEM Images

To better understand the interactions between TiO_2 particles and *C. dubia* and the effects of TiO_2 particles on the death of *C. dubia*, SEM images were obtained. Five *C. dubia* neonates, less than 24 h, were placed in 100 mg/L of P25 solution and another five neonates were placed in dilution water as control group. SEM images were taken after 12 h of exposure. Samples were placed on a 0.22-μ filter paper and vacuum filtration was applied. Phosphate buffer solution was added to remove the slime layer on the sample surface. Samples were dipped in liquid nitrogen and sputtered with gold/palladium to minimize the surface charging effect. Hitachi S4700 filed emission SEM system was used for bioimaging. Acceleration voltage was set at 1 KeV, and 7–15 mm working distance was used for different magnification.

2.8. Other Observations

LC50 is an important endpoint of toxicity test; however, this endpoint is not able to reveal the die-off of *C. dubia* during experiment. Hence, visual observation was taken to gain information on the death of *C. dubia*.

3. RESULTS AND DISCUSSION

3.1. Effect of Particle Size

Twenty-four-hour acute toxicity tests aimed to investigate the effect of particle size and concentration on the survival rate of *C. dubia*. Eleven different particle sizes were applied in this study. Figure 1.1 shows the dose–response curves for the 11 TiO_2 samples. All particle sizes gave similar pattern in dose–response relationship. With doses of TiO_2 particles increasing from 10 to 1000 mg/L, the survival rate of *C. dubia* decreased. LOEC, NOEC, LC50 and standard error values of 11 particles are shown in Table 1.2. Due to data transformation during data analysis, the standard errors were not symmetrical after back transformation. Generally, LOEC and NOEC values increased with increasing size of particles. LOEC value of particles less than 5.3 nm was 30 mg/L. Particle sizes between 13 and 204 nm had LOEC values from 30 to 60 mg/L. The LOEC values of particles greater than 600 nm, i.e. 636 and 1467 nm reached 100 and 400 mg/L, respectively.

Figure 1.2 shows the relationship among LC50s (in log scale) as a function of primary particle size. When particle size ranged approximately 5 nm, the LC50 value varied between 114 and 352 mg/L. When the particle size increased from 13 to 1467 nm, the LC50 value increased logarithmically.

3.2. Effect of Photoperiod

The EPA method suggests a photoperiod of 16-h light and 8-h dark. To address the effect of photoperiod, this procedure was modified. The relationship between LC50 values (in log scale) and the photoperiods is shown in Fig. 1.3. Results show that the LC50 value and the standard deviation remained relatively unchanged with respect to photoperiod. This implies that there was no significant difference in the effect of photoperiod on *C. dubia*

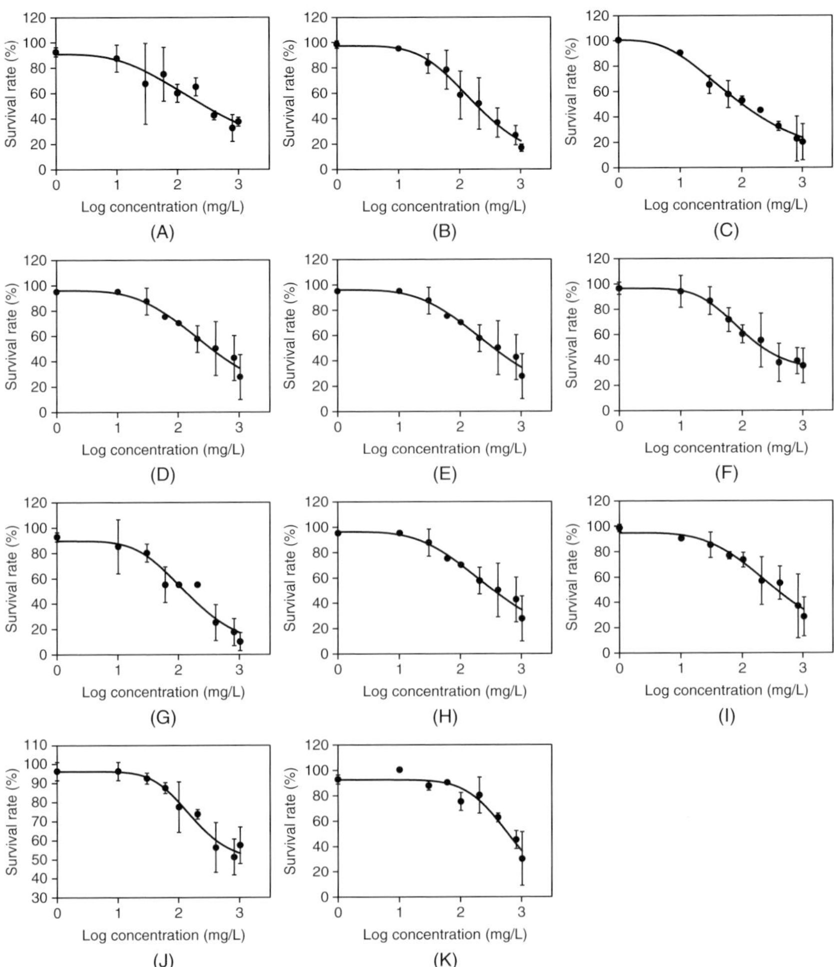

Figure 1.1 Twenty-four-hour acute toxicity test survival dose–response curves of *C. dubia* for different primary particles. *C. dubia* neonates, less than 24 h old were exposed to different concentrations of TiO_2 suspensions and placed in the environmental chamber at 25 °C and under 50–100 ft-c light intensity. Primary particle size: (A): 3 nm (Reade 5); (B): 4.8 nm (ST-01); (C): 6.5 nm (UV-100); (D): 12.9 nm (Y350); (E): 17.9 nm (ST-21); (F): 30 nm (P25); (G): 44 nm (Y660); (H): 73 nm (Y780); (I): 122 nm (Y840); (J): 245 nm (Y970); (K): 387 nm (Y1100). Graph was plot by Sigmaplot.

3.3. Effect of Secondary Particle Size

Table 1.1 shows the steady-state particle size of nano–TiO_2 particles in the growth chamber. The sizes of most steady-state particles fall into a narrow range of approximately 700 to 900 nm, independent of the

Table 1.2 Summary of 24-h acute toxicity test results, including LOEC, NOEC, LC50 values and standard error of LC50 value of *C. dubia* in nine different particle sizes. LC50 and standard error were calculated by Trap. NOEC and LOEC were calculated by Fisher's exact test

Primary particle size (nm)	LC 50-SE (mg/L)	LC 50 (mg/L)	LC50 + SE (mg/L)	NOEC (mg/L)	LOEC (mg/L)
5.2	260	352	476	10	30
5.3	184	209	238	10	30
4.7	95	114	137	10	30
13	97	126	165	30	60
23	183	229	288	30	60
34	293	322	353	10	30
46	129	170	225	30	60
116	292	368	463	30	60
204	350	409	479	10	30
636	868	1100	1400	60	100
1467	508	611	733	200	400

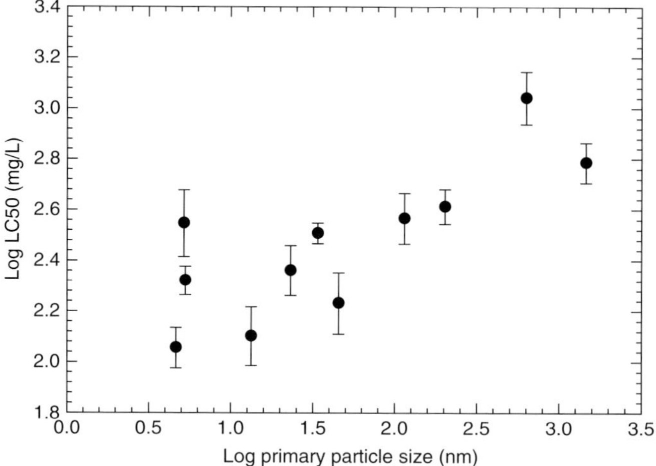

Figure 1.2 Relationship between LC50 and primary particle size (in log scale) from 24 h acute toxicity test of *C. dubia*. Geometric means were used in the measurement of primary particle size. Error bars are standard error.

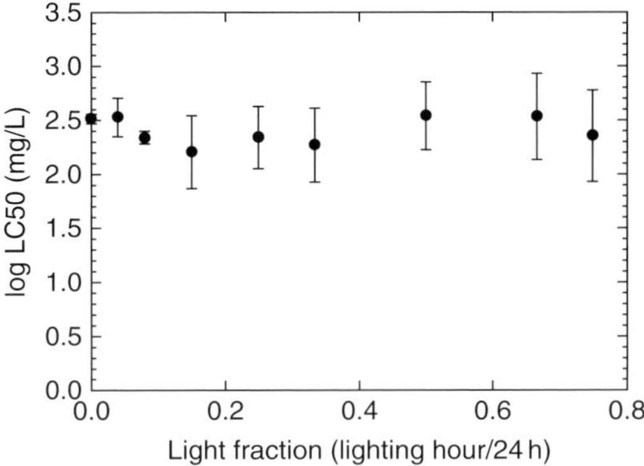

Figure 1.3   Relationship between LC50 and photoperiod. Error bars are standard deviation calculated triplicates.

primary particle size except that of P25 and Y350. The steady-state particle size of P25 and Y350 was approximately 1800 nm. Although the reason remains unknown, the steady-state particle size of P25 and Y350 was greater than that of other particles and the toxicity effect of P25 and Y350 was severe compared to that of Y660, ST-01, ST-21 and Y840. However, the secondary particle size was in the range 700–800 nm, and therefore, secondary particle size cannot be considered an implicit indicator of nanotoxicity.

3.4. Sedimentation Behavior

The kinetics of particle sedimentation in the growth chamber was determined. The particle concentration (concentration of the particles that remained in suspended state) changed with sedimentation time as shown in Fig. 1.4. Y350 and P25 settled rapidly to reach a minimum constant level at sedimentation time of approximately 400 min. A plot of the logarithm of particle concentration with time yielded a straight line in which the slope is the sedimentation rate constant, k (1/min). Figure 1.5 shows the relationship between LC50 and sedimentation rate constant. Results demonstrate that particles with large k value were less toxic to *C. dubia* than those with small k value. Furthermore, particles with similar k values had close LC50 value. Among all particles studied, Y970 (636 nm) had the greatest k value, whereas ST-01 (5.3 nm) had the smallest k value. Y350 (13 nm), UV-100 (4.7 nm), Y660 (46 nm), Y780 (116 nm) and Y840 (204 nm) had very close K value, approximately 2.5×10^{-3} to 4.3×10^{-3} min^{-1}. Generally, the LC50 remained constant, ca. 300 mg/L, when the k value was less than

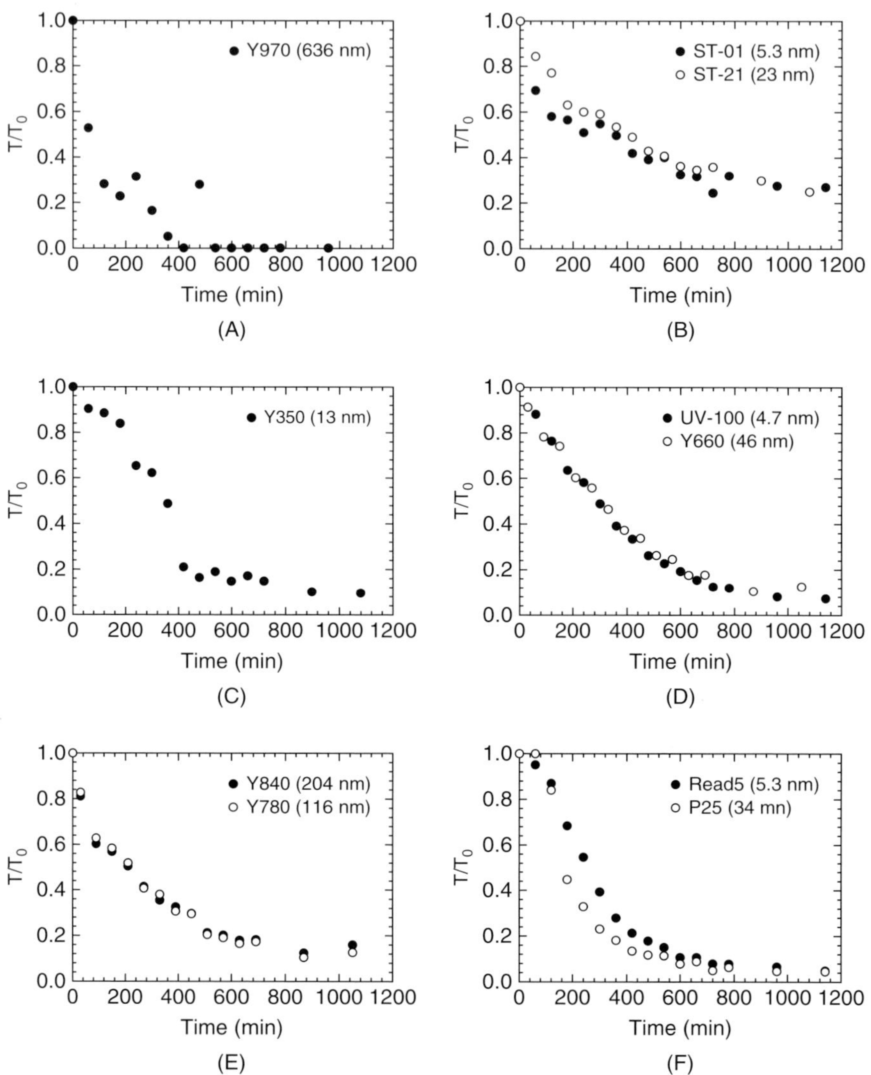

Figure 1.4   24 hr settling curves of eight particle sizes. (A): 636 nm (Y970); (B) 5.3 nm (ST-01) and 23 nm (ST-21); (C) 13 nm (Y350); (D) 4.7 nm (UV-100) and 46 nm (Y660); (E) 116 nm (Y780) and 204 nm (Y840); (F) 5.3 nm (Reade 5) and 34 nm (P25).

4.5×10^{-3} min^{-1}. At k > 4.5×10^{-3} min^{-1}, the LC50 value increased linearly from ca. 300 to 1500 mg/L for Y790, in which the primary particle size (245 nm) was the largest among all the particles studied. It might be that the larger the primary particle size, the smaller the total number of

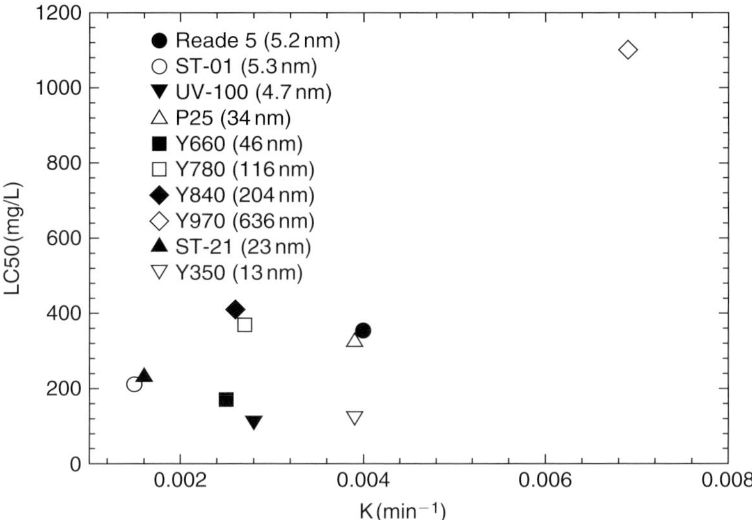

Figure 1.5　Relationship between *k* values and LC50.

particles per unit mass of particles compared with that of smaller primary particle size. Therefore, at given particle mass and at similar *k* value, the total number of smaller particles was large enough to impose effect on *C. dubia*. It is further noted that the *k* value of Reade 5 and P25 were close, so were the LC50 values of these two nanoparticles, i.e. P25 and Reade 5. The result reveals that particles with smaller *k* value can remain suspended in the water column much longer than that with larger *k* value; longer suspension period increased the contact time with *C. dubia*; consequently the toxicity to *C. dubia* was boosted.

3.5.　SEM Images

To gain better understanding of the interaction between *C. dubia* and particles, SEM images were taken. Figure 1.6A1, B1, and C1 shows the images of *C. dubia* in dilution water without exposing to nanoparticles, as control. Figure 1.6A2, B2, and C2 shows images of *C. dubia* in the suspension of P25 for 12 h. Results reveal vividly the adsorption and accumulation of TiO_2 nanoparticle on *C. dubia* carapace surface. The shell of *C. dubia* in suspension (Figure 1.6A2) was not as smooth as that of the control group (Figure 1.6A1); particles were embed into the shell and caused wrinkles on the carapace (Figure 1.6C2). The nano–TiO_2 particles were not only adsorbed on the carapace surface of *C. dubia* but also accumulated around the joints and in the ripples of carapace (Figure 1.7).

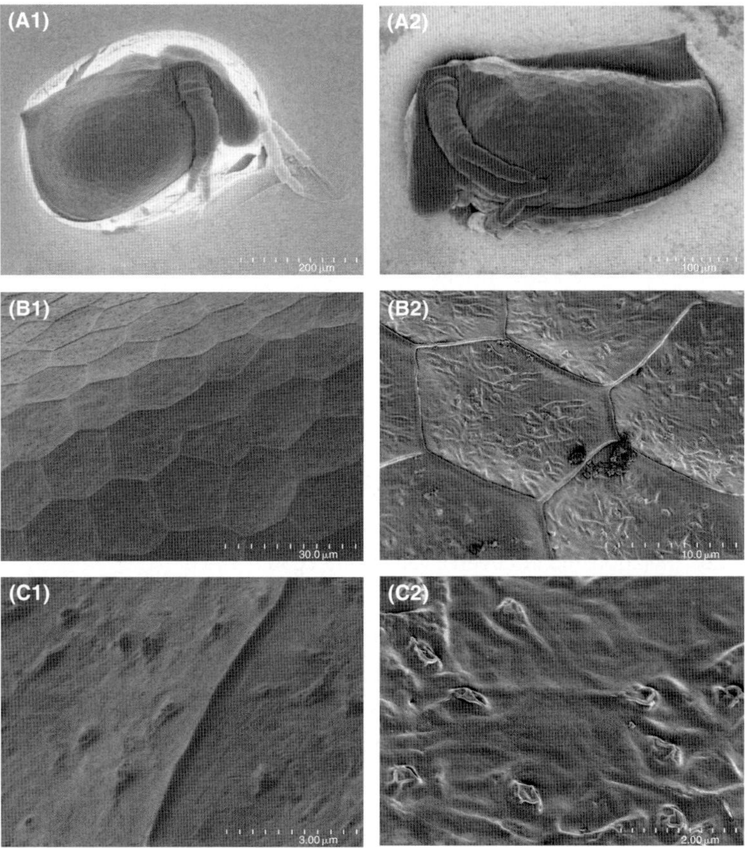

Figure 1.6 SEM images of *C. dubia* neonates, under 24 h of exposure. (1) dilution water after 12 h and (2) 100 mg/L P25 suspension after 12 h. (A) entire body of *C. dubia*; (B) general view of carapace; (C) closer look of carapace.

3.6. General Observations

Although immobility of neonate exposed to nano-TiO_2 particles (e.g., <100 nm) at low concentrations (e.g., 10–30 mg/L) was not statistically affected during experiments to assess the effect of particle size, it was observed that at times neonates exposed to 30 mg/L of TiO_2 suspension were not as energetic as those exposed to 10 mg/L. Moreover, it was observed that *C. dubia* also ingested the nano-TiO_2 particles. After 24 h of exposure, it can be seen that the nano-TiO_2 particles actually were accumulated inside the gut and formed a "white zone" within the neonate's body. While exposure is higher in concentration, the "white zone" in the body

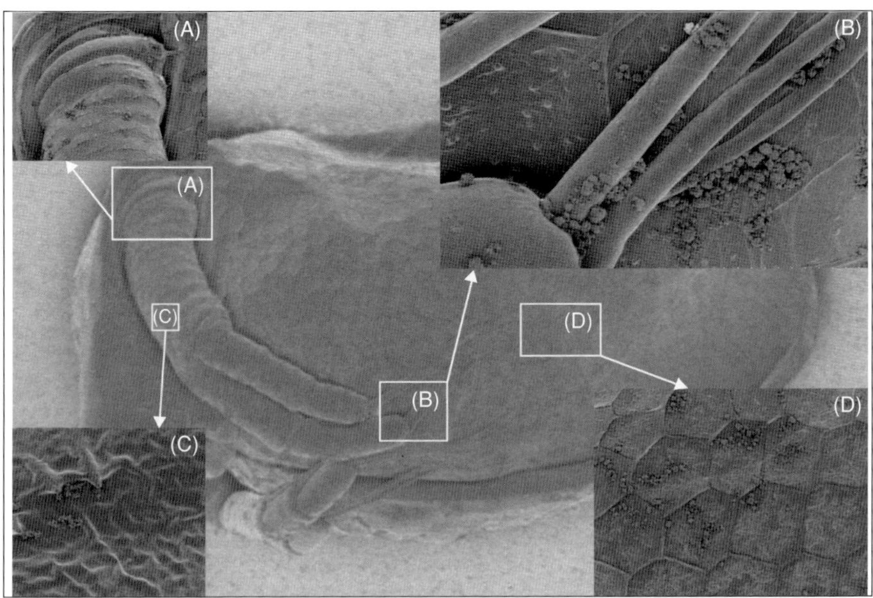

Figure 1.7   SEM images of *C. dubia* neonates, less than 24 h, in 100 mg/L P25 suspension after 12 h (A) The base of antenna; (B) end of antenna; (C) a close look of antenna surface; and (D) a close look of carapace surface.

became thicker and thicker, and the body color of the neonates became less transparent. Eventually, at high nano–TiO_2 concentrations (>100 mg/L) both live and dead neonates showed completely nontransparent (white) color.

Even if the suspensions were sonicated, particles aggregated eventually. *C. dubia* might not be able to ingest aggregated particles; however they could be adsorbed or can "stick" to the carapace surface or on the top of antenna, which causes huge physical body burden to *C. dubia*. Several times, neonates lying on the bottom of the Petri dish above TiO_2 particles tried to wave the antenna loaded with particles laboriously.

Several authors have reviewed nanoparticle toxicity [16, 17]. Due to special physical, chemical properties and high surface area, nano–sized particles could interact with cells more efficiently than bulk material and induced higher toxicity. A number of in vivo and in vitro studies have investigated the toxicity of nanoparticles. Results of many vivo studies had proved that nanoparticles by oral gavages were able to translocate to and remain in systemic organs [18–21, 27], damaged kidney, liver and spleen and caused anemia in the animals. Results of in vitro studies indicated that reactive oxygen species (ROSs), production of malondialdehye (MDA), and leakage of lactate dehydrogenase (LDH) are commonly applied as biomarkers in detecting cytotoxicity and oxidative stress. The nanoparticles that had been studied, including nano-scale silica [22], C_{60} [23], metal oxide [24] and metal particles [25], induced oxidative damage on cell membrane by generating ROS in various cell lines, such as human liver cell, lung cancer cell, and rat liver cell. Study conducted by Witzmann and Monteiro-Riviere [26] also indicated that multiwall carbon nanotubes were able to change protein expression in human epidermal keratinocytes.

Wang et al. [20] showed that after 2 weeks of long oral gavages experiment of TiO_2, female mice had developed hepatic injury and myocardial damage. Particles were found to be retained in liver, spleen, kidney, and lung tissue. In pulmonary toxicity researches, Chen et al. [19] indicated that TiO_2 could induce the change of gene expression that might cause inflammatory response in lung. Furthermore, results of recent study showed that surface-treated ultra fine TiO_2 was actually toxic and lethal. Exposure to overload environment would induce inflammatory responses and lung disease in rat. It was thought that lung and DNA damages could be caused by reactive oxygen species and finally cancers or tumors were formed [28]. Results of recent investigations also showed that TiO_2 could generate ROSs which then brought damages to cell membranes [22, 23].

Although many nano-TiO_2 toxicity studies had been conducted in mammalian vivo, there were very few studies on the impact of nano-TiO_2 on the natural environment. There were two studies on the effect of particle size of nano-TiO_2 on Daphnia [8, 9]. Lovern and Klaper used two particle sizes, i.e. 30 and 100–500 nm and reported that smaller particles had decreased LC50 value (e.g., 5.5 mg/L) and mortality rate increased with concentrations, whereas larger particles showed low mortality ratio over the concentration range studied, namely, from 50 to 500 mg/L. Therefore the LC50 of larger particle size cannot be determined. Kerstin and Markus reported that larger particles (e.g., 100 nm) were less toxic than smaller particles (e.g., 25 nm). Due to narrow range of concentrations (0 to 3 mg/L) studied, full dose–response curves could not be obtained for both the particle sizes. Results of the above studied verified the hypothesis that smaller particles are more toxic than the larger particles.

On the basis of the above results and published literature, it is possible to delineate the possible exposure pathway of TiO_2 on *C. dubia*. The way of exposure of *C. dubia* in TiO_2 suspensions can be divided into two parts:

inside and outside the body. Particles can either be ingested by *C. dubia* or be adsorbed on the carapace surface or can even be embedded into the carapace. Once the particles enter the guts physically, they stick to the gut's surface and block the absorption of nutrients. Particles adsorbed on the surface of carapace or accumulated around the base joints and segment of antenna (important systems for swimming and feeding) can cause body burden to *C. dubia* and further affect the swimming, feeding and even escaping actions of *C. dubia*. Moreover, particles adsorbed on carapace might block the oxygen exchange and respiration processess. Eventually, *C. dubia* dies from lack of oxygen or exhaustion. Chemically, particles in gut can also translocate across intercellar space or transcytosis, before they reach the body chamber. Because the body of *C. dubia* is nearly transparent, lights can easily penetrate the body. Particles are activated to create ROSs that damage the cell membrane, affect mitochondrial function and destroy cell function. ROSs can be generated by activated TiO_2 particles that locate on carapace and damage the carapace. Finally, damage from either physical or chemical stresses caused death or immobility of *C. dubia*.

Eleven different particle sizes and wide range of concentration were studied to assess the effect of particle size on *C. dubia* survival rate. Generally, larger particles were less toxic to *C. dubia* than the smaller ones. However, of the three particle sizes, the smallest nano-TiO_2 particles of approximately 5 nm (e.g., Reade 5, ST-01, and UV-100) exhibited various degrees of toxicity, despite the relatively similar primary particle size.

Although results indicated the undisputable effect of particle size of nanomaterials, with the currently available information it was not possible to assert the different toxicities exhibited by the three smallest particles. The first guest might be the photocatalytic property of TiO_2. Study showed that nano-TiO_2 had the greatest photocatalytic property at 17 nm [29], and incidentally severe impacts on *C. dubia* around this size range were observed. Hence, photocatalytic property of TiO_2 might affect the mortality rate of *C. dubia* via hydroxyl radicals attack. To further verify this hypothesis, *C. dubia* was exposed to nano-TiO_2 for different light periods. However, results indicated that photoperiod had insignificant impact on the survival of *C. dubia*. Results from recent study of the photocatalytic characteristics of nano-TiO_2 showed that the generation of hydroxyl radicals required extreme pH condition, vigorous mixing, and high intensity of UV light and the amount of hydroxyl radical generated was only $0.075\,\mu M\ min^{-1}$ for P25 under best performance conditions

[30]. The experimental conditions of the present study was no where close to the best-performance situation, therefore there might not be sufficient hydroxyl radicals presence. Moreover, the presence of ions and specific chemical impurities can be scavengers for the hydroxyl radicals. Therefore, there was no significant effect of photoperiod on the survival of *C. dubia*.

Several physical and chemical factors, such as secondary particle size, adsorption, and contact time, could govern the toxicity. Nanoparticle aggregation would occur in solution with ionic strength; therefore secondary particle size should be the first item to be considered. The steady-state secondary particle sizes of all nano-TiO_2 particles was in the range of 600–900 nm, except that of P25 and Y350 in which the secondary particle size was ca. 17000 nm. Results also show that the toxicity was not related to the secondary particle size. This could be attributed to the sedimentation of the nano-iO_2 particles on to the bottom of the growth chamber after 24 h of exposure experiments. Thus the contact time between *C. dubia* and nanoparticles could be an important toxicity indicator. On the basis of sedimentation study, particles that exhibit similar toxicity not only share similar settling pattern, but also share rate constant, k values. Generally, when k increased, the LC50 also increased, which means that particles settled faster had shorter contact time with *C. dubia* and subsequently were less lethal to the organisms. Research on cladoceran ecology shows that the major food component of cladoceran is small algae, in the size range from 1 to 25 μm and toxic algae would be rejected during forage [31].

Results of secondary particle size measurements revealed that particles used in particle size effect study reached steady sizes that were less than 25 μm. Particles were retained in midgut by rhythmic contraction of hindgut, which means TiO_2 particles were not rejected by *C. dubia* during forage and were treated as sort of food sources or at least, were not distasteful. While contact time and the total number of particles in suspension increase, the possibility of particle ingestion also increases. The abundance of food also affects the feeding behavior and then changes the respiration rate. While food is superabundant, cladoceran needs to clean thoracic legs by grooming to keep filters clean and furthermore increases the respiration rate [32]. Since *C. dubia* accepted TiO_2 particles as food source, an increase in the total number of particles by either increasing the concentration or decreasing the primary particle sizes can be treated as an increase in food abundance. This might cause *C. dubia* to clean thoracic legs more often and then increase the respiration rate. However, TiO_2 particles are not able to provide energy as real food source, and amount of particle adsorbed on carapace also increased with concentrations and total number of particles. The extra loading on carapace would decrease feeding and swimming efficiency [33]. SEM photos have proved that nano-TiO_2 particles are able to not only adsorb on the surface of carapace but also to accumulate around the joint. Under this situation, *C. dubia* were not able to get energy, moreover body burden also increased by maintaining the efficiency of feeding and swimming.

Besides the total number of particles and contact time, there might be other factors affecting the toxicity of nano-TiO_2, such as surface property. For example, ST-01 and ST-21, from the same manufactured company had different primary and secondary particle sizes. The method of particle production might cause different surface property of particle and further affect the toxicity.

4. CONCLUSION

Results indicated that physical aspects played an important role in the death mechanism with respect to chemical factors; but evidence in hand is not sufficient to show that chemical aspects, such as hydroxyl radicals, do not contribute any to the death of *C. dubia* in this study. In physical terms, the primary and secondary particle sizes are not the most direct factors to cause toxicity difference among various primary particle sizes. On the other hand, results of sedimentation study provided a clue to discover the mechanism of nanotoxicity. Contact time and total number of particles in suspension could be the reasons of toxicity. Furthermore, particle surface property might be another potential factor affecting the degree and extent of interaction between particles and *C. dubia*. Work on chemical impact on the death mechanism will need more effort and long term toxicity effect, such as productivity, is another issue that needs to be addressed in future study.

ACKNOWLEDGMENT

This work was supported by US EPA Science To Achieve Results (STAR) Program grant #R-83172101. Additional support was provided by the Center of Environmental Health, Safety and Engineering, Industrial Technology Research Institute, Hsing Chu, Taiwan, ROC.

References

[1] Bowering N, Walker GS, Harrison PG. Photocatalytic decomposition and reduction reactions of nitric oxide over Degussa P25. Appl Catal B 2006;62:1–10.

[2] Danion A, Disdier J, Chantal G, Paisse O, Jaffrezic-Renault N. Photocatalytic degradation of imidazolinone fugicide in TiO$_2$ -coated optical fiber reactor. Appl Catal B 2006;62:274–81.

[3] Gora A, Toepfer B, Puddu V, Puma GL. Photocatalytic oxidation of herbicides in single-component and multicompinent systems: reaction kinetics analysis. Appl Catal B 2006;65:1–10.

[4] Carp O, Huisman CL, Reller A. Photoinduced reactivity of titanium dioxide. Prog Solid State Chem 2004;32:33–177.

[5] Oberdorster E. Manufactured nanomaterials (fullerenes, C$_{60}$) induce oxidative stress in the brain of juvenile largemouth bass. Environ Health Perspect 2004;112:1058–62.

[6] Oberdorster E, Zhu S, Blickley TM, McClellan-Green P, Hassch ML. Ecotoxicology of carbon-based engineered nanoparticles: effects of fullerene (C$_{60}$) on aquatic organisms. Carbon 2006;44:1112–20.

[7] Zhu S, Oberdorster E, Hassch M L. Toxicity of an engineered nanoparticle (fullerene, C$_{60}$) in two aquatic species, *Daphnia* and fathead minnow. Mar Environ Res 2006;62:S5–9.

[8] Kerstin HR, Markus S. Ecotoxic effect of photocatalytic active nanoparticles (TiO$_2$) on algae and Daphnia. Environ Sci Pollution Res 2006;13(4):1–8.

[9] Lovern SB, Klaper R. Daphnia magna mortality when exposed to titanium dioxide and fullerene (C$_{60}$) nanoparticles. Environ Toxicol Chem 2006;25(4):1132–7.

[10] Gensemer RW, Naddy RB, Stubblefield WA, Hockett JR, Santore R, Paquinc P. Evaluating the role of ion composition on the toxicity of copper to Ceriodaphnia dubia in very hard waters. Comp Biochem Physiol Part C 2002;133:87–97.

[11] Banks KE, Turner PK, Wood SH, Matthewsa C. Increased toxicity to Ceriodaphnia dubia in mixtures of atrazine and diazinon at environmentally realistic concentration. Ecotoxicol Environ Saf 2005;60:28–36.

[12] Lan CH, Lin TS, Peng CP. Aquatic toxicity of nitrogen mustard to Ceriodaphnia dubia, Daphnia magna, and Pimephales poemelas. Ecotoxicol Environ Saf 2005;61:273–9.

[13] Wheelock CE, Miller JL, Miller ML, Phillips BM, Gee JS, Tjeerdema RS. Influence of container adsorption upon observed pyrethroid toxicity to Ceriodaphnia dubia and Hyalella azteca. Aquat Toxicol 2005;74:47–52.

[14] US Environmental Protection Agency. Short-term Methods for Estimating the Chronic Toxicity of Effluents and Receiving Water to Freshwater Organisms. 4th Ed. (2002). Office of Water. 1200 Pennsylvania Ave. NW Washington, DC 20460.

[15] Tseng YH, Lin HY, Kuoc CS, Li YY, Huang CP. Thermostability of nano-TiO$_2$ and its photocatalytic activity. React Kinet Catal Lett 2006;89(1):63–9.

[16] Kipen HM, Laskin DL. Smaller is not always better: nanotechnology yields nanotechnology. Am J Physiol: Lung Cell Mol Physiol 2005;289:L696–7.

[17] Oberdorster G, Oberdorster E, Oberdorster J. Nanotoxoicology: an emerging discipline evolving from studies. Environ Health Perspect 2005;113:823–39.

[18] Jani PU, McCarthy DE, Florenece AT. Titanium dioxide (rutile) particle uptake from the rat GI tract and translocation to systemic organs after oral administration. Int J Pharm 1994;105(2):157–68.

[19] Chen HW, Su SF, Chien CT, Lin WH, Yu SL, Chou CC, Chen JJW, Yang, PC. Titanium dioxide nanoparticles in induce emphysema-like lung injury in mice. FASEB J 2006;20:E1732–41.

[20] Wang B, Feng WY, Wang TC, Jia G, Wang M, Shi JW, Zhang F, Zhao YL, Chai ZF. Acute toxicity of nano- and micro-scale zinc powder in healthy adult mice. Toxicol Lett 2006;161:115–23.

[21] Wang J, Zhou G, Chen C, Yu H, Wand T, Ma Y, Jia G, Gao Y, Li B, Sun J, Li Y, Jia F, Zhao Y, Chai Z. Acute toxicity and biodistribution of different sized titanium dioxide particles in mice after oral administration. Toxicol Lett 2006;168:176–85.

[22] Lin WH, Huang YW, Zhou XD, Ma Y. In vitro toxicity of silica nanoparticles in human lung cancer cell. Toxicol Appl Pharmacol 2006;217:252–9.

[23] Sayes CM, Gobin AM, Ausman KD, Mendez J, West JL, Colvin VL. Nano-C_{60} cytotoxicity is due to lipid peroxidation. Biomaterials 2005;26:7587–95.

[24] Jeng HA, Swanson AJ. Toxicity of metal oxide nanoparticles in mammalian cells. J Environ Sci Health A 2006;41:2699–711.

[25] Hussain SM, Hess KL, Gearhart JM, Geiss KT, Schlager JJ. In vitro toxicity of nanoparticles in BRL 3A rat liver cells. Toxicol In Vitro 2005;19:975–83.

[26] Witzmann FA, Monteiro-Riviere NA. Multi-wall carbon nanotube exposure alters protein expression in human keratinocytes. Nanomedicine 2006;2:158–68.

[27] Chen Z, Meng H, Xing G, Chen C, Zhao Y, Jia G, Wang TC, Yuan H, Ye C, Zhao F, Chai Z, Zhu C, Fang X, Ma B, Wan L. Acute toxicologyical effects of copper nanoparticles in vivo. Toxicol Lett 2006;163:109–20.

[28] Rehn B, Seiler F, Rehn S, Bruch J, Maier M. Investigations on the inflammatory and genotoxic lung effects of two types of titanium dioxide: untreated and surface treated. Toxicol Appl Pharmacol 2003;189:84–95.

[29] Lin H, Huang CP, Li W, Ni C, Shah SI, Tseng YH. Size dependency of nanocrystalline TiO_2 on its optical property and photocatalytic reactivity exemplifi ed by 2-chlorophenol. Appl Catal B 2006;68(1):1–11.

[30] Hirakawa T, Yawata K, Nosaka Y. Photocatalytic reactivity for O_2^- and OH^- radical formation in anatase and rutile TiO_2 suspension as the effect of H_2O_2 addition. Appl Catal A Gen 2007;325:105–11.

[31] Porter KG, Orcutt JJD. Nutritional adequacy, manageability and toxicity as factors that determine the food quality of green and blue-green algae for *Daphnia*. Am Soc Limnol Oceanogr Spec Symp 1980;3:268–81.

[32] Porter KG, Gerritsen J, Orcutt JD. The effect of food concentration on swimming patterns, feeding behavior, ingestion, assimilation, and respiration by Daphnia. Limnol Oceanogr 1982;27(5):935–49.

[33] Allen YC, De Stasio BT. Individual and population level consequences of an algal epibiont on *Daphnia*. Limnol Oceanogr 1993;38(3):592–601.

CHAPTER 2

High Capacity Removal of Mercury(II) Ions by Poly(Hydroxyethyl Methacrylate) Nanoparticles

Deniz Türkmen[*], **Nevra Öztürk**[**], **Sinan Akgöl**[**] *and*
Adil Denizli[*]

[*] Department of Chemistry, Hacettepe University, Ankara, Turkey
[**] Department of Chemistry, Adnan Menderes University, Aydin, Turkey

Contents

1. INTRODUCTION

Nanotechnology is an enabling technology that deals with nanometer-sized objects [1]. It is expected that nanotechnology will be developed at several levels: materials, devices, and systems. The level of nanomaterials is the most advanced at present, both in scientific knowledge and in commercial applications. A decade ago, nanoparticles were studied because of their size-dependent physical and chemical properties [2]. Now, they have entered a commercial exploration period [3]. Many published works focused on the synthesis of micrometer-sized polymer matrices [4]. Only limited work has been published on the application of nanosized particles in the adsorption of heavy metal ions. Nanosized particles can produce larger specific surface area and, therefore, may result in high binding capacity for metal ions. Therefore, it may be useful to synthesize nanosized particles and utilize them for the removal of heavy metal ions [5–10].

Surface modification can be accomplished by physical and chemical binding or surface coating of desired molecules, depending on the specific applications [11]. Surface modification is an active research area in the fields of microelectronics, biotechnology, and material science. Properties such as adhesion, wettability, biocompatibility, and binding affinity of the surface can be altered and tuned to the specific requirements by chemical or physical modification of the surfaces [12]. Nowadays, modified materials are well known and have been investigated intensively due to their potential applications in many areas such as biology, medicine, and environment. Surface modification of particles by organic compounds can be achieved via organic vapor condensation, polymer coating, surfactant binding, and direct silanization. Direct silanization is attractive for improving stability and control of surface properties [13].

Mercury is a common pollutant of water, resulting from the burning of coal by power plants, and in the inappropriate disposal from batteries, paints, lights, and industrial by-products. Mercury poisoning is becoming more important because of the extensive contamination of water and fish and the increasing consumption of fish in the human diet [14]. Mercury is cytotoxic, exerting its effect by depleting the thiol reserves in the mitochondria, resulting in cell death. It is extremely neurotoxic and leads to dizziness, irritability, tremor, depression, and memory loss [15]. It is also toxic to the kidneys and colon, the two main sites of excretion. Mercury is released very slowly from the body with a half-life of at least 60 days, resulting in increasing amounts with chronic consumption of contaminated fish [16].

In the gastrointestinal tract, methylmercury (MeHg) is absorbed to approximately 95%, Hg^{2+} to approximately 7%, and elemental Hg to less than 0.01%. The absorption of elemental mercury (Hg°) in the lung is approximately 80%. Within tissues, MeHg is slowly demethylated to Hg^{2+}. Elemental mercury is rapidly oxidized to the mercurous form (Hg^+) and then to the mercuric form (Hg^{2+}) in blood by catalase [17]. However, the time that this transformation takes is sufficient for Hg° to reach the tissues of the central nervous system, which is its primary target organ [18]. The kidney is considered the target organ for the Hg^+ and Hg^{2+}, but these forms are also known to accumulate readily in virtually all ectodermal and endodermal epithelial cells and glands. At the target tissues, the mercury entities exert various cytotoxic effects as a result of binding to sulfhydryl groups [19].

The goal of this study is to report the synthesis of poly(hydroxyethyl methacrylate) (PHEMA) nanoparticles carrying reactive imidazole containing 3-(2-imidazoline-1-yl)propyl(triethoxysilane) (IMEO) and their use in the adsorptive removal of mercury(II) ions from synthetic solutions by metal chelation. PHEMA nanoparticles (150 nm in diameter) were produced by a surfactant-free emulsion polymerization technique. Then, IMEO was attached to the nanoparticles as a metal complexing agent. PHEMA-IMEO nanoparticles were characterized by transmission electron microscopy (TEM) and Fourier transform infrared spectroscopy (FTIR). Removal studies were conducted to evaluate the binding capacity of Hg^{2+} onto the PHEMA-IMEO nanoparticles. Elution of Hg^{2+} and regeneration of the silanized nanoparticles were also tested.

2. MATERIALS AND METHODS

2.1. Chemicals

Hydroxyethyl methacrylate (HEMA, Sigma Chem. Co., St. Louis, USA) and ethylene glycol dimethacrylate (EGDMA, Aldrich, Munich, Germany) were distilled under vacuum (100 mmHg). 3-(2-imidazoline-1-yl)propyl (triethoxysilane) (IMEO, molecular weight: 274.43 g/mol) was purchased from Sigma. Poly(vinyl alcohol) (molecular weight: 100,000, 98% hydrolyzed) was purchased from Aldrich (Munich, Germany). All other chemicals were of the highest purity commercially available and were used without further purification. All water used in the experiments was purified using a Barnstead (Dubuque, IA, USA) ROpure LP® reverse osmosis unit with a high flow cellulose acetate membrane (Barnstead D2731), followed by a Barnstead D3804 NANOpure® organic/colloid removal and ion exchange packed bed system.

2.2. Synthesis of PHEMA Nanoparticles

Surfactant-free emulsion polymerization was carried out according to the literature procedure with minor modifications as reported elsewhere [20]. Briefly, the stabilizer, PVAL (0.5 g), was dissolved in 50 ml deionized water for the preparation of the continuous phase. Then, the monomer mixture HEMA/ EGDMA (0.6 ml/0.01 ml) was added to the dispersion, which was mixing in an ultrasonic bath for about half an hour. Potassium persulphate (KPS, initiator) concentration in monomer phase was 0.44 mg/ml. Prior to polymerization, an initiator was added to the solution and nitrogen gas was blown through the medium for approximately 1–2 min to remove dissolved oxygen.

Polymerization was carried out in a constant temperature shaking bath at 70 °C under nitrogen athmosphere for 24 h. After the polymerization, the nanoparticles were washed with methanol and water several times to remove the unreacted monomers. For this purpose, the nanoparticles were precipitated and collected with the help of a centrifuge (Zentrifugen, Universal 32 R, Germany) at the rate of 18,000 g for 1 h and resuspended in methanol and water several times. Then, the PHEMA nanoparticles were further washed with deionized water.

2.3. Silanization of PHEMA Particles

Silane is a coupling agent and its bifunctional molecule bonds to both the exposed composite filler particles and the bonding resin [21]. The silane compounds readily react with the surface hydroxyl groups of the different supports [22]. It is assumed in the literature that the silane molecules are first hydrolyzed by the trace quantities of water present either on the surface of the support or in the solvent, followed by the formation of a covalent bond with the surface [23]. For the silanization, PHEMA nanoparticles and IMEO (mol ratio 1:10) were mixed and stirred at 25 °C for about 4 days. At the end of this period, stirring was stopped. The silanization reaction takes place at 25 °C without any catalyst as shown in Fig. 2.1. The resulting silanized nanoparticles were centrifuged and washed with dichloromethane. Then, the nanoparticles were resuspended in distilled water. To evaluate the degree of silanization (i.e., IMEO loading), the PHEMA nanoparticles were subjected to Si analysis using flame atomizer atomic absorption spectrometer (AAS, AAnalyst 800, Perkin Emler, USA). The reaction between the PHEMA and IMEO is given in Fig. 2.1.

2.4. Characterization Experiments

The average particle size and size distribution were determined using Zeta Sizer (Malvern Instruments, Model 3000 HSA, England).

The nanoparticles were imaged in dry state using TEM (FEI Company Tecnai™, G2 Spirit, Biotwin, 20–120 kV). The microscope sample was prepared by placing a drop of the polymer dispersion on a carbon–coated Cu grid, followed by solvent evaporation at room temperature.

FTIR spectra of the IMEO, the PHEMA, and the PHEMA–IMEO nanoparticles were obtained using a FTIR spectrophotometer (Varian FTS 7000, USA). The dry nanoparticles (about 0.1 g) were thoroughly mixed with KBr (0.1 g, IR Grade, Merck, Germany) and pressed into a tablet form and the spectrum was recorded. To prepare a liquid sample (e.g., IMEO)

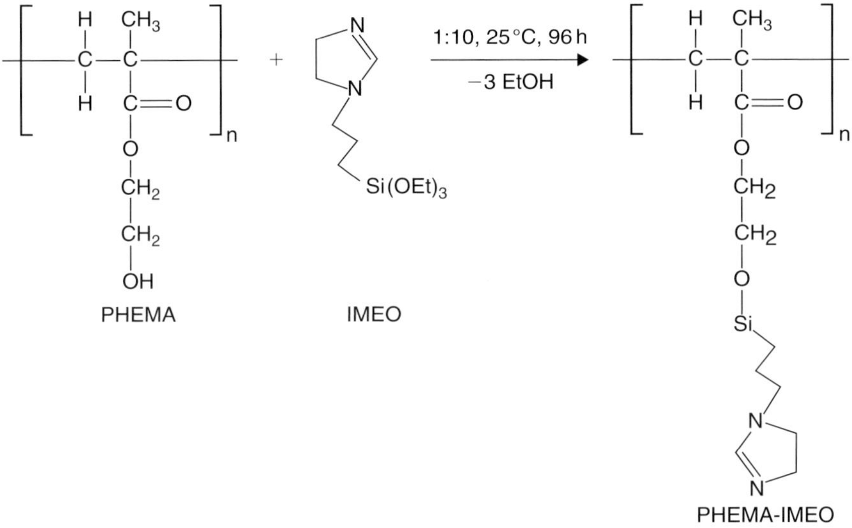

Figure 2.1 Schematic presentation of the reaction between PHEMA and IMEO.

to FTIR analysis, first place a drop of the liquid on the surface of a highly polished KBr plate, then place a second plate on top of the first plate so as to spread the liquid in a thin layer between the plates, and clamps the plates together. Finally, wipe off the liquid out of the edge of plate, and then record the spectrum.

To evaluate the degree of silanization, the PHEMA nanoparticles were subjected to Si analysis using flame atomizer AAS.

The surface area of the PHEMA nanoparticles was calculated using the following expression:

$$N = 6 \cdot 10^{10} \cdot S/\pi \cdot \rho_s \cdot d^3 \tag{1}$$

Here, N is the number of nanoparticles per milliliter; S is the % of solids; ρ_s is the density of bulk polymer (g/mL); d is the diameter (nm). The number of nanoparticles in per milliliter suspension was determined using mass–volume graph of nanoparticles. From all these data, specific surface area of the PHEMA nanoparticles was calculated by multiplying N and surface area of one nanoparticle.

2.5. Hg^{2+} Adsorption Studies

Adsorption of Hg^{2+} from aqueous solutions was investigated in batch experiments. Effects of Hg^{2+} concentration and pH of the medium on the adsorption rate and capacity were studied. 100 mL aliquots of aqueous solutions containing different amounts of Hg^{2+} (in the range of 10–500 mg/L) were treated with the nanoparticles at different pH (in the range of 2.0–7.0) (adjusted with HCl-NaOH). The nanoparticles (100 mg) were stirred with a mercury nitrate salt solution at room temperature for 2 h. All glassware for adsorption experiments were washed with 1 M HNO_3 and rinsed thoroughly with deionized water. The concentration of the Hg^{2+} in the aqueous phase was measured using an AAS. A Shimadzu Model AA-6800 Flame Atomic Absorption Spectrophotometer (Japan) was used. For mercury determinations, Mercury Vapor Unit (MVU)-1A was used. Deuterium background correction was applied throughout the experiments, and the spectral slit width was 0.5 nm. The instrument response was periodically checked with a known Hg^{2+} standard solution. The adsorption experiments were performed in replicates of three and the samples were analyzed in replicates of three as well. For each set of data present, standard statistical methods were used to determine the mean values and standard deviations. Confidence intervals of 95% were calculated for each set of samples to determine the margin of error. The adsorption capacity of the nanoparticles was calculated according to the mass balance on mercury ion.

2.6. Competitive Adsorption

Competitive heavy metal adsorption from aqueous solutions containing Hg^{2+}, Cd^{2+} and Pb^{2+} was also investigated in a batch experimental system. A solution (100 ml) containing 50 mg/L of each metal ion was treated with the nanoparticles (100 mg) at a pH of 5.0 in the flasks and stirred magnetically at 100 rpm. The temperature was maintained at 25 °C. After a sufficient amount of time for equilibration, the solution was centrifuged, and the supernatant was removed and analyzed for remaining metal ions. The amounts of adsorbed heavy metal ions were then determined by difference. Equilibration time was relatively short; the adsorption experiment (from initial contact to final determination) was completed in 30 h.

2.7. Desorption and Repeated Use

Desorption of Hg^{2+} ions were studied with 0.5% thiourea in 0.05 M HCl solution. The nanoparticles were placed in this desorption medium and stirred continuously (at a stirring rate of 600 rpm) for 15 min at room temperature. The desorption ratio was calculated from the amount of Hg^{2+} ions adsorbed on the nanoparticles and the final concentration of Hg^{2+}

ions in the desorption medium. To test the reusability of the nanoparticles, adsorption–desorption procedure of Hg^{2+} ions was repeated 20 times using the same nanoparticles. To regenerate after desorption, the nanoparticles were washed with $0.1\,M$ HNO_3.

3. RESULTS AND DISCUSSION

Nanoparticles can produce larger specific surface area and therefore may result in high metal-complexing ligand loading. Therefore, it may be useful to synthesize nanoparticles with large surface area and utilize them as suitable carriers for the adsorption of metal ions. The specific surface area was calculated as $1779\,m^2/g$. PHEMA nanoparticles with an average size of 150 nm in diameter and with a polydispersity index of 1.171 were produced by surfactant-free emulsion polymerization. It is apparent that the PHEMA nanoparticles are perfectly spherical with a relatively smooth surface and uniform as shown by the TEM images (Fig. 2.2). The small polydispersity index suggests that nucleation is fast compared to particle growth, and also the absence of a secondary nucleation step. In addition, the total monomer conversion was determined as 98.5% (w/w) for PHEMA nanoparticles. PHEMA nanoparticles were highly dispersive in water by ultrasonication due to hydroxyl groups on the surface of nanoparticles. The dispersion state of the particles was confirmed visually by the observed white color of the suspension. The aqueous dispersion of nanoparticles was stable for several days.

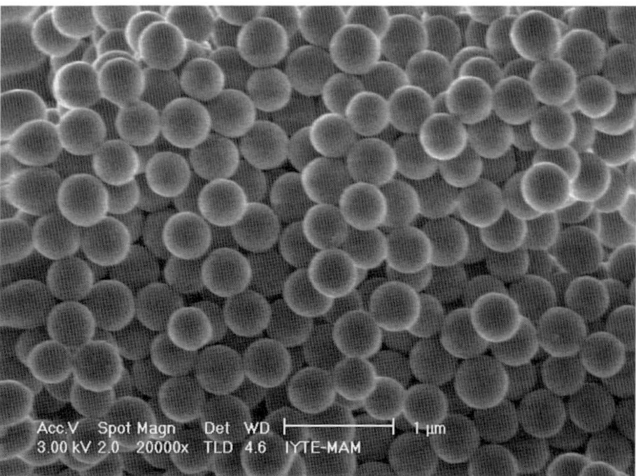

Figure 2.2 Transmission electron microscopy image of PHEMA nanoparticles.

FTIR spectra of the IMEO, the PHEMA, and the PHEMA-IMEO are shown in Fig. 2.3. In the FTIR spectrum of the IMEO, the strong absorption bands at $1605\,\mathrm{cm}^{-1}$, assigned to the characteristic $n(C{=}N)$ vibrations, indicated a strong band at 2928 and $2974\,\mathrm{cm}^{-1}$ $n(C{-}H)$

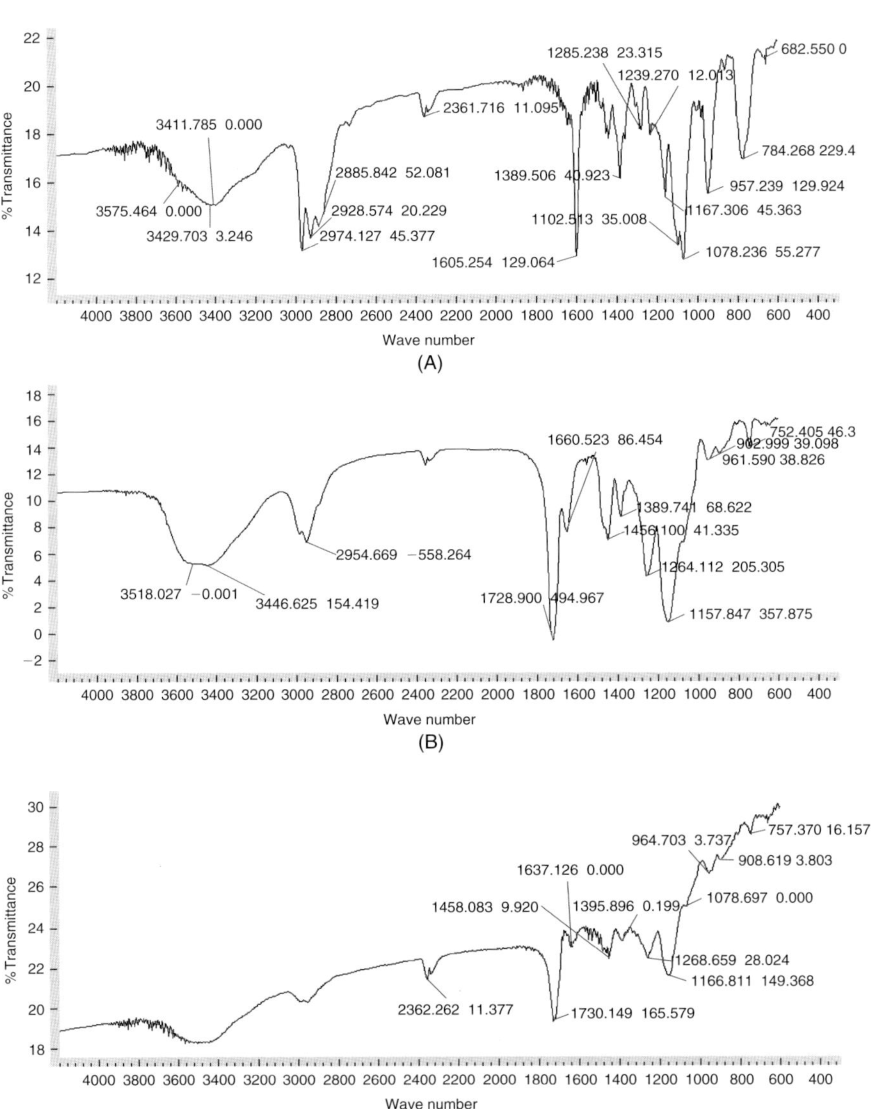

Figure 2.3   FTIR spectra of (A) IMEO; (B) PHEMA; (C) PHEMA-IMEO.

(Fig. 2.3A). The n(O$-$H) stretching vibration in PHEMA, observed in the 3600–3410 cm^{-1} range as broad absorptions, indicated a strong band at 1730 cm^{-1} due to n(C=O) group and the 2954 cm^{-1} n(C$-$H) stretching of CH$_3$ indicated the 1268 cm^{-1} n(C$-$O) stretching vibration (Fig. 2.3B). The characteristic n(C=O), n(C=N), and n(C$-$H) stretching vibration bands of the PHEMA-IMEO are observed at 1728 cm^{-1}, 1660 cm^{-1}, and 2954 cm^{-1}, respectively (Fig. 2.3C). In addition, the n(Si$-$O$-$C) vibration band is observed at 1264 cm^{-1}. As a result, the peak position of 1264 cm^{-1} is related to n(Si$-$O$-$C) and the observance of C=N bands of the PHEMA-IMEO at 1660 cm^{-1} and the shifts of the C=N vibration to higher frequencies of 1660 cm^{-1} due to the silanization of nanoparticles.

3.1. Effect of Hg^{2+} Concentration

Figure 2.4 shows the equilibrium concentration of Hg^{2+} dependence of the adsorbed amount of the Hg^{2+} onto the both PHEMA and PHEMA-IMEO nanoparticles. Adsorption of Hg^{2+} onto the PHEMA nanoparticles was very low (approximately 0.144 mg/g) because PHEMA nanoparticles do not contain any binding sites for complexation of Hg^{2+}. This very low adsorption value of Hg^{2+} may be due to weak interactions between Hg^{2+} and hydroxyl groups on the surface of the PHEMA nanoparticles. However, IMEO incorporation into the polymer structure significantly increased the adsorption capacity to 746 mg/g. The adsorption values increased with increasing equilibrium concentration of Hg^{2+}, and a saturation value is

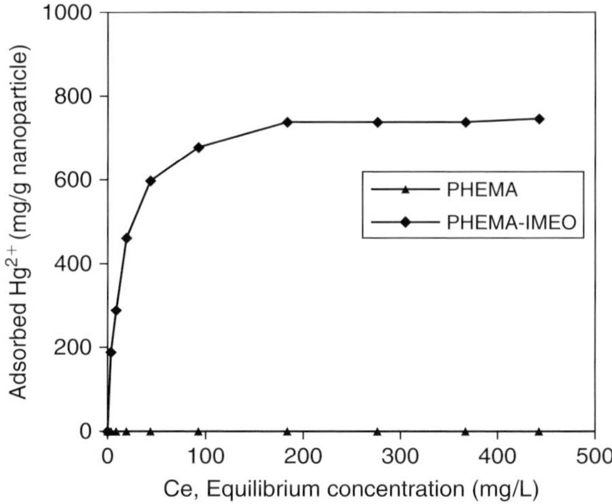

Figure 2.4 Effect of Hg^{2+} concentration on adsorption of Hg^{2+} on the PHEMA and PHEMA-IMEO nanoparticles; IMEO loading: 1021 μmol/g; pH: 5.0.

achieved at an ion concentration of 180 mg/L, which represents saturation of the active binding sites on the PHEMA-IMEO nanoparticles.

IMEO content of the adsorbent nanoparticles used in this group of experiments was 247 μmol/g. The maximum Hg^{2+} adsorption capacity achieved in the studied range is approximately 221.8 mmol per unit mass of the nanoparticles. This seems to give a stochiometry of one IMEO group per one mercury ion.

Different polymeric adsorbents carrying metal-chelating ligands with a wide range of adsorption capacities for mercury ions have been reported (Table 2.1). Comparing the maximum adsorption capacities, it seems that

Table 2.1   Comparison of adsorption capacities of different adsorbents

Adsorbent	Chelating ligand	Capacity (mg/g)	Reference
Styrene-divinylbenzene	Thiol	20	[24]
PMMA	Ethylenediamine	30	[25]
Polystyrene	Dithiocarbamate	32	[26]
Poly(glycidyl methacrylate–divinyl benzene)	Phosphoric acid	40	[27]
PHEMA	Dithizone	42	[28]
Polystyrene	Sulfur-chlorinated jajoba wax	50	[29]
PEGDMA	Acrylamide	54	[30]
Soy protein hydrogel	Ethylenediamine tetraacetic acid	60	[31]
Poly(vinyl alcohol)	Procion Blue MX–3G	69	[32]
N-hydroxymethyl thioamide	Thioamide	72	[33]
Poly(vinyl pyridine)	Dithizone	144	[34]
Silica	3-trimethoxysilyl-1-propanethiol	184	[35]
Silica Gel	Polyethyleneimine	200	[36]

Table 2.1   (*Continued*)

Adsorbent	Chelating ligand	Capacity (mg/g)	Reference
Poly(N-vinylimidazole)	Imidazole	200	[37]
PHEMA	Thiazolidine	222	[38]
Cellulose	Polyethyleneimine	288	[39]
PHEMA	Polyethyleneimine	334	[40]
Amberlite IRC 718	Iminodiacetic acid	360	[41]
Poly(glycidyl methacrylate–divinyl benzene)	Thio	400	[42]
PHEMA	Methacryloylamidocysteine	1018	[43]
PHEMA	N-methacryloyl-histidine	1234	[44]
PHEMA	N-methacryloyl-(L)-alanine	746	In this study

the adsorption capacity achieved with the novel IMEO-incorporated PHEMA nanoparticles is rather satisfactory.

3.2. Effect of pH

The most critical parameter for metal adsorption is pH as it influences both the polymer surface chemistry and the solution chemistry of soluble metal ions. Due to the deprotonation of the acidic groups of the metal complexing ligand (IMEO molecules), its adsorption behavior for metal ions is influenced by the pH value, which affects the surface structure of adsorbents, the formation of metal hydroxides, and the interaction between adsorbents and metal ions. Therefore, to establish the effect of pH on the adsorption of Hg^{2+} onto both the PHEMA and the PHEMA-IMEO nanoparticles, we repeated the batch adsorption equilibrium studies at different pH in the range 2.0–7.0. In this group of experiments, the equilibrium concentration of Hg^{2+} and the adsorption equilibrium time were 43 mg/L and 2 h, respectively. The pH dependence of adsorption values of

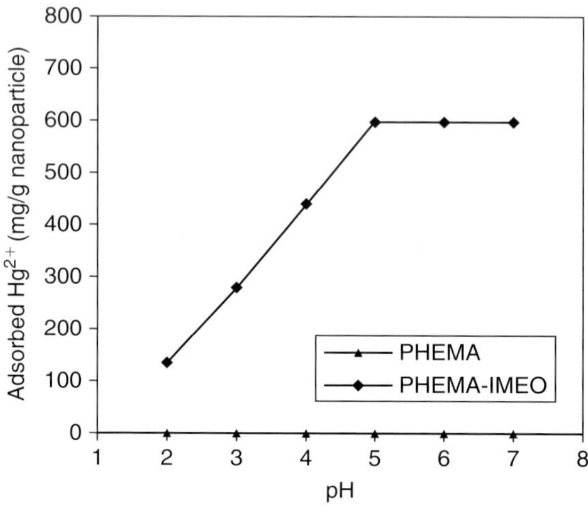

Figure 2.5 Effect of pH on adsorption of Hg^{2+} on the PHEMA and PHEMA-IMEO nano-particles; IMEO loading: 1021 µmol/g; equilibrium concentration of Hg^{2+}: 43 mg/L.

Hg^{2+} is shown in Fig. 2.5. In the case of PHEMA nanoparticles, adsorption is pH independent. But it is indicated that the adsorption of Hg^{2+} onto the PHEMA–IMEO nanoparticles was pH dependent. The results show that mercury adsorption by the PHEMA–IMEO nanoparticles was very low at pH 2.0 but increased rapidly with increasing pH and then reached the maximum at pH 5.0. The competitive adsorption of hydrogen ions with Hg^{2+} ions for imidazole groups at lower pH values accounts for the observed low adsorption capacity. Because the imidazole groups are most likely protonated at a low pH, the nanoparticles are positively charged, resulting in a strong electrostatic repulsive force between the IMEO on the nanoparticles and positively charged metal ions. Hg^{2+} adsorption around pH 3.0–4.0 was also low. It is well known in adsorption mechanisms that a decrease in solubility favors an improvement in adsorption performance.

3.3. Competitive Adsorption

As seen in Table 2.2, adsorbed amounts of Hg^{2+} ions are higher than those obtained for Cd^{2+} and Pb^{2+}, not only in weight basis but also in molar basis. The adsorption capacities are 238.8 mg/g for Hg^{2+}, 45.9 mg/g for Cd^{2+}, and 19.6 mg/g for Pb^{2+}. From these results, the order of affinity is Hg^{2+} > Cd^{2+} > Pb^{2+}. This trend is presented on the basis of mass (mg) metal adsorption per gram adsorbent, and these units are important in quantify-ing respective metal capacities in real terms. However, a more effective

Table 2.2 Competitive adsorption of Hg^{2+}, Cd^{2+}, and Pb^{2+} from their mixture onto PHEMA-IMEO nanoparticles (IMEO loading: 1021 μmol/g; pH: 5.0)

Metal ions	Metal ions adsorbed	
	(mg/g polymer)	(μmol/g polymer)
Hg^{2+}	238.8	1190.5
Cd^{2+}	45.9	408.3
Pb^{2+}	19.6	94.5

approach for this work is to compare metal adsorption on a molar basis; this gives a measure of the total number of metal ions adsorbed, as opposed to total weight, and is an indication of the total number of binding sites available on the adsorbent matrix, to each metal. In addition, the molar basis of measurement is the only accurate way of investigating competition in multicomponent metal mixtures. Molar basis units are measured as μmol per gram of dry adsorbent. It is evident from Table 2.2 that the order of capacity of PHEMA-IMEO nanoparticles is as follows: $Hg^{2+} > Cd^{2+} > Pb^{2+}$. It is clear from Table 2.2 that the PHEMA-IMEO nanoparticles showed more affinity to Hg^{2+} ions.

3.4. Behavior of the Elution

The regeneration of the adsorbent is likely to be a key factor in improving process economics. To be useful in metal remediation processes, metal ions should be easily eluted under suitable conditions. Elution of the Hg^{2+} from the metal-chelating nanoparticles was performed in a batch experimental setup. Various factors are probably involved in determining rates of Hg^{2+} elution, such as the extent of hydration of the metal ions and polymer microstructure. However, an important factor appears to be binding strength. When HNO_3 is used as an elution agent, the coordination spheres of chelated Hg^{2+} ions is disrupted and subsequently Hg^{2+} ions are released from the polymer surface into the desorption medium. In this study, the elution time was found to be 15 min. Elution ratios are very high (up to 99%). The ability to reuse the PHEMA-IMEO nanoparticles was shown in Fig. 2.6. The adsorption behavior is stable for 10 cycles of use and it could be used at least 25 times. The adsorption capacity of the recyled nanoparticles can be maintained at 96% level at the 25th cycle. This means that the newly synthesized nanoparticles have great potential for industrial removal applications.

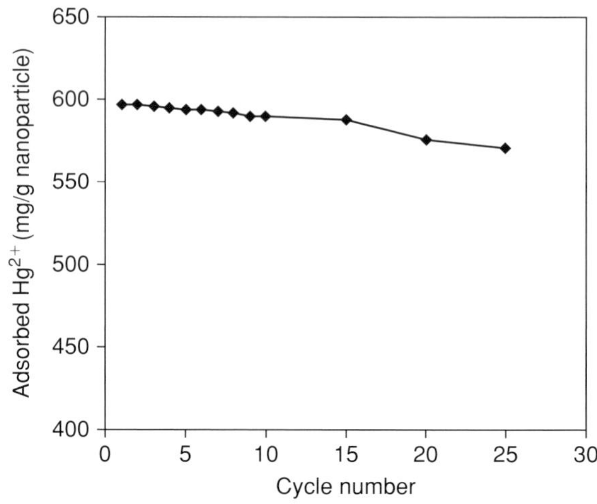

Figure 2.6 Adsorption–elution cycles for Hg^{2+}; Hg^{2+} equilibrium concentration: 43 mg/L; pH: 5.0.

References

[1] Ma ZY, Guan YP, Liu XQ, Liu HZ. Synthesis of magnetic chelator for high-capacity immobilized metal affinity adsorption of protein by cerium initiated graft polymerization. Langmuir 2005;21:6987–94.

[2] Sun L, Dai J, Baker GL, Bruening ML. High-capacity, protein-binding membranes based on polymer brushes grown in porous substrates. Chem Mater 2006;18:4033–39.

[3] Dai J, Bao Z, Sun L, Hong SU, Baker GL, Bruening ML. High-capacity, protein-binding membranes based on polymer brushes grown in porous substrates. Langmuir 2006;22:4274–81.

[4] Peng K, Hidajat K, Udin MS. Adsorption of bovine serum albumin on nanosized magnetic particles. J Colloid Interface Sci 2004;271:277–83.

[5] Choi SW, Kwon HY, Kim WS, Kim JH. Thermodynamic parameters on poly(D,L-lactide-*co*-glycolide) particle size in emulsification–diffusion process. Colloids Surf A 2002;201:283–89.

[6] Lu S, Ramos J, Forcada J. Self-stabilized magnetic polymeric composite nanoparticles by emulsifier-free miniemulsion polymerization. Langmuir 2007;23:12893–900.

[7] Li X, Hernandez JP, Haque SA, Durrant JR, Palomares E. Functionalized titania nanoparticles for mercury scavenging. J Mater Chem 2007;17:2028–32.

[8] Takafuji M, Ide S, Ihara H, Xu Z. Preparation of poly(1-vinylimidazole)-grafted magnetic nanoparticles and their application for removal of metal ions. Chem Mater 2004;16:1977–83.

[9] Lu C, Liu C. J Removal of nickel(II) from aqueous solution by carbon nanotubes. J Chem Technol Biotechnol 2006;81:1932–40.

[10] Qi L, Xu Z. Lead sorption from aqueous solutions on chitosan nanoparticles. Colloids Surf A 2004;251:183–90.

[11] Singh RP, Way JD, Dec SF. Silane modified inorganic membranes: Effects of silane surface structure. J Memb Sci 2005;259:34–46.

[12] Bilkova Z, Slovakova M, Horak D, Lenfeld J, Churacek J. Enzymes immobilized on magnetic carriers: Efficient and selective system for protein modification. J Chromatogr B 2002;770:177–81.

[13] Öztürk N, Günay ME, Akgöl S, Denizli A. Silane-modified magnetic beads: Application to immunoglobulin G separation. Biotechnol Prog 2007;23:1149–56.

[14] Garcia MF, Garcia RP, Garcia NB, Sanz-Medel A. On-line preconcentration of inorganic mercury and methylmercury in sea-water by sorbent-extraction and total mercury determination by cold vapour atomic absorption spectrometry. Talanta 1994;41:1833–39.

[15] D'Itri PA, D'Itri FM. Mercury contamination. New York: Wiley; 1977.

[16] Lai EPC, Wong B, Vandernoot VA. Preservation of solid mercuric dithizonate samples with polyvinyl chloride for determination of mercury(II) in environmental waters by photochromism-induced photoacoustic spectrometry. Talanta 1993;40:1097–105.

[17] Berglund M, Lind B, Björnberg KA, Palm B, Einarsson O, Vahter M. Inter-individual variations of human mercury exposure biomarkers: a cross-sectional assessment. Environmental Health A 2005;4:1–11.

[18] Aschner M, Aschner JL. Mucocutaneous lymph-node syndrome - Is there a relationship to mercury exposure. Am J Dis Child 1989;143:1133–34.

[19] Martin MD, Williams BJ, Charleston JD, Oda D. Spontaneous exfoliation of teeth following severe elemental mercury poisoning: Case report and histological investigation for mechanism. Oral Surg 1997;84:495–501.

[20] Öztürk N, Akgöl S, Arısoy M, Denizli A. Reversible adsorption of lipase on novel hydrophobic nanospheres. Sep Purif Technol 2007;58:83–90.

[21] Lin CT, Lee SY, Keh ES, Dong DR, Huang HM, Shih YH. Influence of silanization and filler fraction on aged dental composites. J Oral Rehabil 2000;27:919–26.

[22] Soares CJ, Giannini M, De Oliveira MT, Paulillo LAMS, Martins LRM. Effect of surface treatments of laboratory-fabricated composites on the microtensile bond strength to a luting resin cement. J Appl Oral Sci 2004;12:45–50.

[23] Silberzan P, Leger L, Ausserre D, Benattar JJ. Silanation of silica surfaces. A new method of constructing pure or mixed monolayers. Langmuir 1991;7:1647–51.

[24] Lezzi A, Cobianco S, Roggrero A. Synthesis of thiol chelating resins and their adsorption properties toward heavy metal ions. J Polym Sci A 1994;32:1877–83.

[25] Denizli A, Özkan G, Arica MY. Preparation and characterization of magnetic polymethylmethacrylate microbeads carrying ethylene diamine for removal of Cu(II), Cd(II), Pb(II), and Hg(II) from aqueous solutions. J Appl Polym Sci 2000;78:81–9.

[26] Denizli A, Kesenci K, Pişkin E. Dithiocarbamate-incorporated monosize polystyrene microspheres for selective removal of mercury ions. React Funct Polym 2000;44:235–43.

[27] Jyo A, Matsufune S, Ono H, Egawa H. Preparation of phosphoric acid resins with large cation exchange capacities from macroreticular poly(glycidyl methacrylate-co-divinylbenzene) beads and their behavior in uptake of metal ions. J Appl Polym Sci 1997;63:1327–34.

[28] Salih B, Say R, Denizli A, Genc O, Piskin E. Determination of inorganic and organic mercury compounds by capillary gas chromatography coupled with atomic absorption spectrometry after preconcentration on dithizone-anchored poly(ethylene glycol dimethacrylate-hydroxyethyl methacrylate) microbeads. Anal Chim Acta 1998;371:177–85.

[29] Binman S, Belfer S, Shani A. Metal sorption properties of sulfur-chlorinated jojoba wax bound to polystyrene beads. J Appl Polym Sci 1997;63:625–33.

[30] Kesenci K, Say R, Denizli A. Removal of heavy metal ions from water by using poly(ethyleneglycol dimethacrylate-co-acrylamide) beads. Eur Polym J 2002;38: 1443–48.

[31] Hwang DC, Damodaran S. Metal-chelating properties and biodegradability of an ethylenediaminetetraacetic acid dianhydride modified soy protein hydrogel. J Appl Polym Sci 1997;64:891–906.

[32] Denizli A, Arpa C, Bektas S, Genc O. Adsorption of mercury(II) ions on Procion Blue MX-3G-attached magnetic poly(vinyl alcohol) gel beads. Ads Sci Technol 2002;20:91–106.

[33] Liu CY, Chang HT, Hu CC. Complexation reactions in a heterogeneous system. Inorg Chim Acta 1990;172:151–58.

[34] Shah S, Devi S. Preconcentration of mercury(II) on dithizone anchored poly(vinyl pyridine) support. React Funct Polym 1996;31:1–9.

[35] Cestari AR, Airoldi C. Chemisorption on thiol–silicas: divalent cations as a function of ph and primary amines on thiol–mercury adsorbed. J Colloid Interf Sci 1997;195:338–42.

[36] Delacour ML, Gailliez E, Bacquet M, Morcellet M. Poly(ethylenimine) coated onto silica gels: Adsorption capacity toward lead and mercury. J Appl Polym Sci 1999;73:899–906.

[37] Rivas BL, Maturana HA, Molina MJ, Gomez-Anton MR, Pierola IF. Metal ion binding properties of poly(N-vinylimidazole) hydrogels. J Appl Polym Sci 1998;67:1109–18.

[38] Arpa C, Saglam A, Bektas S, Patır S, Genc O, Denizli A. Adsorption of mercury(II) ions by oly(hydroxyethyl methacrylate) adsorbents with thiazolidine groups. Ads Sci Technol 2002;20:203–13.

[39] Navarro RR, Sumi K, Fujii N, Matsumara M. Mercury removal from wastewater using porous cellulose carrier modified with poly(ethyleneimine). Water Res 1996;30:2488–94.

[40] Denizli A, Şenel S, Alsancak G, Tuzmen N, Say R. Mercury removal from synthetic solutions using poly (2-hydroxyethyl methacrylate) gel beads modified with poly(ethyleneimine). React Funct Polym 2003;55:121–30.

[41] Becker NSC, Eldridge RJ. Selective recovery of mercury(II) from industrial waste-waters .1. Use of a chelating ion-exchanger regenerated with brine. React Polym 1993;21:5–14.

[42] Atia A, Donia AM, Yousif AM. Synthesis of amine and thio chelating resins and study of their interaction with zinc(II), cadmium(II) and mercury(II) ions in their aqueous solutions. React Funct Polym 2003;56:75–82.

[43] Denizli A, Garipcan B, Emir S, Patir S, Say R. Heavy metal ion adsorption properties of methacrylamidocysteine-containing porous poly(hydroxyethyl methacrylate) chelating beads. Ads Sci Technol 2002;20:607–17.

[44] Say R, Garipcan B, Emir S, Patir S, Denizli A. Preparation and characterization of the newly synthesized metal-complexing-ligand N-methacryloylhistidine having PHEMA beads for heavy metal removal from aqueous solutions. Macromol Mater Eng 2002;287:539–45.

CHAPTER 3

CO$_2$ Response of Nanostructured CoSb$_2$O$_6$ Synthesized by a Nonaqueous Coprecipitation Method

Carlos R. Michel, Alma H. Martínez *and*
Héctor Guillén-Bonilla
Departamento de Física, CUCEI Universidad de Guadalajara, Guadalajara, Jalisco, México

Contents

1. INTRODUCTION

Although CO_2 has played an important role in preserving the average atmospheric temperature over centuries, the emission of this gas in large amounts during recent decades is producing a phenomenon of great concern to mankind, global warming. With the goal of developing reliable solid-state gas sensors, notable scientific research has been made worldwide. As a result, several compounds such as SnO_2, ZnO, WO_3, TiO_2, and other inorganic oxides have been intensively tested for the detection of toxic gases like CO and NO_2 [1–4]. However, less information about materials tested for sensing CO_2 and other greenhouse gases is found in the literature.

In addition to the oxides already mentioned, several oxides possessing different crystal structures have been studied for gas-sensing purposes. Recently, investigating the gas-sensing properties of $CoSb_2O_6$ prepared by an aqueous solution–polymerization method, our group found a good response to CO_2 and O_2 [5]. This oxide has the trirutile type structure, as well as other complex antimony oxides, and has attracted the attention due to their unique physical and chemical properties.

On the other hand, for the preparation of nanostructured materials with micro and nanoporosity, the aqueous solution-polymerization methods usually produce good results [6–8]. In these methods, the homogeneous distribution of metal ions is made using a polymer, such as polyvinyl alcohol (PVA) or polyethylene glycol dissolved in water. Then, during calcination, the polymer decomposes producing a large amount of CO_2 and other gases, which develops extensive porosity and very small particles in the resulting solid material. The synthesis of nanoceramics by these methods largely depends on the solubility of metal nitrates and polymers in water.

In the search of other polymers that facilitate the formation of nanostructured materials, the synthesis of nickel nanoparticles using polyvinyl pyrrolidone (PVP) in ethylene glycol was recently reported [9]. The authors found that the molecular weight of PVP plays an important role in the formation of the nanoparticles and that by increasing the molecular weight of PVP, the growth and agglomeration of the metal particles can be prevented.

The synthesis of ceramic nanoparticles using nonaqueous solution-polymerization methods has been less studied, probably due to the limited null solubility of some reactants in common solvents such as alcohols. For instance, attempts to obtain $CoSb_2O_6$ by a solution-polymerization method using PVA in ethyl alcohol did not produce good results because PVA is not soluble in this alcohol; therefore, other solvents and chemical compounds were tested for optimal results.

In this work, the synthesis of nanostructured $CoSb_2O_6$ by a nonaqueous coprecipitation method using PVP, cobalt nitrate, and antimony trichloride in ethyl alcohol was explored. The relationship between the PVP concentration in solution and the surface morphology was investigated. The potential application of nanostructured $CoSb_2O_6$, as an environmental gas-sensor material, was studied by measuring the dynamic variation of resistance in air, CO_2, and O_2. The CO_2-sensing response was also studied by recording intensity vs. voltage graphs at different CO_2 concentrations.

2. EXPERIMENTAL

For the preparation of $CoSb_2O_6$, stoichiometric amounts of antimony trichloride (Sigma-Aldrich, 99%) and cobalt nitrate hexahydrate (Alfa Aesar, 98–102%) were dissolved separately in 5 ml of anhydrous ethyl alcohol (J.T. Baker), and the resulting transparent solutions were colorless and red, respectively. To test the role of PVP on the morphology of $CoSb_2O_6$, three solutions containing, respectively, 0.05, 0.22, and 0.44 g of PVP (Sigma-Aldrich, average molecular weight = 10,000) in 5 ml of ethyl alcohol were prepared. Because PVP is soluble in ethyl alcohol, clear solutions were obtained. The next step was the mixture of each PVP solution with the solution of cobalt nitrate; this was made under strong stirring, producing transparent red liquids. After 20 min of stirring, the solution of antimony chloride was added, which formed immediately blue–turquoise suspensions, with a pH value of approximately 2; these were stirred for 18 h. The evaporation of the suspensions was made by microwave radiation using a domestic microwave oven, working at low power; the temperature of the beakers was maintained below 50 °C. The resulting precursors were annealed in static air from 200 to 600 °C, using a muffle type furnace with programmable temperature control. A heating rate of 100 °C/h was used in each calcination.

The structural characterization was made by X-ray powder diffraction (XRD) at room temperature, using a Rigaku Miniflex apparatus (Cu K_α radiation). The diffraction angle (2θ) was scanned from 10° to 70°. The surface morphology was observed by scanning electron microscopy (SEM) using a Jeol JSM-5400LV microscope. The observation of nanostructural characteristics was made by transmission electron microscopy (TEM) using a Jeol JEM-1010 microscope (operated at 100 kV). Before the TEM observation, the powders were sonicated in isopropyl alcohol for 5 min. Electrical conductivity and gas-response characterization were performed on thick films made with the powder prepared with 0.22 g of PVP. The films were made by dispersing 0.1 g of $CoSb_2O_6$ in 2 ml of ethyl alcohol using an ultrasonic bath; then the mixture was deposited into alumina substrates forming films with 5 mm diameter and ~300 μm thickness. Silver wires were fixed to the films as electrical contacts. The electrical measurements were carried out using the two-point probe method, from room temperature to 600 °C, using a digital data acquisition unit (Agilent 34970A) and a digital voltmeter (Agilent 34401A). The variation of intensity with voltage was recorded with a potentiostat/galvanostat Solartron 1285A. During the gas-sensing characterization, the gases were supplied by a MKS 647C mass flow controller.

3. RESULTS AND DISCUSSION

Figure 3.1 shows the XRD patterns of precursor powders obtained using the three different concentrations of PVP, after their calcination at 600 °C for 4 h. Clearly, single-phase $CoSb_2O_6$ was produced in each synthesis; the identification of the peaks was made by using the international centre for diffraction data (ICDD) file No. 18-0403, corresponding to $CoSb_2O_6$. The

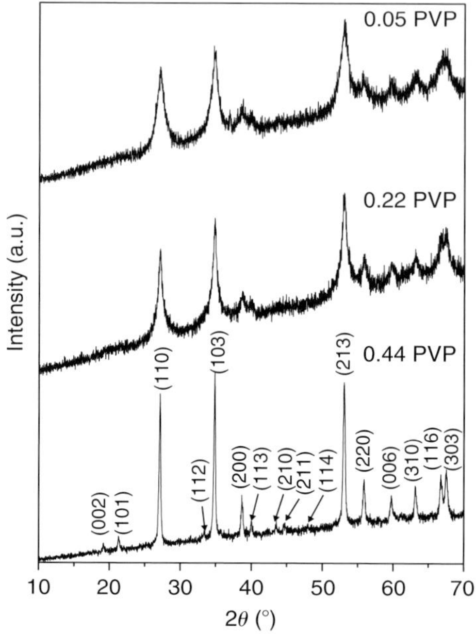

Figure 3.1 XRD patterns of $CoSb_2O_6$ precursor powders prepared with 0.05, 0.22, and 0.44 g of PVP, after a calcination at 600 °C in air.

XRD patterns of samples synthesized with 0.05 and 0.22 g of PVP exhibit wider peaks than the sample synthesized with 0.44 g, indicating a larger average particle size in the latter.

The crystal structural evolution while increasing the calcination temperature was studied using XRD on a sample prepared with 0.22 g of PVP. This sample was selected due to its particular nanostructural characteristics, which will be presented later in the SEM and TEM results. Figure 3.2 shows the XRD patterns of the precursor powder calcined from 200 to 600 °C. For the sample fired at 200 °C, some peaks of low intensity placed in 2θ from 15 to 20° and from 30 to 50° can be observed. The identification of such peaks was not possible using the ICDD database because none of them correspond to crystalline phases containing cobalt or antimony; probably, they are associated with products of decomposition of PVP. At a calcination temperature of 300 °C, the XRD pattern shows mainly an amorphous material; however, the incipient presence of the three main peaks of $CoSb_2O_6$, placed at $2\theta = 27, 34.8$, and 53°, can be observed. From 400 °C, the main peaks of single-phase $CoSb_2O_6$ can be clearly identified, though the calcination at 600 °C produced a better crystallized material.

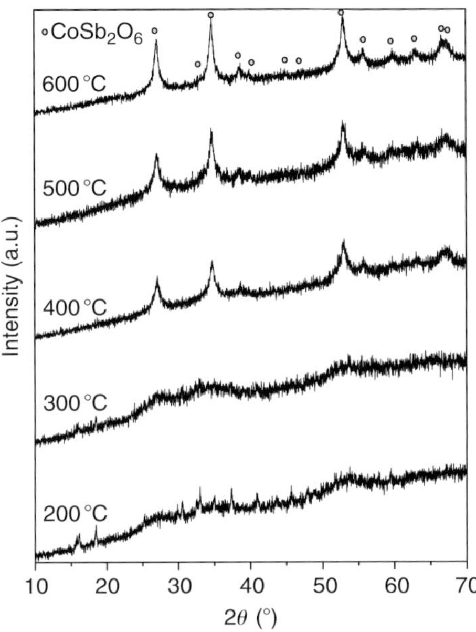

Figure 3.2 Crystal structure evolution as a function of the calcination temperature for a CoSb$_2$O$_6$ precursor powder made with 0.22 g of PVP.

Figure 3.3 shows the typical surface morphology of CoSb$_2$O$_6$ prepared with 0.05 g (A), 0.22 g (B, C), and 0.44 g (D, E) of PVP, calcined at 600 °C. These observations were made by SEM at low magnification on the three samples, but in the case of the last two samples, larger magnifications are also shown, exhibiting their microstructure in more detail. When 0.05 g of PVP was used, a compact agglomeration of very fine particles was obtained; little porosity and few cracks were also observed (Fig. 3.3A). When 0.22 g of PVP was used, large scale of thin wires was developed (Fig. 3.3B and C). The length of these wires was several micrometers and about 100 nm width. When 0.44 g of PVP was used, a similar formation of wires, than the sample prepared with 0.22 g, can be observed in Fig. 3.3D and E; however, the size of the wires is considerably larger. These results show that the amount of PVP used in solution determines the microstructure of CoSb$_2$O$_6$.

To know whether the concentration of PVP solely determines the formation of wires in CoSb$_2$O$_6$, two different parameters were studied. First, the effect of stirring the suspension after their preparation was investigated. Then a new synthesis of CoSb$_2$O$_6$ using 0.22 g of PVP was made. After mixing the cobalt, PVP, and antimony solutions, the resulting suspension

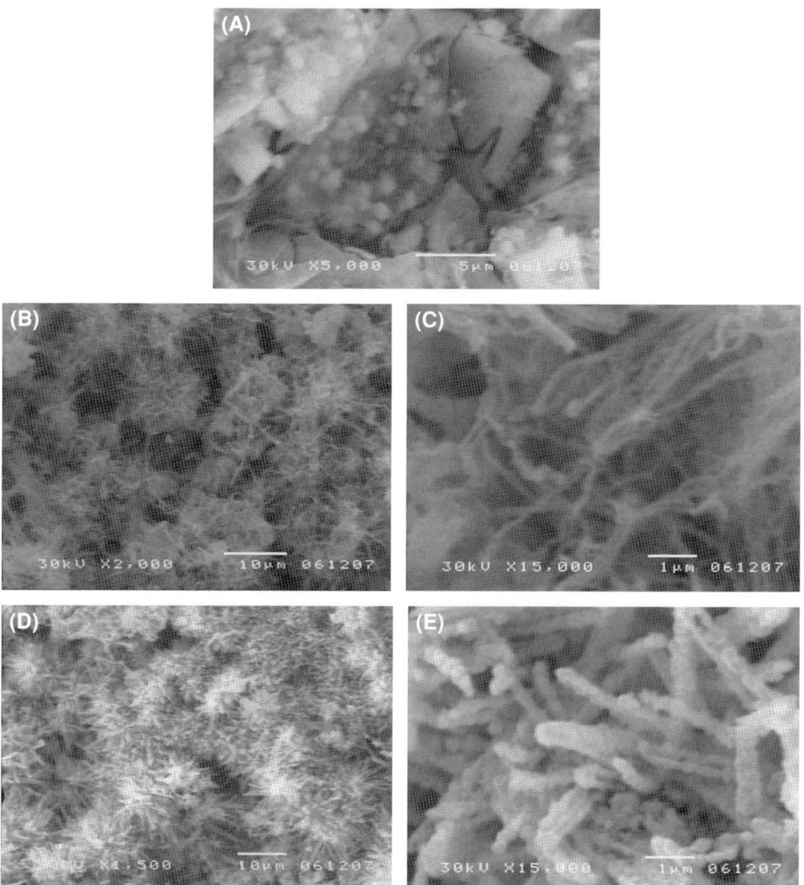

Figure 3.3   SEM images of $CoSb_2O_6$ prepared with 0.05 g (A), 0.22 g (B, C), and 0.44 g (D, E) of PVP.

was under strong stirring for 1 h. Then it was maintained in calm for 18 h. This time was the same used in the previous synthesis under stirring. The evaporation of the solvent and further calcination was made using the same procedure than before. Figure 3.4 shows a typical XRD pattern of the powder obtained and its microstructure observed by SEM (inset). Clearly single-phase $CoSb_2O_6$ was obtained; however, the morphology shows particles with irregular shape and size smaller than ~1.3 μm. Evidently, in this case, the wires were not produced.

The second parameter studied was the effect of the time that the sample is calcined at 600 °C, and this was made with the purpose of investigating a possible growth of the wires. This sample was prepared under stirring

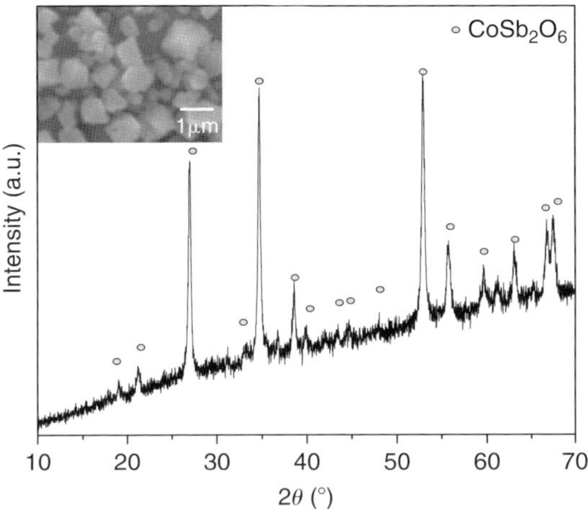

Figure 3.4 XRD pattern and SEM image (inset) of a CoSb$_2$O$_6$ sample prepared with 0.22 g of PVP, without stirring (600 °C).

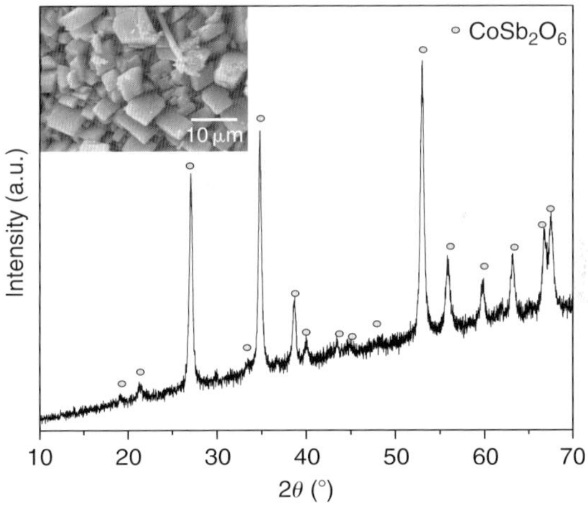

Figure 3.5 XRD pattern and SEM micrograph (inset), corresponding to a CoSb$_2$O$_6$ sample synthesized with 0.22 g of PVP, calcined at 600 °C for 24 h.

and remained at 600 °C for 24 h. Figure 3.5 shows the XRD pattern and its microstructure (inset); as it was expected, single-phase CoSb$_2$O$_6$ was obtained as shown in the XRD pattern. However, the microstructure mainly corresponds to rectangular flat particles, with size in the range 1–10 μm. This result indicates that grain growth dominates during the sintering. Therefore, controlling the PVP concentration, the time of stirring, the time of calcination, and the morphology of CoSb$_2$O$_6$ can be greatly modified using this synthesis method.

On the other hand, Fig. 3.6 shows the typical TEM images, in bright-field mode, of $CoSb_2O_6$ prepared with 0.05 g (A), 0.22 g (B), and 0.44 g (C) of PVP calcined at 600 °C for 4 h. For the synthesis with 0.05 g, a compact arrangement of nanoparticles with few porosity can be observed; this morphology corresponds to the particles observed in Fig. 3.3A. The average particle size was measured from several TEM images resulting in 8 nm; Fig. 3.7A shows the particle size distribution. For $CoSb_2O_6$ prepared with 0.22 g of PVP, Fig. 3.6B exhibits the formation of nanoparticles and nanorods. In the case of nanoparticles, an average particle size of ~10 nm was found, and for nanorods an average diameter of 8 nm and length until 65 nm were observed. Figure 3.7B shows the diameter and length size distributions for this sample. It is noticeable that SEM shows wires for this sample, whereas TEM shows nanoparticles and nanorods. This can be explained by a fragmentation of the wires occurred during the sonication of the sample prior to $CoSb_2O_6$ observation by TEM.

For $CoSb_2O_6$ synthesis using 0.44 g of PVP, Fig. 3.6C shows a single wire exhibiting a bend in one of the ends, which was commonly observed by SEM (Fig. 3.3E). In this sample, the wires are composed of the aggregation of larger particles than those obtained for sample prepared with 0.22 g

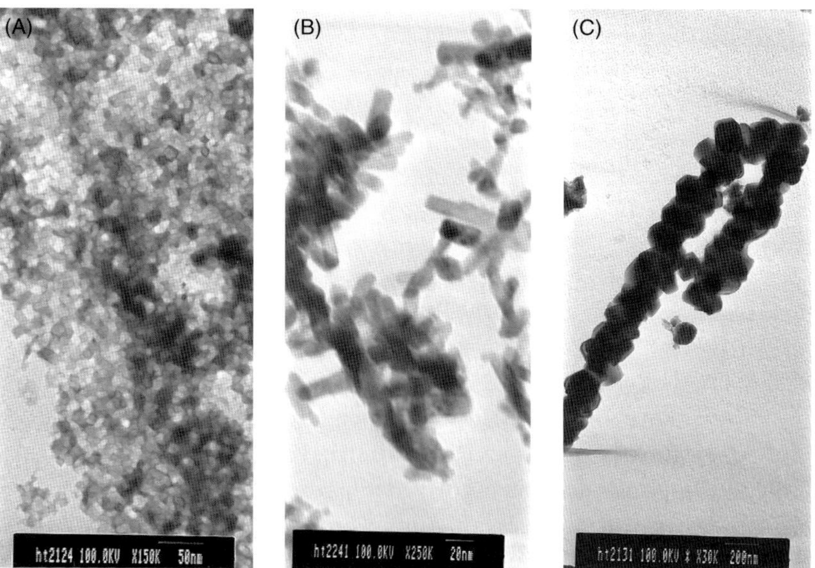

Figure 3.6 TEM images of $CoSb_2O_6$ prepared with 0.05 g (A), 0.22 g (B), and 0.44 g (C) of PVP (calcined at 600 °C for 5 h).

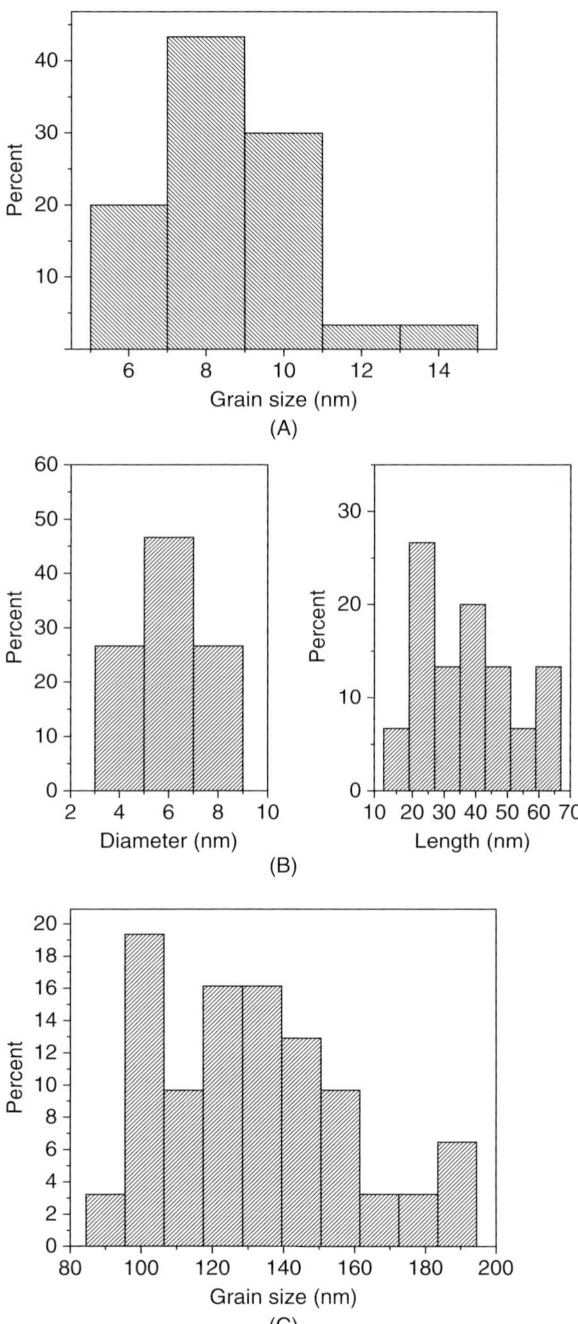

Figure 3.7 Particle size distributions of CoSb$_2$O$_6$ synthesized with 0.05 g (A), 0.22 g (B), and 0.44 g (C) of PVP.

of PVP. The size of individual particles was measured, and Fig. 3.7C shows their corresponding particle size distribution.

To test whether the synthesis of $CoSb_2O_6$ gives reproducible results, several preparations using each PVP concentration were made, obtaining similar microstructural characteristics than those presented in this work.

Comparing the nanostructural characteristics observed by TEM, clearly, an increase in particle size is observed while increasing the amount of PVP, which is in agreement with the XRD and SEM results. On the other side, although the $CoSb_2O_6$ powder made with $0.22\,g$ of PVP did not have the smallest particle size, a larger specific surface area can be qualitatively inferred in this sample, because less agglomeration and plenty of nanoparticles was observed. Due to these characteristics, this powder was used for testing as environmental gas-sensing material.

To investigate the conductivity of the target $CoSb_2O_6$, thick films were heated from room temperature to $600\,°C$ in dynamic air, CO_2, and O_2 while measuring the resistance. The conductivity (σ) was calculated according to the well-known equation [10]:

$$\sigma = \frac{l}{RS} \tag{1}$$

where R is the resistance between the two electrical contacts, separated by a length l, and S is the section of the film. Figure 3.8 shows typical Arrhenius-type plots obtained in the three gases; the increment of the conductivity with temperature observed in this figure is characteristic of the semiconductor behavior. Although the gas type had a little effect on the shape of each curve, the smallest conductivity values were registered in CO_2, in most of the temperature ranges studied. The conductivity goes from $\sim 10^{-6}$ to $\sim 10^{-2}(\Omega\,cm)^{-1}$ and has two branches, with a transition temperature (T_1) at $\sim 350\,°C$ for air and O_2 and at $\sim 460\,°C$ for CO_2. The activation energies for conduction (E_A) were calculated from the slopes of each section of the curves, Table 3.1 shows these results. The transition temperature (T_1) is associated with a change in the conduction mode from a semiconductor to metallic; therefore, E_A values are smaller at higher temperatures.

For the evaluation of the environmental gas-sensing response of $CoSb_2O_6$ thick films, the variation of resistance with time was measured while supplying alternatively air, CO_2, and O_2. To test the response in CO_2, dry synthetic air was supplied for approximately $5\,min$ using a flow rate of $0.22\,sccm$; then CO_2 was delivered for approximately $2.5\,min$ using

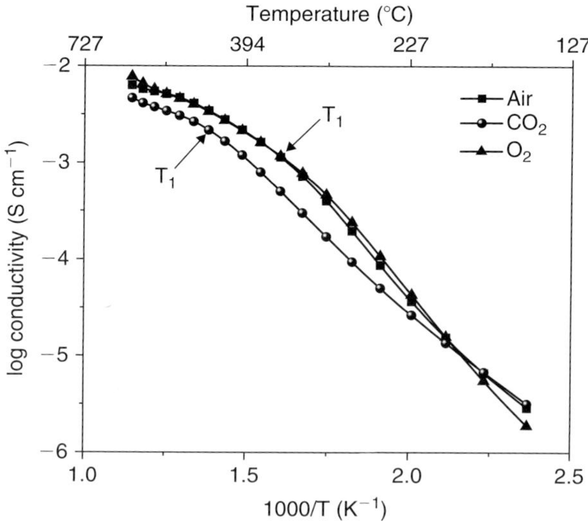

Figure 3.8   Arrhenius plots of CoSb$_2$O$_6$ (0.22 g of PVP) thick films measured in air, O$_2$, and CO$_2$.

Table 3.1   Calculated activation energies of conduction (E_A) for CoSb$_2$O$_6$ prepared with 0.22 g of PVP

Gas	E_A (eV) ($T < T_1$)	E_A (eV) ($T > T_1$)
Air and O$_2$	0.76	0.33
CO$_2$	0.60	0.25

the same flow rate; finally, an air flow was supplied. This operation was performed several times to confirm the reproducibility. To test the films in oxygen, a similar procedure was used. In this case, the CO$_2$ flow was replaced by O$_2$.

Figure 3.9 shows the typical dynamic response in CO$_2$ (A), and O$_2$ (B); these graphs were recorded at temperatures of 400 and 500 °C, respectively, where a larger response was registered. Figure 3.9A shows the introduction of CO$_2$ that produced a variation of resistance (ΔR) of approximately 2.7 kΩ, whereas in O$_2$, the opposite behavior is observed. In this gas, ΔR was approximately -0.5 kΩ. Regardless of the gas used, a reproducible response pattern was recorded in both tests.

Because the resistance of CoSb$_2$O$_6$ decreased in O$_2$, whereas in CO$_2$ has the opposite behavior, this oxide can be considered a P-type semiconductor

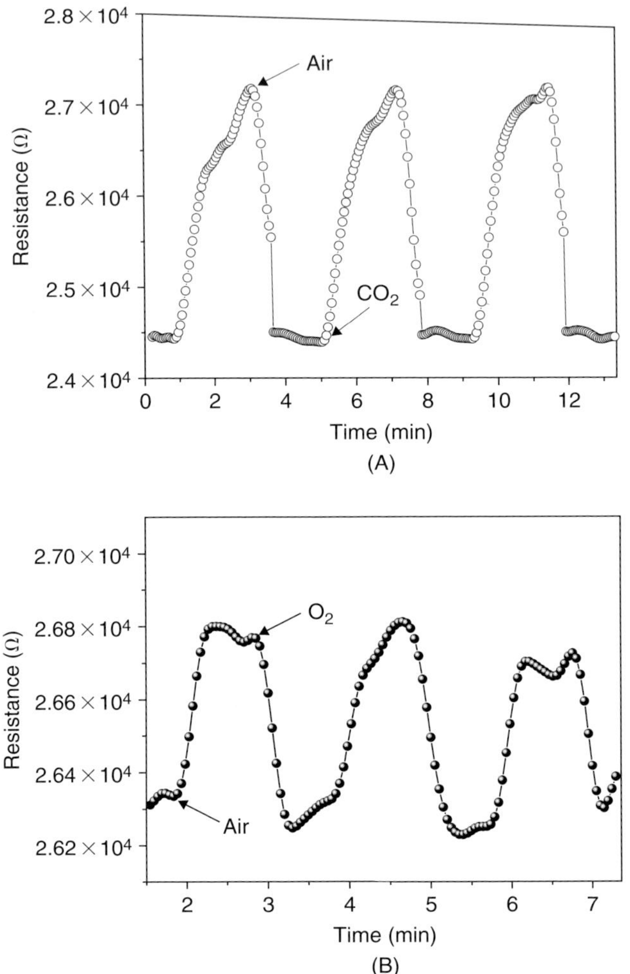

Figure 3.9   Dynamic response of resistance of nanostructured $CoSb_2O_6$ (0.22 g of PVP) thick films in CO_2 (A) and O_2 (B), measured at 400 °C.

material [11]. Besides, when oxygen surrounds the $CoSb_2O_6$ surface, at least two possible adsorption processes can occur:

$$O_2(gas) + 2e^- \rightarrow 2O^-(ads), \tag{2}$$

$$O_2(gas) + 4e^- \rightarrow 2O^{2-}(ads) \tag{3}$$

where (ads) means the adsorbed species [12]. The adsorption of oxygen produces an increase in the number of holes in the valence band, which increases the conductivity. For the detection of CO_2, a rather different mechanism may be involved because the carbon is fully oxidized and the electron transfer is less probable. Ishihara et al., have suggested that a thin carbonation layer may be formed during the CO_2 detection, which changes the electrical conductivity of the material [13]. According to the results shown in Fig. 3.9A, the carbonation layer is removed upon the introduction of air because a full recovery of the resistance is observed in each air$-CO_2$ cycle.

To test the ability of $CoSb_2O_6$ to detect CO_2 at different concentrations, current–voltage $(I-V)$ graphs were recorded from -10 to $10\,V$, at $500\,°C$. Figure 3.10 shows the results obtained for several CO_2:air ratios. In these graphs, for each CO_2:air ratio, a curve can be easily distinguished, and these curves have a nonlinear trend in both positive and negative voltages. When a 100% CO_2 flow was used, the smaller current values were recorded at positive voltages (Fig. 3.10A). When the concentration of CO_2 was decreased, for CO_2:air ratios from 1:1 to 1:3, a clear increase in the current can be observed, and this increment in current is caused by the increment of oxygen contained in the air.

For negative voltages, the inverse behavior was obtained (Fig. 3.10B); however, the magnitude of the current was significantly smaller compared to the results at positive voltages. Moreover, comparing the values of the polarization curves at positive voltages, with those obtained in the tests of dynamic response, a good agreement can be observed.

To test the response of $CoSb_2O_6$ thick films to a variation in O_2 concentration, polarization curves, using several O_2:air flows, were recorded. However, the results show a superposition of the curves, and a poor sensitivity to this gas was concluded.

4. CONCLUSION

A nonaqueous coprecipitation method based on PVP in ethyl alcohol was used for the preparation of $CoSb_2O_6$. Using this simple method, $CoSb_2O_6$ was obtained at a low calcination temperature, with good control on the stoichiometry. By an appropriate control of the amount of PVP, the time of stirring, and calcination, it was possible to prepare nanostructured $CoSb_2O_6$ powders with an average size $\sim10\,nm$, having a narrow particle size distribution. Because PVP is an inexpensive polymer, which has been available for more than five decades, this may be a convenient method for the synthesis of this $CoSb_2O_6$, and other inorganic oxides.

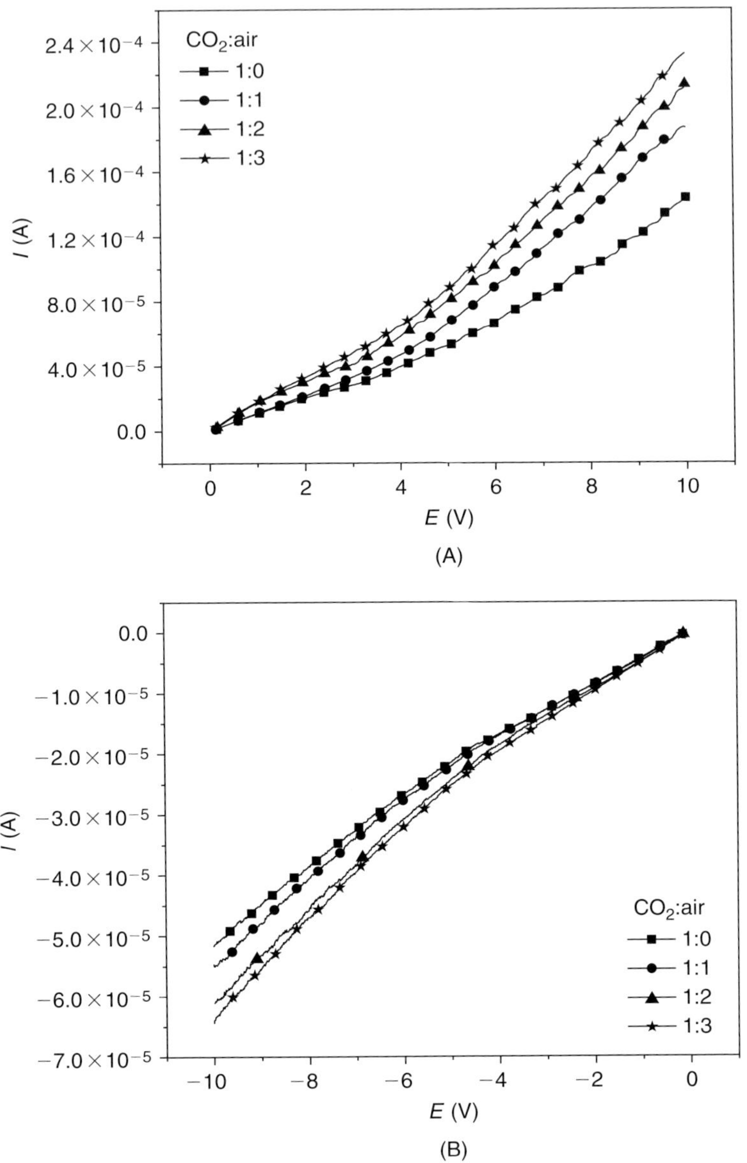

Figure 3.10   Current–voltage curves of $CoSb_2O_6$ (0.22 g of PVP), measured at 400 °C in several CO_2:air proportions.

The gas-sensing properties were studied on the $CoSb_2O_6$ powder with the best nanostructural characteristics for which thick films were prepared with this material, without further modification of its nanostructure. According to the results obtained from the electrical characterization of O_2 and CO_2, $CoSb_2O_6$ can have a potential use as a CO_2 sensor material.

ACKNOWLEDGMENTS

The author is grateful to the Coordinación General Académica of the Universidad de Guadalajara and the National Council of Science and Technology of Mexico (CONACYT) for financial support under the project I-52204. A.H.M. and H.G.B want to thank CONACYT for their doctorate scholarships in the Universidad de Guadalajara.

References

[1] Yamazoe N. Toward innovations of gas sensor technology, Sens Actuators B 2005; 108:2–14.

[2] Shimizu Y, Egashira M. Basic aspects and challenges of semiconductor gas sensors, Mater Res Soc Bull 1999;6:18–24.

[3] Graf M, Barrettino D, Taschini S, Hagleitner C, Hierlemann A, Baltes H. Metal oxide-based monolithic complementary metal oxide semiconductor gas sensor microsystem, Anal Chem 2004;76:4437–4445.

[4] Azad AM, Akbar SA, Mhaisalkar SG, Birkefeld LD, Goto KS. Solid-state gas sensors: A review, J Electrochem Soc 1992;139:3690–3704.

[5] Michel CR, Martínez AH, Jiménez S. Gas sensing response of nanostructured trirutile-type CoSb$_2$O$_6$ synthesized by solution-polymerization method, Sens Actuators B 2008;132:45–51.

[6] Lee SJ, Kriven WM. Crystallization and densification of nano-size amorphous cordier-ite powder prepared by a PVA solution-polymerization route, J Am Ceram Soc 1998;81:2605–2612.

[7] Michel CR, Delgado E, Martínez AH. Evidence of improvement in gas sensing properties of nanostructured bismuth cobaltite prepared by solution-polymerization method, Sens Actuators B 2007;125:389–395.

[8] Michel CR, López-Mena E, Martínez AH. Improvement of the gas sensing behavior in nanostructured Gd$_{0.9}$Sr$_{0.1}$CoO$_3$ by addition of silver, Mater Sci Eng B 2007;141:1–7.

[9] Li D, Komarneni S. Microwave-assisted polyol process for synthesis of Ni nanoparticles, J Am Ceram Soc 2006;89:1510–1517.

[10] Schroder DK. Semiconductor material and device characterization. New York: Wiley; 2006. p. 11.

[11] Moseley PT, Stoneham AM, Williams DE. Oxide semiconductors: patterns of response behaviour according to materials type, In: Moseley PT, Norris J, Williams DE, editors. Techniques and mechanisms in gas sensing. Bristol: Adam Hilger; 1991. p. 108–38.

[12] Bochenkov VE, Sergeev GB. Preparation and chemiresistive properties of nanostruc-tured materials, Adv Colloid Interface Sci 2005;116:245–254.

[13] Ishihara T, Kometani K, Mizuhara Y, Takita Y. Mixed oxide capacitor of CuO-BaTiO$_3$ as a new type CO$_2$ gas sensor, J Am Ceram Soc 1992;75:613–618.

CHAPTER 4

Capture of Carbon Dioxide by Modified Multiwalled Carbon Nanotubes

Chungsying Lu*, Bilen Wu*, Wenfa Chen*, Yu-Kuan Lin* *and* **Hsunling Bai****

* Department of Environmental Engineering, National Chung Hsing University, 250 Kuo Kuang, Taichung, Taiwan
** Institute of Environmental Engineering, National Chiao Tung University, Hsinchu, Taiwan

Contents

1. INTRODUCTION

The greenhouse gas carbon dioxide (CO_2) capture and storage (CCS) technologies from fossil fuel–fired power plant have received significant attention after the Kyoto Protocol came into force on February 16, 2005. Various CO_2 capture technologies including absorption, adsorption, cryogenics, membranes, etc. have been investigated [1, 2]. Among these, the absorption–regeneration technology has been recognized as the most developed process so far, with the amine-based or ammonia-based absorption processes having received the greatest attention [3–5].

However, as the energy penalty of the absorption process is still too high, other technologies are being investigated throughout the world. The Intergovernmental Panel on Climate Change (IPCC) special report concluded that the design of a full-scale adsorption process might be feasible and the development of a new generation of materials that would efficiently adsorb CO_2 will undoubtedly enhance the competitiveness of adsorptive separation in a flue gas application [6]. Possible adsorbents include activated carbon [7, 8], zeolites [9, 10], silica adsorbents [11, 12], single-walled carbon nanotubes [13], and nanoporous silica-based molecular basket [14, 15].

Carbon nanotubes (CNTs) are unique and one-dimensional macro-molecules that have outstanding thermal and chemical stability [16]. These nanomaterials have been proven to possess good potential as superior adsorbents for removing many kinds of organic and inorganic pollutants in air streams [17–19] or in aqueous environments [20–22]. The large adsorption capacity of pollutants by CNTs is mainly attributable to their pore structure and the existence of a wide spectrum of surface functional groups, which can be achieved by chemical modification or thermal treatment to make CNTs that possess optimum performance for particular purposes. Thus, a chemical modification of CNTs would be also expected to have good potential for greenhouse gas CO_2 capture. However, such studies are still limited in the literature.

This study investigates the physicochemical properties of raw and 3-aminopropyl-triethoxysilane (APTS)-modified CNTs and their adsorption properties of CO_2. Effects of temperature and moisture on the CO_2 adsorption in modified CNTs are also given.

2. MATERIALS AND METHODS

2.1. Adsorbents

Commercially available multiwalled CNTs with inner diameter less than 10 nm (L-type, Nanotech Port Co., Shenzhen, China) were selected as adsorbents in this study. The length of CNTs was in the range 5–15 μm and the amorphous carbon content in the CNTs was less than 5 wt%. These data were provided by the manufacturer.

Raw CNTs were thermally pretreated using an oven at 300 °C for 60 min. After thermal treatment, the CNTs were dispersed into flasks containing various kinds of chemical agents, including N-[3-(trimethoxysilyl)propyl] ethylenediamine (EDA), polyethylenimine (PEI; Acros Organics, analytical reagent, NJ, USA), and APTS (Riedel-de Häen, analytical reagent, Seelze, Germany), to determine the optimum modification of CNTs for enhancing CO_2 capture. Literature screening indicates that these chemical agents have been used to modify carbon adsorbents [13], zeolites [14, 23], or silica adsorbents [12, 24, 25]. The mixture was then shaken in an ultrasonic cleaning bath (model D400H, Delta Instruments Co., USA) for 20 min and was refluxed at 100 °C for 20 h to remove metal catalysts. After cooling to room temperature, the mixture was filtered through a 0.45-μm mylon fiber filter, and the solid was washed with deionized water until the pH of the filtrate was 7. The filtered solid was then dried at 70 °C for 2 h.

2.2. Adsorption Experiment

The experimental setup for CO_2 capture by CNTs is shown in Fig. 4.1. The adsorption column was made of Pyrex tube of length 20 cm and internal diameter 1.5 cm. The adsorption column was filled with 1.0 g of CNTs (packing height = 3.5 cm) and was placed within a temperature control box (model CH-502, Chin Hsin, Taipei, Taiwan) to maintain a temperature

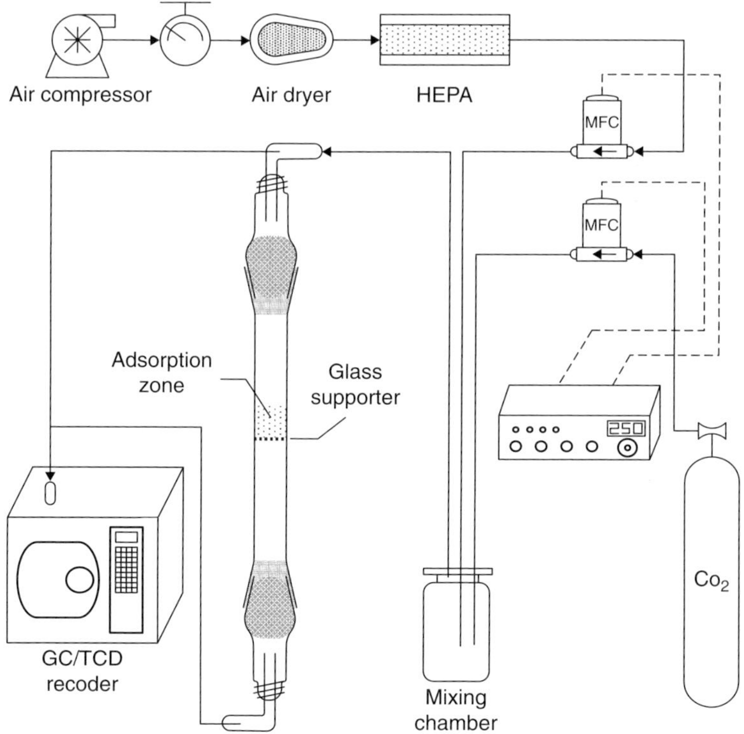

Figure 4.1   A diagram of the experimental setup.

at 25 °C with the exception of temperature effect study, in which the temperature range 5–45 °C (in a 10 °C increment) was tested.

Compressed air was passed through a silica-gel air dryer to remove moisture and oil and then was passed through a high-efficiency particulate air (HEPA) filter (Gelman Science, Ann Arbor, MI, USA) to remove particulates. The clean air was then served as a diluting gas and was mixed with pure CO_2 gas that was obtained from a pure CO_2 cylinder (99.9% purity) before entering the absorption column. The influent CO_2 concentration and the system flow rate were controlled using mass flow controllers (model 247C four-channel readout and model 1179A, MKS instruments Inc., MA, USA). The mixed air was then passed downward into the adsorption column. The influent and effluent air streams were flowed into a gas chromatograph (GC) equipped with a thermal conductivity detector (TCD) by an auto sampling system. The variations in the influent CO_2 concentration were below 0.2%, and the system flow rate was controlled at 0.08 lpm that is equal to an empty-bed retention time of 4.6 s.

The relative humidity was kept at 0% with the exception of moisture effect study, in which the relative humidity range 0–100% was tested. Moisture was introduced into the air stream by dispersing the diluting gas through a water bath before being mixed with pure CO_2 gas.

2.3. Adsorption Capacity

The amount of CO_2 adsorbed on CNTs (q, mg/g) was calculated as

$$q = \frac{1}{m} \int_0^t Q \times (C_{\mathrm{in}} - C_{\mathrm{eff}}) dt \tag{1}$$

where m is the mass of virgin adsorbents (g), t is the contact time (min), Q is the influent flow rate (lpm), and C_{in} and C_{eff} are the influent and effluent CO_2 concentrations (mg/l).

2.4. Analytical Methods

CO_2 concentration in the air stream was determined using a GC-TCD (model GC-2010, Shimadzu Instruments, Japan). A 30-m fused silica capillary column with inner diameter 0.32 mm and film thickness 5.0 μm (AB-PLOT GasPro, USA) was used for CO_2 analysis. The GC-TCD was operated at an injection temperature of 50 °C, a detector temperature of 100 °C, and an oven temperature of 55 °C.

The structural information of CNTs was evaluated by a Raman spectrometer (model Nanofinder 30 R., Tokyo Instruments Inc., Japan). The carbon content in CNTs was determined using a thermogravimetric analyzer (model TG209 F1 Iris, NETZSCH, Bavaria, Germany).

The physical properties of CNTs were determined by N_2 adsorption at 77 K using ASAP 2010 surface area and porosimetry analyzer (Micromeritics Inc., Norcross, GA, USA). N_2 adsorption isotherms were measured at a relative pressure range 0.0001–0.99. The adsorption data were then used to determine the surface area of CNTs using the Brunauer–Emmett–Teller (BET) equation, whereas the pore size distributions (PSDs) of CNTs were determined from the N_2 adsorption data using the Barrett–Johner–Halenda (BJH) equation.

The chemical properties of CNTs were determined using a Fourier transform infrared spectroscopy equipped with an attenuated total reflectance (FTIR/ATR; model FTIR-SP-1 Spectrum One, Perkin Elmer, Japan).

3. RESULTS AND DISCUSSION

3.1. Adsorption of CO_2 onto Various Modified Adsorbents

Figure 4.2 shows the q_e of 10% CO_2 with various modified CNTs. The q_e increased after the CNTs were modified by EDA, PEI, and APTS solutions. Because these solutions are polymers with many amine groups, which can react with CO_2 to form carbamate in the absence of water [15], and therefore the q_e increases. The APTS-modified CNTs have greater

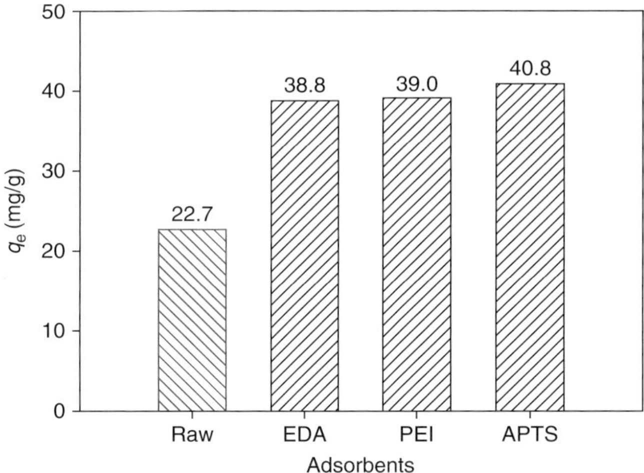

Figure 4.2 Equilibrium amount of 10% CO_2 adsorbed on raw and various modified CNTs.

enhancement in q_e than the EDA- and PEI-modified CNTs. The raw and APTS-modified CNTs were thus selected as adsorbents to study their physicochemical characterizations and adsorption properties of CO_2.

3.2. Characterizations of Raw and Modified CNTs

Figure 4.3 shows the adsorption and desorption isotherms of N_2 onto raw and modified CNTs. It is apparent that the raw CNTs have a more adsorption capacity of N_2 than the modified CNTs, indicating that a less amount of porosity within modified CNTs due to attachment of APTS solutions on the surface of modified CNTs. The adsorption isotherms for raw and modified CNTs is type IV, showing an increase in N_2 adsorption capacity with increasing relative pressure. This reflects that the raw and modified CNTs have a broad pore size range.

Figure 4.4 shows the PSDs of raw and modified CNTs. It is noted that the PSDs of raw CNTs display a bimodal distribution, including a fine fraction (2–4 nm width range) and a coarse fraction (10–30 nm width range). The pores in the fine fraction are the CNT inner cavities, close to the inner CNT diameter, while the pores in the coarse fraction are likely to be contributed by aggregated pores that are formed within the confined space among the isolated CNTs. The PSDs of modified CNTs only show a coarse fraction in the 7–12 nm width range, and the pore volumes of CNTs become small after modification. The low detection limit for the pore size of employed BET analyzer is approximately 2 nm. Therefore, a

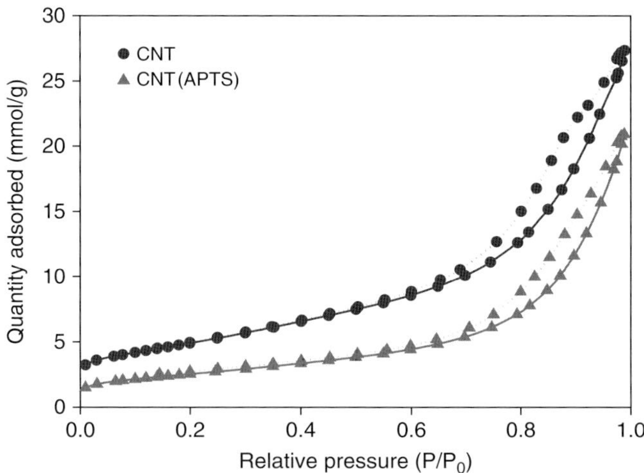

Figure 4.3   Adsorption (solid line) and desorption (dash line) isotherms of N_2 on raw and APTS-modified CNTs.

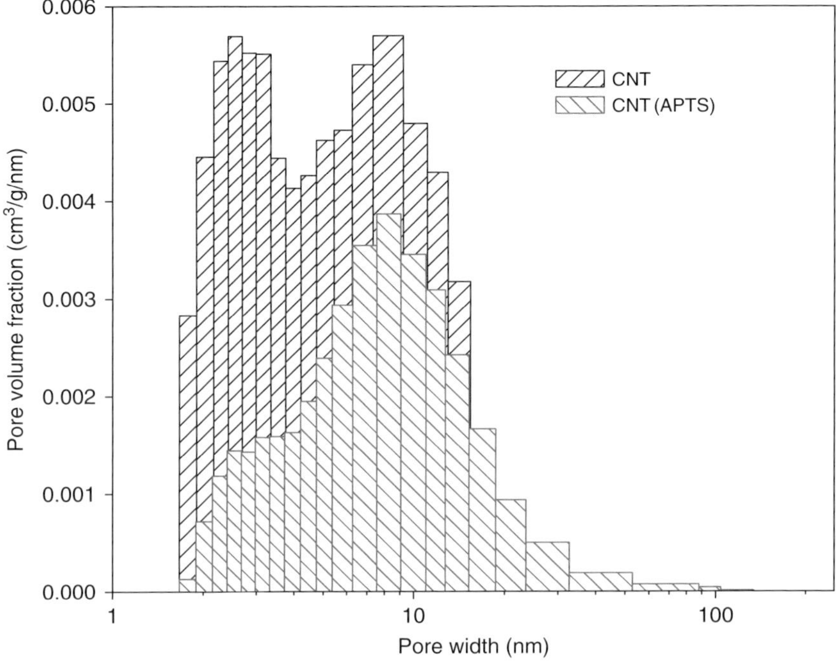

Figure 4.4 PSDs of raw and APTS-modified CNTs.

quantification of the pore size of adsorbents less than 2 nm is not possible in this study.

The physical properties of adsorbents are given in Table 4.1. The modified CNTs have smaller BET surface area and pore volume but larger average pore diameter than the raw adsorbents. Because the entrance of small pores of modified CNTs is covered with APTS solutions and thus leads to a smaller surface and pore volume but a larger average pore diameter. Most pore volumes of modified CNTs are in the 5–20 nm pore size range.

Figure 4.5 shows the Raman spectra of raw and modified CNTs. It is clear that there are two sharp peaks. The peak located between 1330 and 1360 cm^{-1} is the D band that is related to disordered sp^2-hybridized carbon atoms of nanotubes containing vacancies, impurities, or other symmetry-breaking defects. The peak near 1580 cm^{-1} is the G band that is related to graphite E$_{2g}$ symmetry of the interlayer mode, reflecting structural integrity of sp^2-hybridized carbon atoms of the nanotubes. Hence, the extent of carbon–containing defects of adsorbents can be evaluated by intensity ratios of D band to G band (I_D/I_G) [26]. The I_D/I_G ratios of raw and modified CNTs are 0.461 and 0.402, respectively. The I_D/I_G ratio of raw CNTs is slightly lower than that of

Table 4.1 Physical properties of adsorbents

Adsorbents	BET SA (m²/g)	APD (nm)	PV cm³/g	% of total Pore volume in stated pore size (nm) range		
				<5	5–20	>20
CNT	394	8.9	0.91	54.2	43.9	1.9
CNT(APTS)	198	12.2	0.63	31.7	63.5	4.8

Note: SA = surface area; APD = average pore diameter; PV = pore volume.

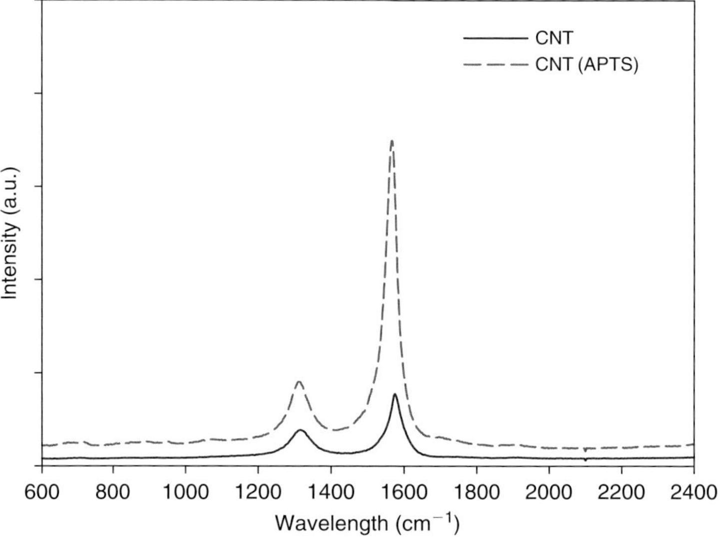

Figure 4.5 Raman spectra of raw and APTS-modified CNTs.

modified CNTs, indicating that the CNTs possess less graphitized structures and more carbon-containing defects after modification.

Figure 4.6 shows the TG curves of raw and modified CNTs. It is obvious that the TG curve of raw CNTs is considerably stable and shows a little weight loss close to 1% below 450 °C. After that, a significant weight loss begins and ends at 670 °C, in which 4.29% remaining weight was found. The modified CNTs have a broader temperature range for weight loss and exhibit three main weight loss regions. The first weight loss region (<550 °C) can be attributed to the loss of various kinds of surface

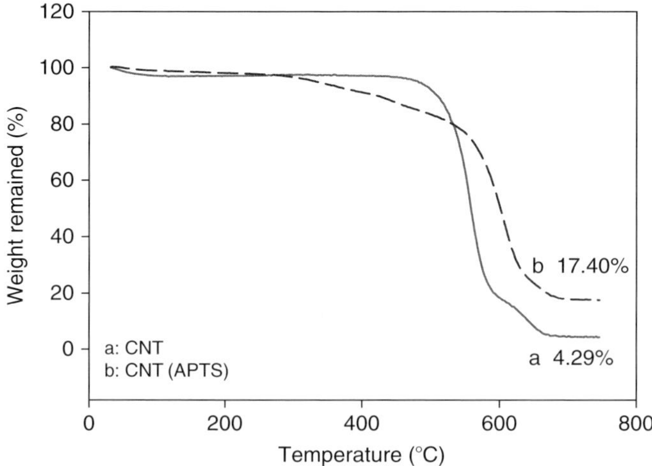

Figure 4.6   TG curves of raw and APTS-modified CNTs.

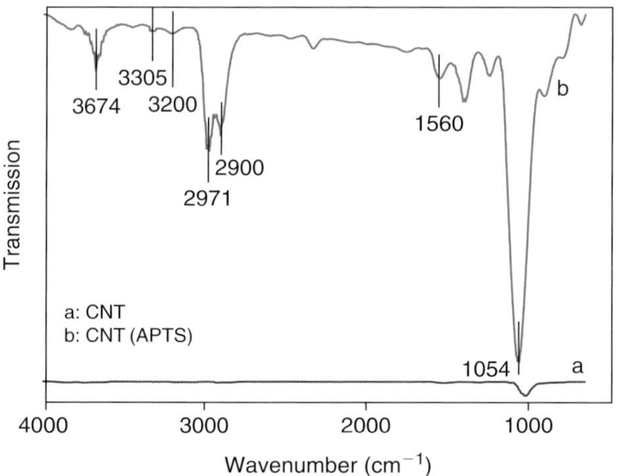

Figure 4.7   IR spectra of raw and APTS-modified CNTs.

functional groups. The rapid weight loss region (550–620 °C) can be attributed to the decomposition of CNTs. The third region only shows a very little weight loss close to 1%, in which 17.4% remaining weight was observed. These results showed that after modification, the carbon content in CNTs decreased from 95.71 to 82.6% due to attachment of APTS solutions on the surface of CNTs.

Figure 4.7 shows the IR spectra of raw and modified CNTs. It is observed that the IR spectra of raw CNTs show no significant peaks. In contrast, the

IR spectra of modified CNTs show several significant bands at 3674–3702, 3101–3305, 2900–2971, 1530–1560, 1240 and 1020–1081 cm^{-1}, which are associated with bridge Si–OH acidic groups [27], asymmetric and symmetric NH$_2$ stretching –OH, CH stretching from CH$_2$CH$_2$CH$_2$–NH$_2$ groups, NH$_2$ deformation of hydrogen-bonded amino group, CH$_3$ asymmetric deformation of Si–CH$_3$, and Si–O–Si lattice vibrations [28], respectively. These functional groups are abundant on the surface of adsorbents, which can provide many chemical adsorption sites for CO$_2$ capture.

3.3. Breakthrough Curves

Figure 4.8 shows the breakthrough curves of 50% CO$_2$ adsorption on raw and modified CNTs. CO$_2$ gas can be efficiently adsorbed on CNTs. The breakpoint time, which is defined as the time at which the breakthrough curve first begins to rise appreciably, and the breakthrough time of CO$_2$ adsorption, respectively, are 1 and 2 min for the raw CNTs and 2 and 4 min for the modified CNTs. Both breakpoint time and breakthrough time become longer after modification by APTS solutions.

3.4. Adsorption Isotherms

Figure 4.9 shows the adsorption isotherms of 5–50% CO$_2$ with raw and modified CNTs. It is seen that the equilibrium adsorption capacity (q_e) increased with an increase in C_{in} and was greatly enhanced after the

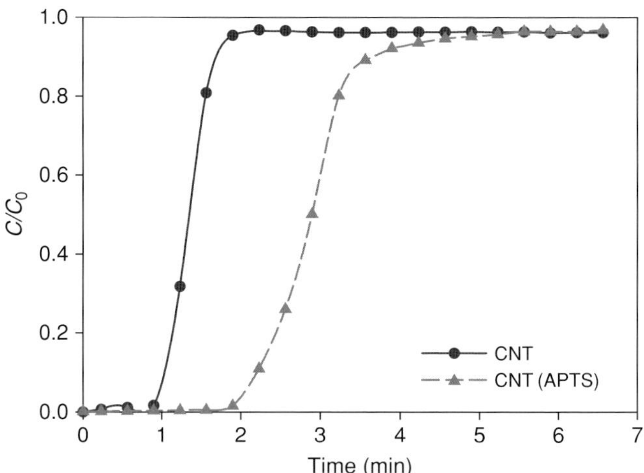

Figure 4.8   Breakthrough curves of CO$_2$ with raw and APTS-modified CNTs.

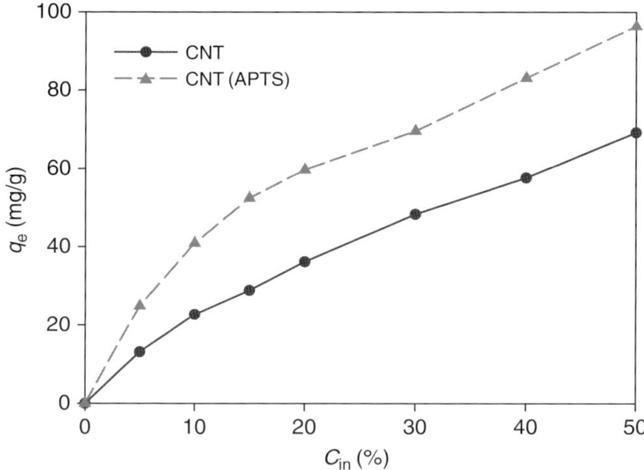

Figure 4.9   Adsorption isotherms of CO_2 with raw and APTS-modified CNTs.

CNTs were modified by APTS solutions. With the 50% CO_2 inlet, the q_e of raw and modified CNTs is 69.2 and 96.3 mg/g, respectively, in which 30.1 mg/g enhancement in CO_2 adsorption was obtained.

3.5. Temperature Effect

Figure 4.10 shows the adsorption isotherms of 5–50% CO_2 with modified CNTs at various temperatures. The q_e decreased with an increase in temperature and increased q_e was found at 5 °C, indicating the exothermic nature of adsorption process. As the temperature increased from 5 to 45 °C, the q_e decreased from 101 to 80.9 mg/g. The adsorption isotherm curves at 5–25 °C are relatively close, reflecting that the effects of temperature change on the q_e are less significant from 5 to 25 °C than from 25 to 45 °C.

3.6. Moisture Effect

Figure 4.11 shows the q_e of 50% CO_2 with modified CNTs at various relative humidities (RHs). It is apparent that the q_e increased with an increase in relative humidity. There are two possible reasons to explain the increase in q_e in the presence of moisture [15]. First, with the adsorption of water on the surface of modified CNTs, CO_2 gas may dissolve into water. Second, the chemical interaction between CO_2 and APTS may be further enhanced to form bicarbonate in the presence of water. As the RH increased from 0 to 100%, the q_e increased from 96.3 to 100 mg/g.

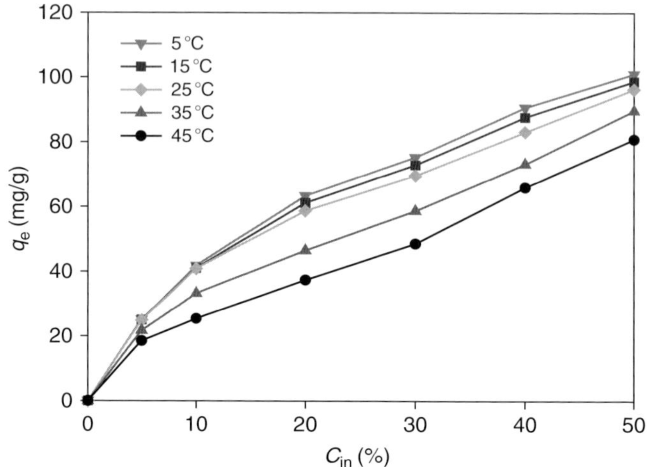

Figure 4.10   Effect of temperature on the q_e of APTS-modified CNTs.

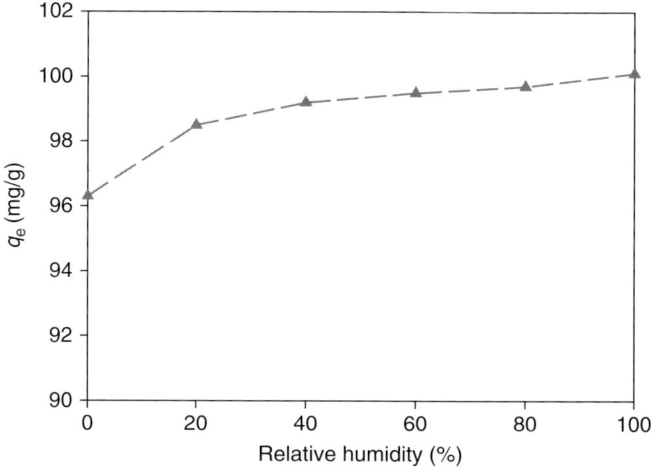

Figure 4.11   Effect of relative humidity on the q_e of APTS-modified CNTs.

3.7. Comparisons with Literature Results

The comparisons of q_e with various APTS-modified adsorbents are given in Table 4.2. It is apparent that the q_e was greatly enhanced after these adsorbents were modified by APTS solutions. The q_e of APTS-modified CNTs is much higher than those of many types of APTS-modified silica adsorbents reported in the literature. These comparisons suggest that the APTS-modified CNTs are promising adsorbents for the capture of CO_2 from air streams.

Table 4.2 Comparisons of CO_2 adsorption performance with various APTS-modified adsorbents

Adsorbents	Modified chemicals	q_e (mg/g)	Conditions	References
CNTs	APTS	96.3	C_{in}:50%, T:25 °C	This study
SBA–15	APTS	6.6	C_{in}:15%, T:60 °C	[12]
SBA–15	APTS	9.9	C_{in}:4%, T:25 °C	[24]
Silica xerogel	APTS	25	C_{in}:5%, T:25 °C	[29]
Silica gel–40	APTS	30	C_{in}:90%, T:20 °C	[30]

4. CONCLUSIONS

The raw and APTS-modified CNTs were selected as adsorbents to study their characterizations and adsorption properties of CO_2 from air streams. The physicochemical properties of CNTs were improved after modification, including the increase in defective structure and surface functional groups, which made CNTs adsorb more CO_2. With the 50% CO_2 inlet, the amount of adsorbed CO_2 on raw and modified CNTs was 69.2 and 96.3 mg/g, respectively. The CO_2 adsorption performance of modified CNTs decreased with an increase in temperature but increased with an increase in relative humidity. The modified CNTs possess higher CO_2 adsorption capacity than many types of APTS-modified silica adsorbents reported in the literature, suggesting that the APTS-modified CNTs are promising adsorbents for the capture of CO_2 from air streams.

ACKNOWLEDGMENT

This work was supported by the National Science Council, Taiwan, under a contract no. NSC95-2623-7-009-014.

References

[1] White CM, Strazisar BR, Granite EJ, Hoffman JS, Pennline HW. Separation and capture of CO_2 from large stationary sources and sequestration in geological formations-coalbeds and deep saline aquifers. J Air Waste Manage Assoc 2003;53:645–715.
[2] Aaron D, Tsouris C. Separation of CO_2 from flue gas: a review. Sep Sci Technol 2005; 40:321–48.

[3] Bai HL, Yeh AC. Removal of CO_2 greenhouse gas by ammonia scrubbing. Ind Eng Chem Res 1997;36:2490–3.

[4] Yeh AC, Bai HL. Comparison of ammonia and monoethanolamine solvents to reduce CO_2 greenhouse gas emissions. Sci Total Environ 1999;228:121–33.

[5] Rao AB, Rubin ES. A technical, economic, and environmental assessment of amine-based CO_2 capture technology for power plant greenhouse gas control. Environ Sci Technol 2002;36:4467–75.

[6] Intergovernmental Panel on Climate Change (IPCC). Special report on carbon dioxide capture and storage. <http://www.ipcc.ch/activity/srccs/index.htm>; Sep. 22 2005.

[7] Siriwardane RV, Shen MS, Fisher EP, Losch J. Adsorption of CO_2 on molecular sieves and activated carbon. Energy Fuels 2001;15:279–84.

[8] Prezepiórski J, Skrodzewicz M, Morawski AW. High temperature ammonia treatment of activated carbon for enhancement of CO_2 adsorption. Appl Surf Sci 2004;225:235–42.

[9] Gao W, Butler D, Tomasko DL. High-pressure adsorption of CO_2 on NaY zeolite and model prediction of adsorption isotherms. Langmuir 2004;20:8083–9.

[10] Siriwardane RV, Shen MS, Fisher EP, Poston JA. Adsorption of CO_2 on zeolites at moderate temperatures. Energy Fuels 2005;19:1153–9.

[11] Gray ML, Soong Y, Champagne KJ, Pennline H, Baltrus JP, Stevens Jr RW, Khatri R, Chuang SSC, Filburn T. Improved immobilized carbon dioxide capture sorbents. Fuel Process Technol 2005;86:1449–55.

[12] Hiyoshi N, Yogo K, Yashima T. Adsorption characteristics of carbon dioxide on organically functionalized SBA-15. Microporous Mesoporous Mater 2005;84:357–65.

[13] Cinke M, Li J, Bauschlicher Jr CW, Ricca A, Meyyappan M. CO_2 adsorption in single-walled carbon nanotubes. Chem Phys Lett 2003;376:761–6.

[14] Xu XC, Song C, Andresen JM, Miller BG, Scaroni AW. Novel polyethylenimine-modified mesoporous molecular sieve of MCM-41 type as high-capacity adsorbent for CO_2 capture. Energy Fuels 2002; 16:1463–9.

[15] Xu XC, Song CS, Miller BG, Scaroni AW. Influence of moisture on CO_2 separation from gas mixture by a nanoporous adsorbent based on polyethylenimine-modified molecular sieve MCM-41. Ind Eng Chem Res 2005;44:8113–9.

[16] Smart SK, Cassady AI, Lu GQ, Martin DJ. The biocompatibility of carbon nanotubes. Carbon 2006;44:1034–47.

[17] Long RQ, Yang RT. Carbon nanotubes as superior sorbent for dioxin removal. J Am Chem Soc 2001;123:2058–9.

[18] Agnihotri S, Rood MJ, Rostam-Abadi M. Adsorption equilibrium of organic vapors on single-walled carbon nanotubes. Carbon 2005;43:2379–88.

[19] Hsu S, Lu C. Modification of single-walled carbon nanotubes for enhancing isopropyl alcohol vapor adsorption from air streams. Sep Sci Technol 2007;42:2751–66.

[20] Lu C, Chung YL, Chang KF. Adsorption of trihalomethanes from water with carbon nanotubes. Wat Res 2005;39:1183–9.

[21] Peng X, Li Y, Luan Z, Di Z, Wang H, Tian B, Jia Z. Adsorption of 1,2-dichlorobenzene from water to carbon nanotubes. Chem Phys Lett 2003;376:154–8.

[22] Rao GP, Lu C, Su F. Adsorption of divalent heavy metal ions from aqueous solutions by carbon nanotubes: a review. Sep Purif Technol 2007;58:224–31.

[23] Xu X, Song C, Andresen JM, Miller BG, Scaroni AW. Preparation and characterization of novel CO_2 "molecular basket" adsorbents based on polymer-modified mesoporous molecular sieve MCM-41. Microporous Mesoporous Mater 2003;62:29–45.

[24] Chang ACC, Chuang SSC, Gray M, Soong Y. In-situ infrared study of CO_2 adsorption on SBA-15 grafted with r-(aminopropyl)triethoxysilane. Energy Fuels 2003;17: 468–73.

[25] Zheng F, Tran DN, Busche BJ, Fryxell GE, Addleman RS, Zemanian TS, Aardahl CL. Ethylenediamine-modified SBA-15 as regenerable CO_2 sorbent. Ind Eng Chem Res 2005;44:3099–105.

[26] Tsai CL, Chen CF. Characterisation of bias-controlled carbon nanotubes. Diamond Relat Mater 2003;12:1615–20.

[27] Ha VTT, Tiep LV, Meriaudeau P, Naccache C. Aromatization of methane over zeolite supported molybdenum: active sites and reaction mechanism. J Mol Catal A: Chem 2002;181:283–90.

[28] Jing SY, Lee HJ, Choi CK. Chemical bond structure on Si-O-C composite films with a low dielectric constant deposited by using inductively coupled plasma chemical vapor deposition. J Korean Phys Soc 2002;41:769–73.

[29] Huang HY, Yang RT, Chinn D, Munson CL. Amine-grafted MCM-48 and silica xerogel as superior sorbents for acidic gas removal from natural gas. Ind Eng Chem Res 2003;42:2427–33.

[30] Knowles GP, Graham JV, Delaney SW, Chaffee AL. Aminopropyl-functionalized mesoporous silicas as CO_2 adsorbents. Fuel Process Technol 2005;86:1435–48.

Kinetics, Thermodynamics, and Regeneration of BTEX Adsorption in Aqueous Solutions via NaOCl-Oxidized Carbon Nanotubes

Fengsheng Su, Chungsying Lu, Kelvin R. Johnston *and* Sukai Hu
Department of Environmental Engineering, National Chung Hsing University, Taichung, Taiwan

Contents

1. INTRODUCTION

The BTEX volatile organic compounds benzene (B), toluene (T), ethyl-benzene (E), and p-xylene (X) are important industrial solvents frequently encountered as industrial contaminants. Every year, these volatile organic compounds, used in gasoline, printing industry, leather industries, and rubber manufacture [1], produce a large amount of BTEX contaminated wastewater that is discharged into the aqueous environment. According to the US Environmental Protection Agency (EPA), the most likely source of BTEX pollution in the environment is due to leakage from underground gasoline storage tanks into ground water [2], leaching from landfills, and discharge from factories and refineries [1]. Even though BTEX is relatively insoluble in water, levels have been recorded at up to 1000 mg/L [3]. This is much higher than the allowed maximum contaminant level of 0.005 mg/L for B, 1 mg/L for T, 0.7 mg/L for E, and a total of 10 mg/L for all three forms of X [1–4]. Since BTEX solutions are flammable, toxic, and carcinogenic substances, the presence of excessive amounts in aqueous surroundings may lead to an adverse impact on water quality and thus endanger public health and welfare [5]. Therefore, the removal of BTEX before releasing into the environment is necessary to meet the growing demand for cleaner water, which has necessitated the development of innovative, cost-effective alternatives for treatment.

There are many conventional methods that are being used to remove the BTEX in wastewater treatment including adsorption, aeration, biological oxidation, and chemical oxidation. Among them, the promising process for the removal of BTEX from wastewater is adsorption, because the used adsorbent can be regenerated by suitable desorption process and it is highly effective and economical. The most widely used adsorbents for BTEX removal are activated carbon [6–8] and zeolite [9]. However, these adsorbents suffer from slow kinetics and low adsorption capacities of BTEX. Therefore, researchers carried out investigation for new promising adsorbents.

Carbon nanotubes (CNTs), a new and exciting part of nanotechnology, are unique and one-dimensional macromolecules that possess thermal and chemical stability [10]. They have been proven to possess great potential as superior adsorbents for removing many kinds of organic pollutants such as dioxin [11] and volatile organic compounds [12–16] from air streams or natural organic matters [17–21] and trihalomethanes [22, 23], xylene [24], resorcinol [25], dyes [26–28], 1,2-dichlorobenzene [29, 30], chlorophenol [31], poly aromatic hydrocarbons [32–42], and herbicides [43] from aqueous solutions, which are summarized in Table 5.1. Compared with other adsorbents such as activated carbon by the foregoing researchers, it is suggested that the CNTs are a promising adsorbent

Table 5.1   Adsorption of organic compounds on CNTs

	Adsorbates	Adsorbents	Adsorption capacity (mg/g)	References
Potable water	Fulvic	MWCNTs	24.57	[17]
	NOM	MWCNTs	26.1	[18, 19]
	Assimilable organic carbon		0.54	
	Humic acid	SWCNTs	121.95	[20]
		MWCNTs	107.53	
	THMs	MWCNTs		[22, 23]
	$CHCl_3$		2.41	
	$CHBrCl_2$		1.23	
	$CHBr_2Cl$		1.08	
	$CHBr_3$		0.92	
Wastewater	o-xylene	SWCNTs	68	[24]
	p-xylene		85	
	Resorcinol	MWCNTs		[25]
	Phenol		15.88	
	Catechol		14.20	
	Pyrogallol		33.61	
	Procion Red MX-5B	MWCNTs	44.64	[26]
	C.I. Reactive Red 2	MWCNTs	26.5	[27]
	1,2-Dichloro-benzene	CNTs	30.8	[29]
		Thermal-treated CNTs	28.7	
	Chlorophenol	NH_3-MWCNTs	110.3	[31]

Continued

Table 5.1   *Continued*

	Adsorbates	Adsorbents	Adsorption capacity (mg/g)	References
	Pyrene ($C_{16}H_{10}$)	MWCNTs	40	[32–34]
	Phenanthrene		43	
	Naphthalene		55	
	Cyclohexanol	MWCNTs	1.580 mmol/g	[35]
	Phenol		1.770 mmol/g	
	Catechol		1.850 mmol/g	
	Pyrogallol		1.780 mmol/g	
	2-Phenylphenol		1.290 mmol/g	
	1-Naphthol		1.500 mmol/g	
	Microcystins	MWCNTs		[36]
	MC–RR		8.1	
	MC–LR		4.5	
	Uraemic toxic	MWCNTs	45	[37]
Herbicides	Atrazine	SWCNTs	36	[43]
		MWCNTs	142	

Note: MWCNTs = multiwalled carbon nanotubes; NOM = natural organic matter; SWCNTs = single-walled carbon nanotubes; THMs = trihalomethanes.

for the removal of organic compounds. Recently, Lu et al. [44] conducted the surface modification of CNTs by NaOCl, HNO_3, H_2SO_4, and HCl solutions and found that the NaOCl-oxidized CNTs are efficient BTEX adsorbents and that they possess good potential applications in wastewater treatment.

Although the NaOCl-oxidized CNTs show promising for BTEX removal in wastewater treatment, the relatively high unit cost of CNTs currently restricts their practical use. The production and the use of nonfunctionalized multiwalled CNTs have now reached industrial levels, but the price of CNTs is still too high to be used in the field. Therefore, if the regeneration of CNTs is not carried out, the practical use of CNTs in wastewater treatment is not feasible. If the regeneration of CNTs is performed, the CNTs should be successfully regenerated and reused through a number of adsorption and regeneration cycles and then can be cost-effective adsorbents.

In this article, NaOCl-oxidized CNTs were used as novel adsorbents to study kinetics, thermodynamics, and regeneration of BTEX adsorption in aqueous solutions. A comparative look at cyclic BTEX adsorption between CNTs and granular activated carbon (GAC) was conducted to evaluate their repeated availability in wastewater treatment. A statistic analysis on the replacement cost of these adsorbents was also given.

2. EXPERIMENTAL/MATERIALS AND METHODS

2.1. Adsorbents

CNTs were fabricated by the catalytic chemical vapor deposition method [45]. The catalyst was prepared by dissolving 2.5 wt% of iron acetate ($Fe(CH_3COO)_2$) and 2.5 wt% of cobalt nitrate ($Co(NO_3)_2$) into 25 mL of deionized water and then mixed with a commercially available zeolite (CVB100, Zeolyst International, Vally Forge, USA). The mixture was constantly stirred to obtain the resulting semisolid mixture, which was subsequently dried overnight at 140 °C and grounded into a fine powder (as-prepared catalyst).

The physical properties of as-prepared catalyst revealed that surface area, pore volume, and average pore diameter of micropores (MPs) are 1458 m^2/g, 0.332 cm^3/g, and 0.23 nm, respectively. Most pore volumes are in the size range of 0.2–0.3 nm. Because of its high surface area, small pore diameter, and narrow pore size distribution, the as-prepared catalyst appears beneficial to the production of good-quality CNTs.

The as-prepared catalyst (~1.5 g) was evenly distributed over quartz boats (3.5 cm i.d. and 16.5 cm in length), which were placed in a quartz tube and located in the central position of a horizontal furnace. The quartz boat was activated by passing N_2 gas through the reaction chamber at 120 mL/min for 15 min. The reaction chamber was heated to 600 °C at a heating rate of 10 °C/min and then followed by passing acetylene (C_2H_2) gas at 40 mL/min and N_2 gas at 295 mL/min through the reaction chamber for 1 h.

After the thermal reaction, the furnace was cooled to room temperature by the passage of N_2 gas at 295 mL/min. The black powders (as-grown CNTs) were collected from the quartz boat. In this way, the CNTs can be fabricated in gram quantities (~3 g of CNTs per hour).

The as-grown CNTs were treated by 30% NaOCl (70 mL of H_2O + 3 mL of NaOCl) solution, which has been found to possess better performance for BTEX adsorption than those oxidized by HCl, HNO_3, and H_2SO_4 solutions in a previous study [44]. The mixture was heated at 100 °C and refluxed for 2 h to remove metal catalysts (Fe and Co nanoparticles). After cooling to room temperature, the mixture was filtered through a 0.45 μm fiber filter. The filtered solid was washed with deionized water until the pH of filtrate was 7 and then dried at 200 °C for 2 h. The weight loss of CNTs after the treatment was <5%.

In order to compare the performance of BTEX adsorption on CNTs with other adsorbents, a commercially available GAC (Filtrasorb 400, Calgon Carbon Co., Tianjia, China) was chosen because of its wide use for removal of organic compounds in wastewater treatment.

2.2. Adsorbates

The used B, T, E, and X were analytical grade with >99% purity and purchased from Merck (Darmstadt, Germany, for B and T; Hohenbrunn, Germany, for E and X). These chemicals were diluted using deionized H_2O to the desired concentrations.

2.3. Batch Adsorption Experiments

The batch adsorption experiments were conducted using 110 mL glass bottles with addition of 60 mg of adsorbents and 100 mL of BTEX solutions at initial concentrations (C_0) of 20–200 mg/L, which were chosen to be representative of BTEX concentration range in an industrial wastewater. The glass bottles were sealed with 20 mm stopper and mounted on a shaker, which was operated at 160 rpm. The experiments were carried out at 5, 25, and 45 °C in a temperature-controlled box (CH-502, Chin Hsin, Taipei, Taiwan). The choice of temperature range is to simulate possible water temperature in the field. All the experiments are repeated two times, and only the mean values were reported. The maximum deviation was <3%. Blank experiments, without the addition of adsorbents, were also conducted to ensure that the decrease in BTEX concentration was actually caused by the adsorption on adsorbents rather than by the adsorption on glass bottle wall or via volatilization.

The adsorption capacity of BTEX on adsorbents (q, mg/g) was calculated as follows:

$$q = \left(C_0 - C_t \right) \times \frac{V}{m} \tag{1}$$

where C_0 and C_t are the BTEX concentrations at the beginning and after a certain period of time (mg/L), V is the initial solution volume (L), and m is the adsorbent weight (g).

2.4. Adsorption/Desorption Experiments

To evaluate the repeated availability of BTEX adsorption on adsorbents, 60 mg of adsorbents was added to 100 mL of solution at a C_0 of 200 mg/L. As adsorption reached equilibrium, the adsorption capacity of BTEX on adsorbents was measured (q_e), and then, the solution was filtered using a 0.45 μm nylon fiber filter to regenerate adsorbents. The filtered adsorbents were added into 50 mL of 10–50% NaOCl solution and 0.01–0.05 M HNO_3 solution to determine the optimum regeneration agent, strength, and time for effective BTEX desorption. The adsorption/desorption process was repeated for 10 cycles. The weight loss of CNTs and GAC after regeneration in each cycle was measured, while the pore structure and the surface functional groups of CNTs before and after 10 cycles of adsorption and regeneration were characterized by their physical properties and infrared (IR) spectra. The recovery is defined as the percentage ratio of the q_e of the regenerated adsorbents to that of the virgin adsorbents.

2.5. Analytical Methods
2.5.1. Benzene, Toluene, Ethylbenzene, and *p*-Xylene

BTEX samples were analyzed by headspace solid phase microextraction [46] and quantified via gas chromatograph (GC) (SRI 8610C, SRI Instruments, CA, USA) with flame ionization detection (FID). Microextraction fiber (Supelco, Bellefonte, USA), which was coated with polydimethylsiloxane/divinylbenzene/carboxen film, was pushed out and exposed directly to the headspace above the sample for 20 min. After extraction, the fiber was immediately inserted into GC injector for desorption and analysis. The GC-FID was operated at injection temperature of 260 °C and detector temperature of 200 °C. The following temperature program was used: 50 °C for 8 min and 4 °C/min to 100 °C for 5 min.

2.5.2. Characterization of Adsorbents

The physical properties of adsorbents were determined by N_2 adsorption/desorption at 77 K using Micromeritics ASAP 2020 volumetric sorption analyzer (Norcross, GA, USA). N_2 adsorption/desorption isotherms were measured at a relative pressure range (P/P_0) of 0.0001–0.99 and then used to determine surface area, pore volume, and average pore diameter of adsorbents via the Barrett–Johner–Halenda equation for pore size range of 1.7–100 nm and the MP equation for pore size below 1.7 nm. The surface functional groups of adsorbents were detected by a Fourier transform infrared ray spectrometer (Spectrum One, Perkin Elmer, MA, USA).

3. RESULTS AND DISCUSSION

3.1. Effect of Contact Time and Temperature

Figure 5.1 shows the effect of contact time and temperature on the BTEX adsorption at a C_0 of 200 mg/L. For all experiments, the adsorption capacity

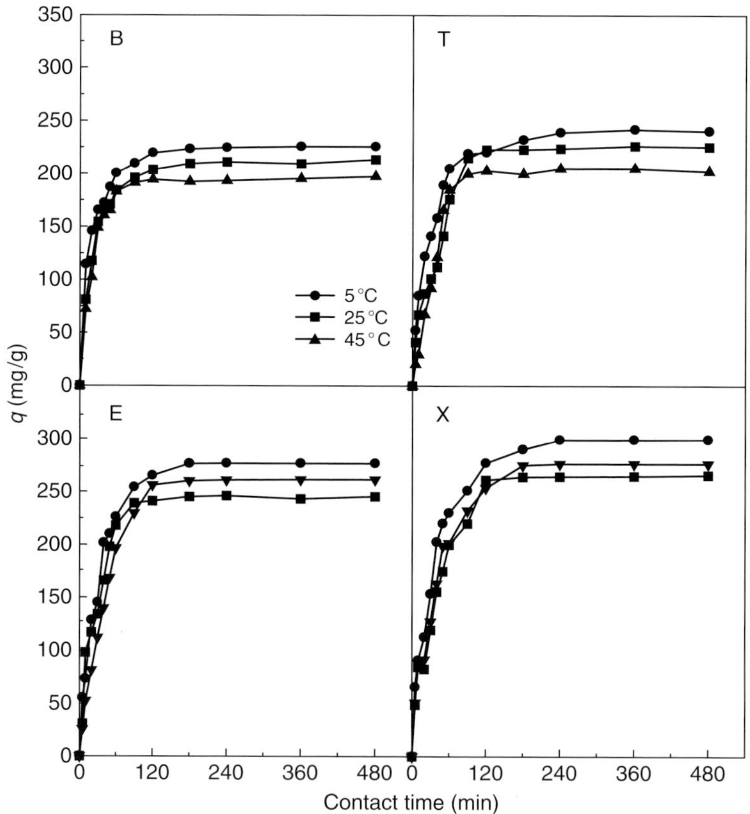

Figure 5.1   Effects of contact time and temperature on BTEX adsorption on CNTs.

of BTEX (q) increased quickly with time and then slowly reached equilibrium. The equilibrium times at 5, 25, and 45 °C are, respectively, 180, 180, and 120 min for B; 240, 120, and 90 min for T; 180, 120, and 90 min for E; and 240, 180, and 120 min for X. It is observed that the equilibrium would be reached faster at higher temperatures, which could be explained by the fact that increasing temperature results in a rise in the diffusion rate of the adsorbates across the external boundary layer and within the CNT pores due to the result of decreasing solution viscosity. A contact time of 240 min, which can assure the attainment of adsorption equilibrium for all the tests, was thus selected in the following tests.

The equilibrium adsorption capacities (q_e) at 5, 25, and 45 °C are, respectively, 226, 213, and 197 mg/g for B; 240, 225, and 205 mg/g for T; 276, 256, and 240 mg/g for E; and 299, 275, and 261 mg/g for X. The q_e decreased with temperature, indicating the exothermic nature of adsorption process. Favorable adsorption of order of X > E > T > B was observed in all tested temperatures which may be due to the increase in molecular weight (B, 78 g < T, 92 g < E, X, 106 g), the decrease in solubility (B, 790 mg/L > T, 530 mg/L > E, 152 mg/L > X, insoluble) [47], and the increase in boiling point (B, 80.1 °C < T, 110.7 °C < E, 136.2 °C < X, 144 °C) [25].

3.2. Adsorption Kinetics

To analyze the rate of BTEX adsorption on CNTs, Lagergren's first-order rate equation was used [48]:

$$\ln \frac{q_e - q_t}{q_e} = -k_1 t \qquad (2)$$

where q_t is the adsorption capacity of BTEX on CNTs at time t (mg/g) and k_1 is the first-order rate constant (1/min).

The k_1 at 5, 25, and 45 °C, determined from the slope of a linear plot of $\ln[(q_e - q_t)/q_e]$ versus t, are, respectively, 0.029, 0.034, and 0.046 min^{-1} for B; 0.022, 0.031, and 0.041 min^{-1} for T; 0.027, 0.034, and 0.038 min^{-1} for E; and 0.022, 0.031, and 0.032 min^{-1} for X. The correlation coefficients (r^2) are in the range of 0.972–0.987 for B, 0.919–0.953 for T, 0.973–0.995 for E, and 0.926–0.962 for X, reflecting that the kinetics of BTEX adsorption on CNTs follows the first-order rate law.

The k_1 increased with temperature which is inconsistent with the temperature dependence of the q_e. This could be explained by the fact that the adsorption rate is faster than desorption rate at lower temperatures, but desorption rate is more sensitivity to temperature change, and it becomes greater at high temperatures. Thus, adsorption would dominate at lower temperatures, while desorption would dominate at higher temperatures. Similar results have been reported in the literature for adsorption of trihalomethanes [23] and natural organic matters [19] on CNTs.

The temperature dependence of the k_1 has been found in practically all cases to be well-represented by the Arrhenius equation [49]

$$\ln k_1 = \ln k_0 - \frac{E_a}{RT} \tag{3}$$

where k_0 is the frequency of adsorption (1/min), E_a is the activation energy of the reaction (J/mol), R is the universal gas constant (8.314 J/mol/K), and T is the absolute temperature (K). A plot of $\ln k_1$ versus $1/T$ as a straight line, the corresponding k_0 and E_a are determined from the intercept and the slope, respectively.

The k_0 of BTEX are 0.911, 1.473, 0.408 and 0.494 min^{-1}, respectively. The E_a of BTEX are 8.018, 9.701, 6.229 and 7.07 kJ/mol, respectively. Notably the E_a are all less than 20 kJ/mol, showing that adsorption of BTEX on CNTs is controlled by diffusion mechanism [50].

3.3. Adsorption Isotherms

The q_es of BTEX at multiple temperatures are correlated with the Langmuir model (Eq. 4) and Freundlich model (Eq. 5):

$$q_e = \frac{abC_e}{1 + bC_e}, \tag{4}$$

$$q_e = K_f C_e^{1/n} \tag{5}$$

where C_e is the equilibrium concentration (mg/L), a and b are the Langmuir constants, and K_f and n are the Freundlich constants. The constants of Langmuir and Freundlich models were obtained from fitting the isotherm model to the q_e of BTEX and are given in Table 5.2. The correlation coefficients (r^2) are in the range of 0.703–0.975 for the Langmuir model and 0.861–0.999 for the Freundlich model, indicating that the q_e is better correlated with the Freundlich model than with the Langmuir model. The constants a and K_f, which represent the maximum adsorption capacity of BTEX, decreased with temperature.

Table 5.2 Constants of Langmuir and Freundlich models

Adsorbates	Temperature (°C)	Langmuir model			Freundlich model		
		a (mg/g)	b (L/mg)	r^2 (dimensionless)	K_f [(mg/g) (L/mg)$^{1/n}$]	$1/n$ (dimensionless)	r^2 (dimensionless)
B	5	476.34	0.014	0.887	29.56	0.49	0.925
	25	259.96	0.030	0.713	29.50	0.45	0.884
	45	236.91	0.030	0.878	24.77	0.46	0.920
T	5	509.90	0.186	0.849	45.21	0.431	0.999
	25	285.03	0.036	0.731	36.39	0.441	0.977
	45	248.00	0.265	0.703	29.64	0.409	0.996
E	5	528.05	0.023	0.942	50.27	0.431	0.989
	25	340.42	0.050	0.975	44.56	0.430	0.993
	45	323.69	0.041	0.965	40.29	0.428	0.986
X	5	599.54	0.035	0.917	58.33	0.466	0.920
	25	414.74	0.048	0.928	57.98	0.442	0.908
	45	390.61	0.049	0.810	49.49	0.436	0.861

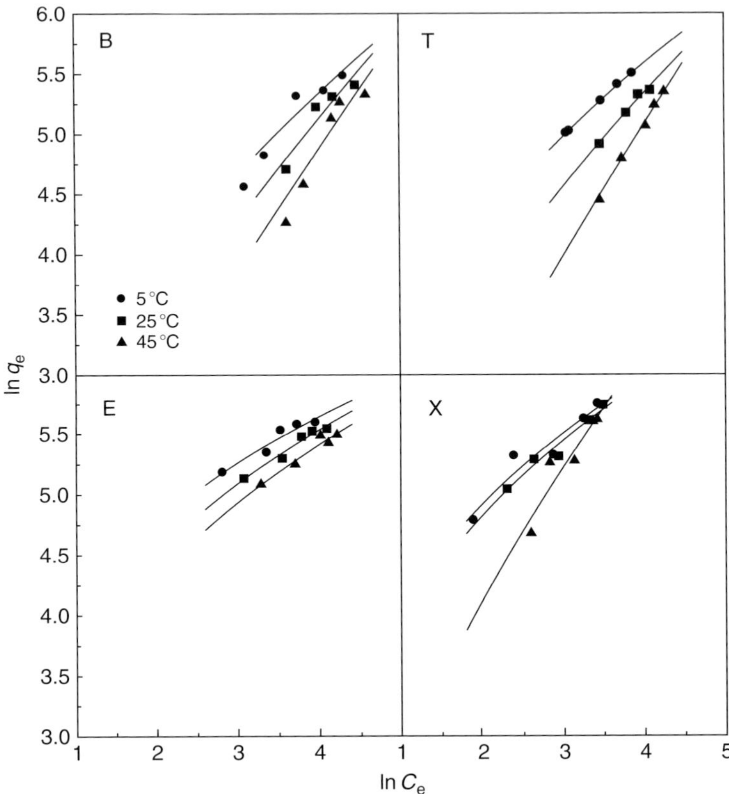

Figure 5.2   Freundlich isotherms of BTEX adsorption on CNTs at multiple temperatures.

Freundlich isotherms (K_f) of BTEX adsorption on CNTs at multiple temperatures are presented in Fig. 5.2. It is evident that the q_e increased with equilibrium concentration (C_e) but decreased with temperature. As the temperature increased from 5 to 45 °C, the K_f decreased, going from 29.56 to 24.77 $(mg/g)(L/mg)^{1/n}$ for B, 45.21 to 29.64 $(mg/g)(L/mg)^{1/n}$ for T, 50.27 to 40.29 $(mg/g)(L/mg)^{1/n}$ for E, and 58.33 to 49.49 $(mg/g)(L/mg)^{1/n}$ for X.

3.4. Comparison with Literature Results

The comparisons of q_e via CNTs in this study, with those conducted via many types of raw and oxidized adsorbents such as GAC, single-walled carbon nanotube (SWCNT), powdered activated carbon, montmorillonite, and zeolite, are summarized in Table 5.3. The present CNTs show a better BTEX adsorption performance than other carbon or silica adsorbents, suggesting that the present CNTs are promising BTEX adsorbents in wastewater treatment.

Table 5.3 Comparisons of q_e via various adsorbents

Adsorbents	q_e, mg/g				Conditions	References
	B	T	E	X		
CNT (NaOCl)[a]	260	285	340	415	pH:7, T:25, S/L:0.06/100, C_0 = 20–200	This study
SWCNT (NaOCl)	60	103			pH:7, T:25, S/L:0.001/40, C_0 = 200	[30]
MWCNT (NaOCl)	36					
SWCNT[a]				78	pH:5.4, T:25, S/L:0.05/140, C_0 = 7–107	[49]
SWCNT (HNO$_3$)[a]				86		
GAC[a]	183	194			pH:7, T:30, S/L:0.15/100, C_0 = 35–442	[8]
GAC (HNO$_3$)[a]	114	122				
PAC	40	40				
Montmorillonite	28	27			S/L:0.1/40, C_0 = 100	[9]
Zeolite	27	20				

Note: CNT = carbon nanotube; T = temperature (°C); S/L = solid/liquid (g/mL); C_0 = initial concentration (mg/L); SWCNT = single-walled carbon nanotube; MWCNT = multiwalled carbon nanotube; GAC = granular activated carbon; PAC = powdered activated carbon.
[a] Maximum adsorption capacity calculated from Langmuir model.

3.5. Adsorption Thermodynamics

The thermodynamic parameters, free energy change (ΔG^0), enthalpy change (ΔH^0), and entropy change (ΔS^0) for BTEX adsorption were calculated using the following equations [51]:

$$\Delta G^0 = -RT\ln K_a, \tag{6}$$

$$\Delta S^0 = \frac{\Delta H^0 - \Delta G^0}{T} \tag{7}$$

where K_a is the thermodynamic equilibrium constant. As the BTEX concentrations in the solution decrease and approach 0, the value of the

Table 5.4 Thermodynamic parameters

Adsorbates	Temperature (°C)	K_o	ΔG^o (kJ/mol)	ΔH^o (kJ/mol)	ΔS^o (J/mol/K)
	5	4.22	−3.33		10.36
B	25	3.76	−3.28	−0.45	9.50
	45	3.45	−3.27		8.87
	5	5.09	−3.76		11.79
T	25	4.54	−3.75	−0.48	10.95
	45	4.09	−3.73		10.20
	5	5.21	−3.82		11.95
E	25	4.61	−3.79	−0.50	11.04
	45	4.17	−3.77		10.31
	5	6.85	−4.45		13.88
X	25	5.97	−4.43	−0.59	12.87
	45	5.25	−4.38		11.92

constant K_a is obtained by plotting a straight line of capacity divided by the equilibrium concentration (q_e/C_e) versus q_e based on a least-square analysis and extrapolating q_e to 0. The intercept of vertical axis gives K_a. The enthalpy change ΔH^0 is determined from the slope of the regression line after plotting $\ln K_a$ against $1/T$. The free energy ΔG^0 and entropy ΔS^0 change are determined from Eqs. 6 and 7. Table 5.4 summaries the values of these thermodynamic parameters. Negative ΔH^0 indicates the exothermic nature of adsorption process. This is supported by the decrease in q_e with temperature as shown in Fig. 5.2. Negative ΔG^0 suggests that the adsorption process is spontaneous with a high preference for BTEX molecules. Positive ΔS^0 reflects the affinity of CNTs for BTEX and the increase of randomness at the solid/liquid interface during adsorption process [52].

3.6. Adsorption/Desorption Study

To test the repeated availability of BTEX adsorption on CNTs, the optimum conditions to reach effective BTEX desorption, such as the regeneration agents, the strength of regeneration solution, and the regeneration time, must be determined.

Figure 5.3 shows the BTEX recoveries of CNTs under various strengths of NaOCl solution. Desorption experiments were conducted for 12 h to assure the attainment of a desorption equilibrium. As the NaOCl strength increased from 10 to 30%, the BTEX recoveries increased from 67.40 to 88.81% for B, 52.17 to 88.82% for T, 40.3 to 88.15% for E, and 40.1 to 89.53% for X, respectively. Figure 5.4 shows the BTEX recoveries of CNTs under various strengths of HNO_3 solution. It is observed that as the HNO_3 strength increased

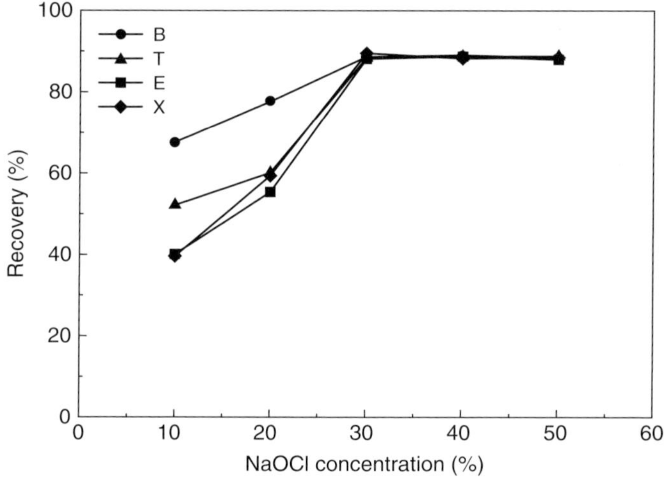

Figure 5.3   BTEX recoveries of CNTs under various NaOCl strengths.

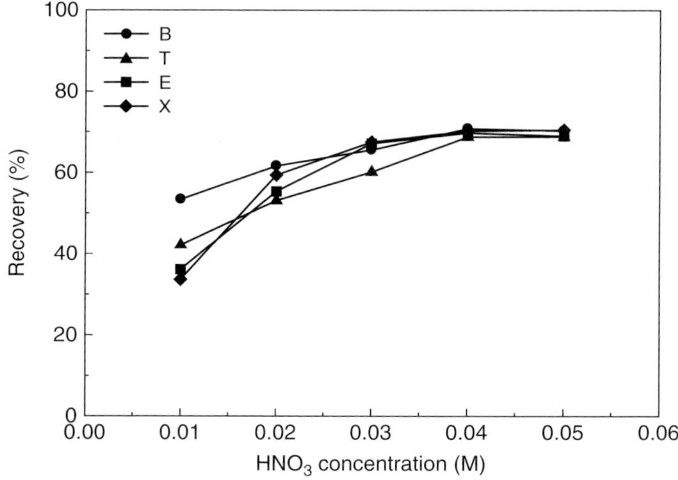

Figure 5.4   BTEX recoveries of CNTs under various HNO_3 strengths.

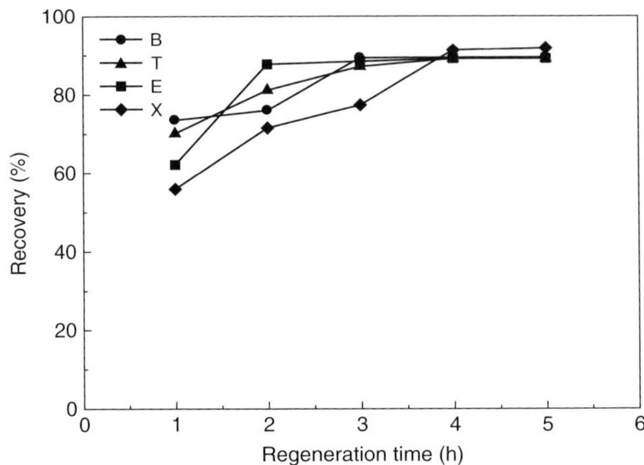

Figure 5.5   BTEX recoveries of CNTs under various regeneration times.

from 0.01 to 0.04 M, the BTEX recoveries increased from 53.40 to 70.28% for B, 42.17 to 68.77% for T, 36.47 to 69.76% for E, and 33.53 to 70.24% for X, respectively. This could be explained by the degradation of more BTEX molecules on the CNT surface at higher regeneration concentrations where more BTEX molecules can be adsorbed into aqueous solutions again and thus lead to higher BTEX recoveries. The BTEX recoveries via NaOCl solution are much higher than those via HNO_3 solution and reached a maximum for NaOCl solution of greater or equal to 30%. Thus, a 30% NaOCl solution was chosen to evaluate the optimum regeneration time.

Figure 5.5 displays the BTEX recoveries of CNTs under various regeneration times. It is evident that the BTEX recoveries increased with regeneration time and reached maximum for 3, 4, 2, and 4 h of operation, respectively. Thus, a regeneration time of 4 h was chosen in the cyclic BTEX adsorption.

It should be noted that the optimum strength of regeneration agent and the regeneration time employed in the desorption experiment depend on the initial BTEX concentration (C_0). A larger q_e was obtained with a higher C_0, which makes desorption of BTEX molecules from the CNT surface more difficult. Thus, a higher strength of regeneration agent or a longer regeneration time is needed with a higher C_0 to reach effective BTEX desorption.

3.7. Cyclic BTEX Adsorption on CNTs and GAC

A comparison of cyclic BTEX adsorption on CNTs and GAC was conducted. The physical properties of CNTs and GAC are given in Table 5.5. The surface area, pore volume, and average pore diameter for pore size

Table 5.5 Physical properties of CNTs and GAC

| Adsorbents | <1.7 nm | | | 1.7–100 nm | | | | |
| | SA (m²/g) | PV (cm³/g) | APD (nm) | 1.7–5 nm | | 5–100 nm | | APD (nm) |
				SA (m²/g)	PV (cm³/g)	SA (m²/g)	PV (cm³/g)	
CNTs	133.33	0.076	0.30	16.74	0.014	149.82	1.106	23.75
GAC	1292.1	0.538	0.42	567.61	1.360	6.91	0.011	2.31
CNTs (after $n = 10$)	198.75	0.054	0.27	25.53	0.022	147.44	0.940	22.23

Note: SA = surface area; PV = pore volume; APD = average pore diameter.

below 5 nm are larger for GAC, whereas the surface area, pore volume, and average pore diameter for pore size above 5 nm are larger for CNTs.

Figures 5.6 and 5.7 present the q_e and the BTEX recoveries of CNTs and GAC under various cycles of adsorption and regeneration (n). It is apparent that as the n increased, the q_e and the BTEX recovery slightly decreased for CNTs but sharply decreased for GAC. This could be explained by the fact that the GAC has a porous structure in which BTEX molecules have to move from the inner surface to the exterior surface of the pores and thus makes desorption of BTEX more difficult.

As n increased from 1 to 10, the q_e of BTEX via CNTs decreased from 199.71 to 172.67 mg/g, 208.43 to 184.17 mg/g, 229.2 to 200.66 mg/g, and 263.30 to 232.93 mg/g; the q_e of BTEX via GAC decreased from 108.16 to 36.68 mg/g, 143.24 to 50.23 mg/g, 147.88 to 50.63 mg/g, and 159.78 to 54.88 mg/g; the BTEX recoveries of CNTs decreased from 89.49 to

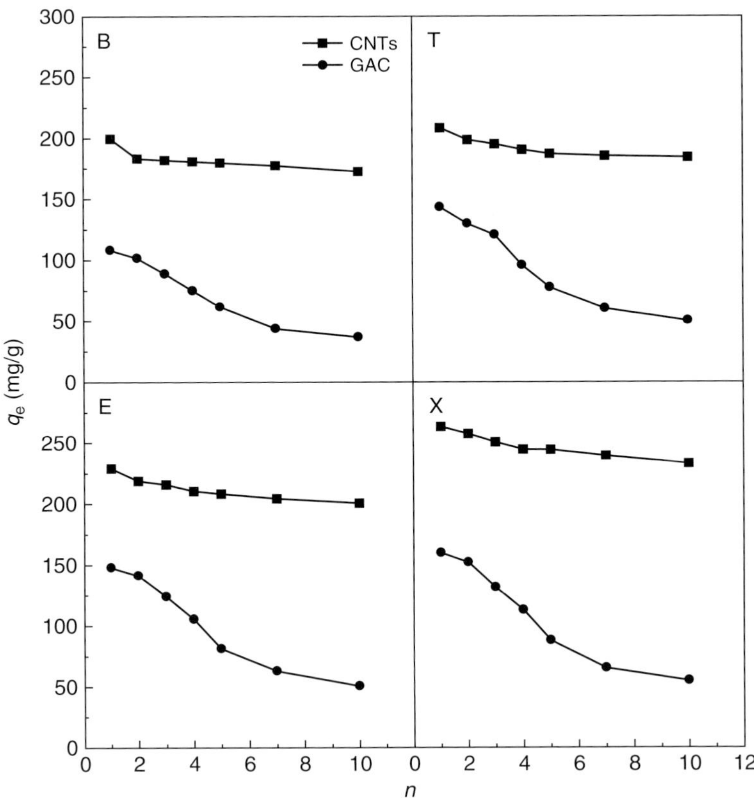

Figure 5.6 Equilibrium capacities of BTEX adsorption on CNTs and GAC under various *n*.

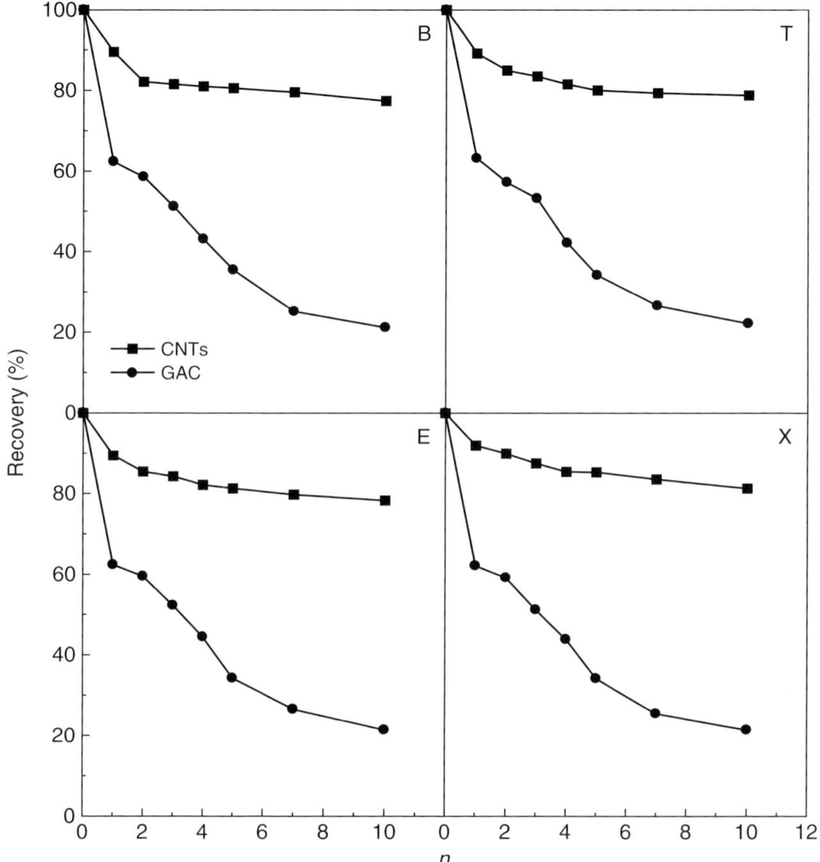

Figure 5.7   BTEX recoveries of CNTs and GAC under various n.

77.38%, 89.16 to 78.78%, 89.4 to 78.3%, and 91.6 to 80.7%; the BTEX recoveries of GAC decreased from 62.35 to 21.15%, 63.15 to 22.15%, 61.4 to 22.3%, and 60.8 to 21.7%. It is apparent that the CNTs not only possess higher adsorption capacity of BTEX but also show better BTEX recovery through 10 cycles of adsorption and regeneration, suggesting that the CNTs are promising BTEX adsorbents in wastewater treatment.

3.8. Stability of Adsorbents

Figure 5.8 shows the percentage weight loss ratios of CNTs and GAC during the regeneration process of various n. As can be seen, the CNTs have less weight loss in each cycle of adsorption and regeneration than the GAC, reflecting that the CNTs appear rather stable after being repeated use than

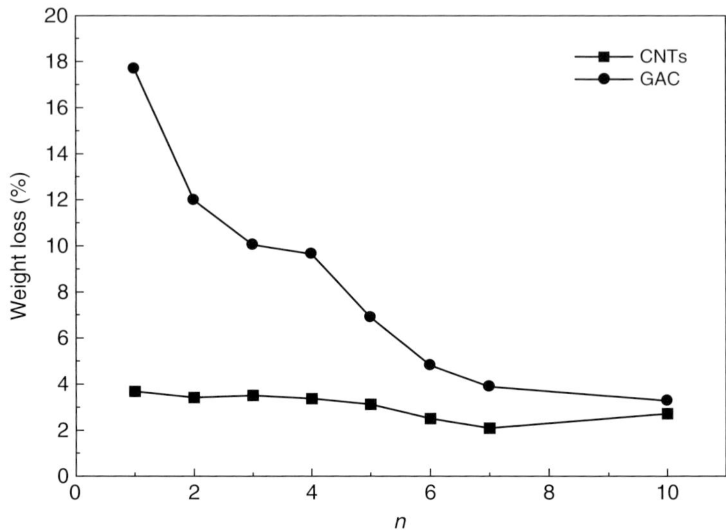

Figure 5.8   Percentage weight loss ratios during regeneration of CNTs and GAC under various *n*.

the GAC. This could be due to a high degree of graphitization of CNTs. The average percentage weight loss ratios of CNTs and GAC are 3.05 and 8.53%, respectively.

The pore structure and the surface functional groups of CNTs before and after 10 cycles of adsorption and regeneration were characterized by their physical properties and IR spectra, respectively. Figure 5.9 shows the N_2 adsorption/desorption isotherms of virgin and regenerated CNTs. Both samples have very similar N_2 isotherms that exhibit a type II shape [53], with a rounded knee at a very low P/P_0 (about 0.01) representing some MPs in CNTs. After a very slow increase up to a P/P_0 0.9, the isotherms display a sharp increment with P/P_0 showing largely mesoporous nature of CNTs. A small closed adsorption/desorption hysteresis loop is also observed with a P/P_0 above 0.7 probably due to the mesopores with a capillary condensation.

Figure 5.10 shows the pore size distributions of virgin and regenerated CNTs. It is noted that the MP size distribution of virgin CNTs only appears in the pore size range of 0.26–0.32 nm which become board (0.22–0.36 nm) after regeneration. Both samples have a similar distribution for pore size range of 1.7–100 nm. The pore volume below pore size 20 nm is larger for regenerated CNTs probably due to the fact that more end caps of

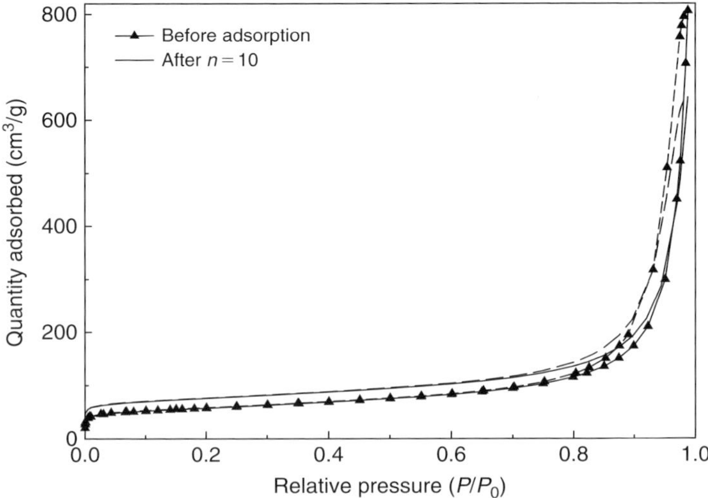

Figure 5.9 N_2 adsorption (solid line) and desorption (dash line) isotherms of virgin and regenerated CNTs.

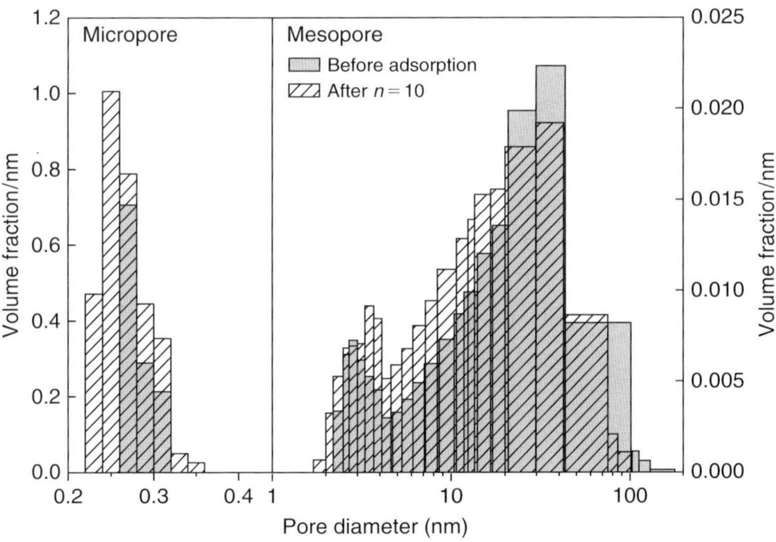

Figure 5.10   Pore size distribution of virgin and regenerated CNTs.

CNTs were opened up during repeated regeneration via NaOCl solution, whereas the pore volume above pore size 20 nm is larger for virgin CNTs. The physical properties of regenerated CNTs are also given in Table 5.5.

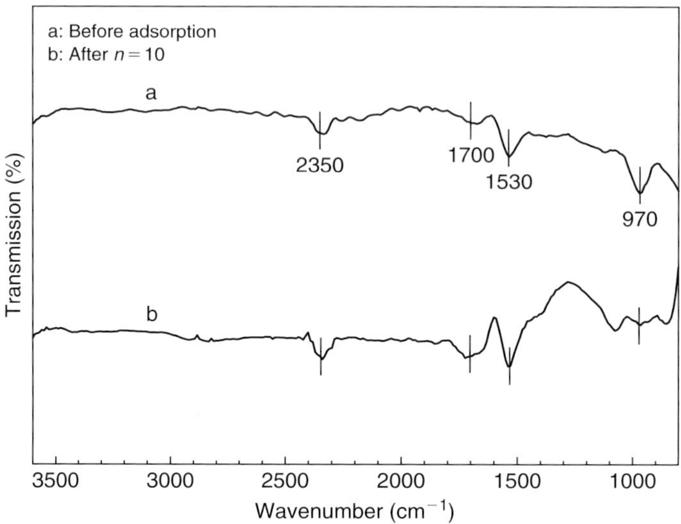

Figure 5.11 IR spectra of virgin and regenerated CNTs.

The surface area, pore volume, and average pore diameter of CNTs have no significant changes before and after regeneration.

Figure 5.11 shows the IR spectra of virgin and regenerated CNTs. It is evident that both samples have similar IR spectra. The band at 2350 cm^{-1} is assigned to $-$OH stretch from strongly H-bonded COOH [54]. The bands at 1700 and 1530 cm^{-1} are related to the stretching variations of C=O groups [6] and carboxylate anion stretch mode [54], respectively. The band at ~970 is associated to C–O stretching of alcoholic compounds [55]. The abundance of these surface oxygen groups on the external and internal surface of CNTs provides numerous chemical sites for BTEX adsorption.

It is apparent that the pore structure and the surface oxygen groups of CNTs were preserved through 10 cycles of adsorption and regeneration, suggesting that the CNTs can be used in the prolonged cyclic operation.

3.9. Cost-Effective Analysis

To evaluate the replacement cost of CNTs and GAC in wastewater treatment, a statistical analysis based on the best-fit regression of q_e versus n was conducted (Fig. 5.6), and the regression equations are given in Table 5.6. The r^2 of BTEX is 0.913–0.995 for CNTs and 0.898–0.918 for GAC.

Figure 5.12 shows the predicted n of BTEX adsorption via CNTs and GAC at a q_e range of 60–200 mg/g. As can be seen, the predicted n rapidly decreased with q_e. At a q_e of 60 mg/g, the predicted n's of BTEX are, respectively, 2.46×10^9, 4.03×10^9, 1.19×10^{10}, and 1.09×10^{12} for

Table 5.6   Regression equations for cost-effective analysis

Adsorbates	CNTs		GAC	
B	$q_e = 195.39 \times n^{-0.0546}$	$r^2 = 0.913$	$q_e = 131.65 \times n^{-0.5044}$	$r^2 = 0.905$
T	$q_e = 207.05 \times n^{-0.056}$	$r^2 = 0.969$	$q_e = 170.15 \times n^{-0.4881}$	$r^2 = 0.918$
E	$q_e = 228.83 \times n^{-0.0577}$	$r^2 = 0.995$	$q_e = 180.99 \times n^{-0.4981}$	$r^2 = 0.898$
X	$q_e = 265.12 \times n^{-0.0536}$	$r^2 = 0.979$	$q_e = 194.99 \times n^{-0.5037}$	$r^2 = 0.902$

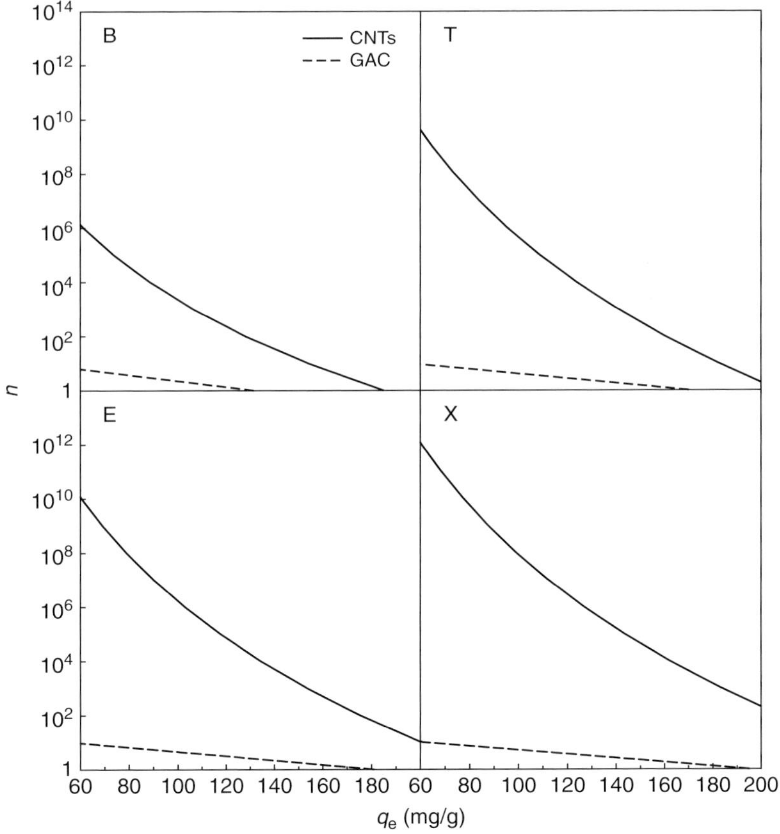

Figure 5.12   Predicted n of BTEX adsorption on CNTs and GAC under various q_e.

CNTs and 6, 12, 9, and 10 for GAC. At a q_e of 120 mg/g, the predicted n's of BTEX are, respectively, 7545, 1.71×10^4, 7.2×10^4, and 2.65×10^6 for CNTs and 1, 2, 2, and 2 for GAC.

It is clear that the CNTs can be reused for BTEX removal through a number of adsorption and regeneration cycles. A significant cost for the replacement of adsorbents can be thus reduced. This is the key factor whether a novel but expensive adsorbent can be accepted by the field or not. It is expected that the unit cost of CNTs can be further reduced in the foreseeable future. Therefore, the NaOCl-oxidized CNTs possess good potential for BTEX removal in wastewater treatment in the near future.

It should be noted that the predicted number of cycles (n) was estimated based on the q_e of 10 adsorption and regeneration cycles since it is quite time-consuming to regenerate adsorbents for over thousands of tests. Furthermore, the adsorbent weight loss was neglected in the estimation of n. Thus, a departure of predicted results from real conditions may possibly occur, and a prolonged test on the adsorption performance of regenerated CNTs is required.

4. CONCLUSIONS

The CNTs were oxidized by NaOCl solution and were used as novel adsorbents to study kinetics, thermodynamics, and regeneration of BTEX adsorption in aqueous solutions. The adsorption kinetics follows the first-order rate law, while the adsorption thermodynamics indicates the exothermic and spontaneous nature. The CNTs show better BTEX adsorption performance than other carbon or silica adsorbents documented in the literature. A comparative study on cyclic BTEX adsorption on CNTs and GAC revealed that the CNTs have a better adsorption performance and show less weight loss through 10 cycles of adsorption and regeneration. This suggests that the CNTs are promising BTEX adsorbents in wastewater treatment. A statistical analysis on the replacement cost of CNTs and GAC revealed that the CNTs can be reused through a number of adsorption/regeneration cycles and thus be possibly a cost-effective BTEX adsorbent despite their initial high unit cost at the present time.

ACKNOWLEDGMENT

Support from the National Science Council, Taiwan, under a contract no. NSC-97-2221-E-005-036-MY3 is gratefully acknowledged.

References

[1] EPA National Primary Drinking Water Standards. 816-F-03-016. Office of Water (4606M). June 2003.

[2] Hazardous Substance Research Centers. BTEX contamination. US EPA funded (GLMAC) for Hazardous Substance Research. http://www.grassrootscoalition.org/gasoilfields/btexGC.pdf; July 2000.

[3] Sawyer CN, McCarty PL, Parkin GF. Chemistry for environmental engineering. 4th ed. Singapore: McGraw-Hill; 1994. p. 224.

[4] Oregon Department of Human Services Environmental Toxicology Section, BTEX. Technical Bulletin - Health Effects Information. 1994.

[5] Purdom PW. Environmental health. 2nd ed. New York: Academic Press; 1980.

[6] Daifullah AAM, Girgis BS. Impact of surface characteristics of activated carbon on adsorption of BTEX. Colloids Surf A 2003;214:181–93.

[7] Lillo-Ródenas MA, Cazoria-Amorós D, Linares-Solano A. Behaviour of activated carbons with different pore size distributions and surface oxygen groups for benzene and toluene adsorption at low concentrations. Carbon 2005;43:1758–67.

[8] Wibowo N, Setyadhi L, Wibowo D, Setiawan J, Ismadji S. Adsorption of benzene and toluene from aqueous solutions onto activated carbon and its acid and heat treated forms: influence of surface chemistry on adsorption. J Hazard Mater 2007;146: 237–42.

[9] Koh SM, Dixon JB. Preparation and application of organo-minerals as sorbents of phenol, benzene and toluene. Appl Clay Sci 2001;18:111–2.

[10] Smart SK, Cassady AI, Lu GQ, Martin DJ. The biocompatibility of carbon nanotubes. Carbon 2006;44:1034–47.

[11] Long RQ, Yang RT. Carbon nanotubes as superior sorbent for dioxin removal. J Am Chem Soc 2001;123:2058–9.

[12] Agnihotri S, Rood MJ, Rostam-Abadi M. Adsorption equilibrium of organic vapors on single-walled carbon nanotubes. Carbon 2005;43:2379–88.

[13] Hsu S, Lu C. Modification of single-walled carbon nanotubes for enhancing isopropyl alcohol vapor adsorption form air stream. Sep Sci Technol 2007;42:2751–66.

[14] Díaz E, Ordóñez S, Vega A. Adsorption of volatile organic compounds onto carbon nanotubes, carbon nanofibers, and high-surface-area graphites. J Colloid Interface Sci 2007;305:7–16.

[15] Liu JM, Li L, Fan HL, Ning ZW, Zhao P. Evaluation of single-walled carbon nanotubes as novel adsorbent for volatile organic compounds. Chin J Anal Chem 2007;35: 830–4.

[16] Shih Y-H, Li M-S. Adsorption of selected volatile organic vapors on multiwall carbon nanotubes. J Hazard Mater 2008;154:21–8.

[17] Wang SG, Liu XW, Gong WX, Nie W, Gao BY, Yue QY. Adsorption of fulvic acids from aqueous solutions by carbon nanotubes. J Chem Technol Biotechnol 2007;82: 698–704.

[18] Lu C, Su F. Adsorption of natural organic matter by carbon nanotubes. Sep Purif Technol 2007;58:113–21.

[19] Su F, Lu C. Adsorption kinetics, thermodynamics and desorption of natural dissolved organic matter by multiwalled carbon nanotubes. J Environ Sci Health A 2007;42: 1543–52.

[20] Chung M, Lu C, Su F. Adsorption of humic acid from aqueous solution by carbon nanotubes. 12th Mainland-Taiwan Environmental Protection Academic Conference, Taichung County, Taiwan, 2005.

[21] Hyung H, Kim J. Natural organic matter (NOM) adsorption to multi-walled carbon nanotubes: effect of nom characteristics and water quality parameters. Environ Sci Technol 2008;42:4416–21.

[22] Lu C, Chung YL, Chang KF. Adsorption of trihalomethanes from water with carbon nanotubes. Water Res 2005;39:1183–9.

[23] Lu C, Chung YL, Chung KF. Adsorption thermodynamic and kinetic studies of trihalomethanes on multiwalled carbon nanotubes. J Hazard Mater 2006; B138:304–10.

[24] Chin CJ, Shih LC, Tsai HJ, Liu TK. Adsorption of *o*-xylene and *p*-xylene from water by SWCNTs. Carbon 2007;45:1254–60.

[25] Liao Q, Sun J, Gao L. The adsorption of resorcinol from water using multi-walled carbon nanotubes. Colloids Surf A 2008;312:160–5.

[26] Wu CH. Adsorption of reactive dye onto carbon nanotubes: equilibrium, kinetics and thermodynamics. J Hazard Mater 2007;144:93–100.

[27] Kuo C-Y. Desorption and re-adsorption of carbon nanotubes: comparisons of sodium hydroxide and microwave irradiation processes. J Hazard Mater 2008;152:949–54.

[28] Liu C-H, Li J-J, Zhang H-L, Li B-R, Guo Y. Structure dependent interaction between organic dyes and carbon nanotubes. Colloids Surf A 2008;313–314:9–12.

[29] Peng X, Li Y, Luan Z, Di Z, Wang H, Tian B, et al. Adsorption of 1,2-dichlorobenzene from water to carbon nanotubes. Chem Phys Lett 2003;376:154–8.

[30] Chen W, Duan L, Zhu D. Adsorption of polar and nonpolar organic chemicals to carbon nanotubes. Environ Sci Technol 2007;41:8295–300.

[31] Liao Q, Sun J, Gao L. Adsorption of chlorophenols by multi-walled carbon nanotubes treated with HNO_3 and NH_3. Carbon 2008;46:553–5.

[32] Yang K, Xing B. Desorption of polycyclic aromatic hydrocarbons from carbon nanomaterials in water. Environ Pollut 2007;145:529–37.

[33] Yang K, Wang X, Zhu L, Xing B. Competitive sorption of pyrene, phenanthrene, and naphthalene on multiwalled carbon nanotubes. Environ Sci Technol 2006;40:5804–10.

[34] Yang K, Zhu L, Xing B. Adsorption of polycyclic aromatic hydrocarbons by carbon nanomaterials. Environ Sci Technol 2006;40:1855–61.

[35] Lin D, Xing B. Adsorption of phenolic compounds by carbon nanotubes: role of aromaticity and substitution of hydroxyl groups. Environ Sci Technol 2008;42:7254–9.

[36] Yan H, Gong A, He H, Zhou J, Wei Y, Lv L. Adsorption of microcystins by carbon nanotubes. Chemosphere 2006;62:142–8.

[37] Ye C, Gong Q-M, Lu F-P, Liang J. Adsorption of uraemic toxins on carbon nanotubes. Sep Purif Technol 2007;58:2–6.

[38] Gotovac S, Yang C-M, Hattori Y, Takahashi K, Kanoh H, Kaneko K. Adsorption of polyaromatic hydrocarbons on single wall carbon nanotubes of different functionalities and diameters. J Colloid Interface Sci 2007;314:18–24.

[39] Gotovac S, Song L, Kanoh H, Kaneko K. Assembly structure control of single wall carbon nanotubes with liquid phase naphthalene adsorption. Colloids Surf A 2007;300: 117–21.

[40] Cho HH, Smith BA, Wnuk JD, Fairbrother DH, Ball WP. Influence of surface oxides on the adsorption of naphthalene onto multiwalled carbon nanotubes. Environ Sci Technol 2008;42:2899–905.

[41] Chen W, Duan L, Wang L, Zhu DQ. Adsorption of hydroxyl-and amino-substituted aromatics to carbon nanotubes. Environ Sci Technol 2008;42:6862–8.

[42] Chen J, Chen W, Zhu DQ. Adsorption of nonionic aromatic compounds to single-walled carbon nanotubes: effects of aqueous solution chemistry. Environ Sci Technol 2008;42:7225–30.

[43] Yan XM, Shi BY, Lu JJ, Feng CH, Wang DS, Tang HX. Adsorption and desorption of atrazine on carbon nanotubes. J Colloid Interface Sci 2008;321:30–8.

[44] Lu C, Su F, Hu S. Surface modification of carbon nanotubes for enhancing BTEX adsorption from aqueous solutions. Appl Surf Sci 2008;254:7035–41.

[45] Mukhopadhyay K, Koshio A, Sugai T, Tanaka N, Shinohara H, Konya Z, et al. Bulk production of quasi-aligned carbon nanotube bundles by the catalytic chemical vapour deposition (CCVD) method. Chem Phys Lett 1999;303:117–24.

[46] Cristina M, Almeida M, Boas LV. Analysis of BTEX and other substituted benzenes in water using headspace SPME-GC-FID: method validation. J Environ Monit 2004;6:80–8.
[47] Eckenfelder Jr WW. Industrial water pollution control. Singapore: McGraw-Hill; 1989.
[48] Tütem E, Apak R, Ünal CF. Adsorptive removal of chlorophenols from water by bituminous shale. Water Res 1998;32:2315–24.
[49] Mangun CL, Yue Z, Economy J. Adsorption of organic contaminants from water using tailored ACFs. Chem Mater 2001;13:2356–60.
[50] Atun G, Sismanoglu T. Adsorption of 4,4'-isopropylidene diphenol and diphenylolpropane 4 4'dioxyacetic acid from aqueous solution on kaolinite. J Environ Sci Health A 1996;31:2055–69.
[51] Niwas R, Gupta U, Kha AA, Varshney KG. The adsorption of phosphamidon on the surface of styrene supported zirconium (IV) tungstophosphate: a thermodynamic study. Colloids Surf A 2000;164:115–9.
[52] Yavuz Ö, Altunkaynak Y, Güzel F. Removal of copper, nickel, cobate and manganese from aqueous solution by kaolinite. Water Res 2003;37:948–52.
[53] Gregg SJ, Sing KSW. Adsorption, surface area and porosity. New York: Academic Press; 1982.
[54] Yue Z, Economy J. Nanoparticle and nanoporous carbon adsorbents for removal of trace organic contaminants from water. J Nanopart Res 2005;7:477–87.
[55] Wang Y, Wu J, Wei F. A treatment method to give separated multi-walled carbon nanotubes with high purity, high crystallization and a large aspect ratio. Carbon 2003;41:2939–48.

CHAPTER 6

Nanostructured Metal Oxide Gas Sensors for Air-Quality Monitoring

David G. Rickerby* *and* **Alessandra M. Serventi****

* European Commission Joint Research Centre, Institute for Environment and Sustainability, 21020 Ispra VA, Italy

** Institut de Recherche d'Hydro-Québec, Varennes, Québec J3X 1S1 Canada

Contents

1. INTRODUCTION

The development of chemical gas sensors has been the subject of intense academic and industrial research activity for the past 40 years. The increasing need to protect the environment and for more efficient control of industrial processes has stimulated the development of various types of gas sensors. These utilize thick or thin films of metal oxides for the detection of inflammable and toxic gases, solid electrolytes for oxygen sensing and to monitor vehicle emissions and production processes, ceramic materials and organic polymers for humidity detection, and artificial olfactory systems with a wide range of applications for the food industry, and these gas sensors are also utilized for medical diagnosis [1]. Another rapidly growing application area is the use of metal oxide sensors for monitoring air pollutants, such as carbon monoxide, nitrogen dioxide, and ozone.

The first thick-film tin dioxide gas sensors were constructed by Taguchi at the beginning of the 1960s [2] and various types were subsequently marketed commercially. In the following years, research was carried out to improve the performance of these sensors whose operating principle is based on measuring the variation of the electrical resistance of semiconductor metal oxides due to the adsorption of gas molecules on the surface. The materials used in fabrication include zinc oxide for the detection of hydrogen, alcohol, carbon monoxide, and various hydrocarbons; tin dioxide for the detection of inflammable and oxidizing or reducing toxic gases; tungsten trioxide for the detection of hydrogen, hydrazine, ammonia, and hydrogen sulfide; titanium dioxide for the detection of oxygen. Polycrystalline tin dioxide is the material most commonly used in commercial gas-sensor applications because it is stable enough to allow reproducible measurements.

Whether for thick-film or thin-film sensors, the response has generally been optimized empirically, without systematic studies of the parameters that influence performance. Various techniques have been used for the deposition of the gas-sensitive layer. These include ultrahigh vacuum and reactive electron beam evaporation [3–10], chemical vapor deposition (CVD) by thermal, plasma or laser technique [11–13], reactive radio-frequency (RF) or magnetron sputtering [14–26], laser ablation [27–32], the sol–gel technique [32–36], screen printing technology [37, 38], electrochemical deposition [7, 39], and pyrosol methods [40–43]. Various noble metals, such as platinum, palladium, and silver, which due to their catalytic properties improve the sensitivity and selectivity, may in addition be incorporated into the active layer [44, 45].

Future improvements in the response of gas sensors will depend on developing a better understanding of the principles governing their operation and, in particular, the influence of the microstructure and morphology of the semiconducting metal oxide layer on the mechanism of adsorption of gas molecules. The requirement to detect simultaneously the component gases contributing to air pollution has stimulated the integration of multiple sensors in single devices, as well as attempts to improve stability and selectivity. The use of nanocrystalline materials [46] allows further improvement in performance compared with conventional polycrystalline material. Reduction of the grain diameter to values comparable to the Debye length and the extremely high surface-to-volume ratios of the nanocrystalline structure significantly enhance the sensor response.

Nanocrystalline materials are commonly defined as having a mean grain size of less than 100 nm. More generally, nanostructured materials are those, such as a thin film, fiber, or a particle, with at least one dimension below 100 nm or containing atomic domains less than this diameter. Although conventional polycrystalline materials are characterized by grains with dimensions of the order of micrometers containing millions of atoms, a grain of a nanocrystalline material typically contains only a few thousand atoms. Because the surface-to-volume ratio increases rapidly with decreasing diameter, a reduction in grain size results in an increase in the density of grain boundaries and triple points and thus an increase in the fraction of atoms lying in interfaces compared to those at regular lattice positions. For example, in a polycrystalline material with a grain size of approximately 100 nm, only a small percentage of atoms are in grain boundaries, whereas if the grain diameter is reduced to 5 nm, the small percentage of atoms adjacent to grain boundaries becomes similar to that within the grain interior [47]. Such drastic reduction of the grain size causes changes in the physical and chemical properties, such as electrical and electronic, thermodynamic and mechanical behaviors [48, 49].

The sensing mechanism is based on the interaction of gas molecules in the air with the surface grains of the metal oxide layer. The size and the interface structure of the grains influence the sensitivity of the sensor. Detection occurs as a result of complex physical and chemical interactions of the gas molecules with ionized oxygen species adsorbed at the surface of the metal oxide grains. The use of a nanocrystalline material increases the efficiency of detection because of the greater specific surface area available to interact with the gas molecules and the higher density of grain boundaries that provide a network of diffusion paths into the interior of the material.

2. THE GAS-SENSING MECHANISM

A sensor is a device that can detect information from the environment and, by converting the energy associated with that information into another form, provide a usable signal. In the case of metal oxide gas sensors, the presence of a relatively small concentration of gas molecules in the air causes a change in the electrical resistivity of the active layer due to modification of the free carrier density due to exchanges between the conduction band and adsorbed species. For an n-type semiconductor, the presence of a reducing gas, such as carbon monoxide, leads to an increase in the electrical conductivity, while an oxidizing gas, such as nitrogen dioxide, leads to a decrease [2].

Tin dioxide is an n–type wide band gap (3.6 eV) semiconductor with rutile type structure, in which the oxygen vacancies act as free electron donors. The gas detection mechanism is governed by two reactions. The first reaction occurs when oxygen present in the air is chemisorbed on the surface of the oxide as molecular O_2^-, atomic O^-, or as hydroxide $(OH)^-$ species [3]. The adsorption of oxygen on the surface of a grain causes formation of a depleted region (the space charge region) as a result of transfer of electrons from the conduction band of tin dioxide to the adsorbed oxygen species [1, 4, 50]. This depleted region has a width L, which is dependent on the oxygen species and the electron density in the conduction band and can be determined from [51]

$$L = L_D \left(2eV_s / kT \right)^{1/2} \tag{1}$$

where $L_D = (\varepsilon kT/ne^2)^{1/2}$ is the Debye length and ε is the permittivity, n is the electron density, e is the electronic charge, k is the Boltzmann constant, T is the temperature, and V_s is the potential at the surface.

The second reaction occurs between the pollutant gas molecules and the adsorbed oxygen. In the case of a reducing gas like carbon monoxide, this leads to a reduction in the height of the potential barrier at the grain boundaries and a consequent decrease in the electrical resistance of an n–type semiconductor due to neutralization of the oxygen ions adsorbed at the surface. This reaction takes place according to the equations [52]

$$2CO + O_2^- \rightarrow 2CO_2 + e^- \tag{2a}$$

$$CO + O^- \rightarrow CO_2 + e^- \tag{2b}$$

In the case of an oxidizing gas like nitrogen dioxide, the corresponding equations are

$$NO_2 + e^- \rightarrow NO_2^- \tag{3a}$$

$$NO_2 + O_2^- + 2e^- \rightarrow NO_2^- + 2O^- \tag{3b}$$

resulting in an increase in the height of the potential barrier and the electrical resistance.

Changes in the resistance of the film due to chemisorption can be directly correlated with the concentration of the gas. The sensor response is defined as R_{gas}/R_{air} for an oxidizing gas and R_{air}/R_{gas} for a reducing gas, where R_{gas} is the resistance in the presence of the gas and R_{air} is the resistance in pure air [53]. The sensitivity S is defined as the percentage change in the resistance so that for an oxidizing gas,

$$S = 100\left(R_{gas} - R_{air}\right)/R_{air} \qquad (4a)$$

while for a reducing gas,

$$S = 100\left(R_{air} - R_{gas}\right)/R_{gas} \qquad (4b)$$

Both the sensitivity and the selectivity are dependent on the operating temperature and can be modified by doping with a noble metal. A significant increase in sensitivity is obtained when the grain size is reduced to approximately twice the Debye length (~3 nm for tin dioxide) so that the width of the space charge layer is comparable to the grain diameter [6, 54, 55].

The microstructure and the surface structure of the sensing layer exert a significant influence on the sensitivity because they influence the rate of surface reactions and the diffusion of gas into the interior of the material. Göpel and Schierbaum [55] developed models of the dependence of electrical conduction on the grain size of metal oxides by representing the band structure as an electronic equivalent circuit. The grain diameter of polycrystalline tin dioxide is generally greater than the Debye length [6, 45, 56], and if it is more than twice the width of the space charge region, the interior of the grain will make a negligible contribution to the variation in the electrical conductivity of the material, which will be dominated by the grain boundary effects (Fig. 6.1). However, if the grain diameter is comparable or less than twice the Debye length, the space charge region will extend over almost the entire grain and the electrical resistance will, therefore, be more sensitive to the presence of the gas [57].

To be useful, a sensor needs to have a selective response to different gases. The required selectivity can be obtained by varying the operating temperature or the use of catalyst [58]. It must be considered that if the temperature is too low, the reaction between the gas molecules and the oxygen species adsorbed at the surface may be too slow to give an adequate sensitivity. On the other hand, if the temperature is too high, the oxidation reaction will take place so rapidly that the concentration of the gas molecules at the surface becomes diffusion limited. For each individual gas, there is a temperature at which the variation of the resistance is maximized with respect to the resistance measured in pure air. It is thus possible to optimize the sensitivity to a specific gas in a mixture by selecting the appropriate operating temperature.

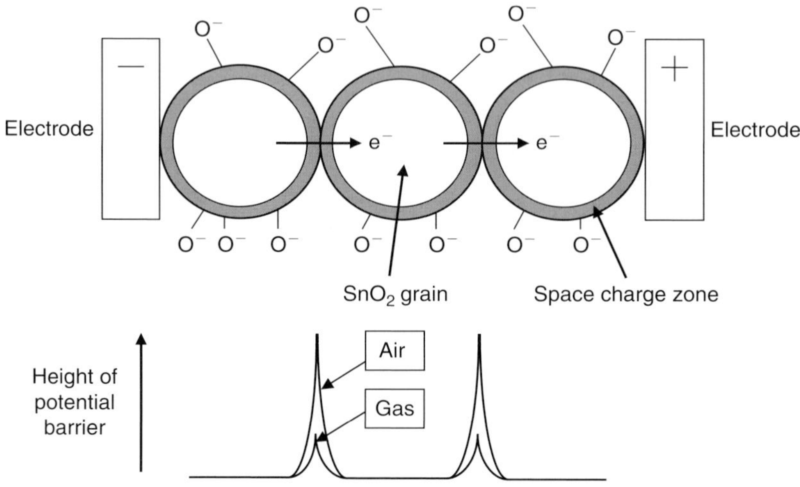

Figure 6.1   Adsorption of oxygen ions on the surface of SnO_2 grains results in the formation of a space charge zone. Interaction of the adsorbed species with gas molecules changes the height of the potential barrier.

3. EFFECT OF CATALYST AND ELECTRICAL CONTACT MATERIALS

The addition of a noble metal, such as platinum, palladium, gold, or silver, to the structure of the metal oxide increases the sensitivity and the selectivity of the sensor and reduces the optimal operating temperature [58]. Yamazoe [45] has proposed a model to describe the effect of a noble metal catalyst in contact with a semiconducting metal oxide. Both chemical and electronic mechanisms have to be considered. The chemical contribution is due to the so-called spillover effect of the gas molecules on the surface of the oxide. This increases the rate of oxidation of the gas and reduces the concentration of adsorbed oxygen species, while the catalyst dissociates the gas molecules enabling them to react with the adsorbed oxygen species.

An electronic interaction occurs between a noble metal particle and the surface of the tin dioxide grain as a result of the difference between the work function of the metal and the electronic affinity of the semiconductor [59]. This leads to oxidation of the metal and hence a variation in the oxidation state of the tin dioxide and the formation of a depleted region inside the grain. The effect is usually associated with platinum and gold, whereas palladium and silver tend more readily to form the stable compounds palladium oxide and silver oxide [4, 7, 45]. In the case of carbon monoxide, the presence of both platinum and palladium increases the sensitivity, particularly at low temperature (275 °C), and reduces the temperature at the maximum sensitivity from over 350 °C to below 300 °C [7, 33].

The total electrical resistance is the result of the contributions from several components: the semiconductor layer, comprising the surface, bulk, and grain boundary components; the substrate on which the metal oxide film was deposited; the electrical contacts; and any catalyst added to increase the sensitivity and the selectivity. A typical sensor can be represented as an ohmic resistance for which the relation between current and potential is linear. The electrical contacts are metallic and are usually made of platinum or gold. The presence of a metallic contact on the tin dioxide layer causes the formation of a Schottky barrier at the interface between the metal and the semiconductor due to the higher work function of the metal compared to the oxide. The contribution of the contacts can, thus, be represented by a diode with nonlinear current–voltage characteristics.

4. THIN-FILM DEPOSITION METHODS

Two commonly used techniques for the deposition of thin films will be considered: reactive RF sputtering and laser ablation. Reactive sputtering is a method that can be used for many different types of metallic and nonmetallic material [60]. It ensures adhesion between the thin film and substrate and a high degree of control of the composition of the deposited material to alter the stoichiometry as required. The surface of a target (the cathode) is bombarded by argon ions in a vacuum of $\sim 10^8$ torr, forming a plasma containing particles of the target material that are deposited on the substrate (the anode). The energy of the incident ions is transferred to the atoms in the surface of the target by atomic collisions [61, 62]. Ejection of atoms or clusters of atoms from the surface occurs when the energy transferred is greater than the binding energy of the material [63, 64].

The rate of sputtering from the target is proportional to the current for a constant applied potential. The power is typically between 50 and 500 W for a potential of 0.5–5 kV. A unique feature of this technique is that the rate of deposition can be kept constant by controlling the argon pressure and the power [65]. There are three different modes of operation: diode sputtering, triode sputtering, and magnetron sputtering, determined by the way in which the potential is applied between the cathode and the anode. However, the present discussion will deal exclusively with magnetron sputtering, which is the system most commonly used in the production of semiconductor devices, integrated circuits, and computer memory chips.

Magnetron sputtering involves the application of a magnetic field perpendicular to the electrical field generated between the anode and the cathode. This magnetic field confines the electrons in the plasma within a region close to the surface of the target. Consequently, the ionic current and the rate of sputtering from the target are increased. A direct current is generally used for conducting materials, while RF sputtering is used for nonconductors and semiconductors. Reactive sputtering occurs when the target material reacts chemically in the presence of a gas in the vacuum chamber. The introduction of a reactive gas, such as oxygen or nitrogen, allows the deposition of thin films of oxides or nitrides. Varying the concentration of the reactive gas allows the stoichiometry of the deposited film to be maintained or modified as desired.

Pulsed laser deposition (PLD) can be used to deposit a wide range of materials [66]. The technique was initially used to produce thin films with CO_2 or Nd:YAG lasers [67], but the rapid development of alternative methods, such as RF sputtering and molecular beam epitaxy (MBE), slowed down further progress in this field. Sputtering became the method of choice for the deposition of metallic and dielectric films, while MBE was extensively applied in the production of semiconductor films for electronic and optical devices. The PLD technique increased in popularity when it was possible to obtain semiconductor thin films of quality comparable to those obtained by MBE. After this, its application for the synthesis of high-temperature superconductors with complex crystal structures marked the beginning of the widespread use of PLD for many different types of metallic, dielectric, semiconductor, ferroelectric, and piezoelectric materials.

The principle of operation of PLD is relatively simple [67, 68]: a laser pulse enters the vacuum chamber via a transparent window and impinges on the material to be deposited. The pulse duration is typically 10–30 nanoseconds with an energy density of 1–10 J/cm. The spot size generally corresponds to an area of approximately $1\,mm^2$, and a few tens of nanometers of the surface material are vaporized after each pulse, forming a plume of neutral atoms and molecules, which are deposited on the substrate located at a distance of several centimeters from the target. The deposition rate is typically approximately 0.1 nm per pulse, with a repetition frequency of 1–100 Hz. The plume is highly supersaturated, with a high level of ionization (~50%) and particle energies ranging from a few eV to several keV. This results in a high rate of deposition and a high nucleation density [69].

One of the most important features of the PLD technique is the possibility of depositing films of mixed composition with the same stoichiometry as the target material. Heating of the material surface due to the laser beam causes melting and vaporization, and all the components of the target are simultaneously vaporized [70]. Deposition can be carried out under vacuum at 10^{-6}–10^{-7} torr or in the presence of a reactive gas such as oxygen, which allows the stoichiometry of the deposited film to be precisely controlled and also has an influence on the grain size.

The fundamental parameters for all thin-film production methods are the temperature of the substrate during deposition, the pressure inside the vacuum chamber, and the rate of deposition. Formation of the film takes place by nucleation and grain growth, involving adsorption, surface diffusion, and chemical bonding phenomena. The film is generated from a number of discrete nuclei that grow at the expense of smaller ones. For both RF sputtering and PLD, film formation is due to three-dimensional island growth [70]. Following nucleation on the substrate surface, growth will occur when the arriving atoms or molecules have a greater propensity to form bonds between themselves than with the substrate.

The substrate temperature is extremely important because it has a strong influence on the microstructure of the film. Thornton [71] generalized the models proposed by Sanders [72] and Movchan and Demschishin [73] to relate thin-film microstructures to the substrate temperature and the melting point of the material. The substrate temperature influences the variation of the free energy associated with the formation of nuclei and the surface diffusion of the deposited atoms. A reduction in this temperature leads to a lower bulk free energy with the result that the critical radius needed to obtain a stable nucleus is smaller. A reduction occurs in the coefficient of surface diffusion of the atoms adsorbed on the substrate that affects the rate of nucleation.

The rate of deposition, as well as the temperature, has a strong influence on the film microstructure [74] because it determines the rate of nucleation. For a given substrate temperature, a lower deposition rate will result in a lower rate of nucleation and a longer mean free diffusion time of mobile atoms on the surface. These conditions favor the formation of well-ordered structures that often exhibit preferential growth directions determined by the orientations of the low-energy crystal surfaces. The films are generally not very dense and show pronounced surface roughness. In the case of a higher rate of deposition, growth takes place rapidly by agglomeration at the initial binding sites due to the continual arrival of new atoms. The atoms do not have sufficient time to migrate to occupy the most energetically favorable sites, and the resulting film microstructures are characterized by a very small grain size and a greater number of lattice defects. The films tend to be denser and their surfaces are generally smoother than those deposited at lower deposition rates.

The presence of a reactive gas in the vacuum chamber influences the rate of deposition and hence the grain size. These gas molecules reduce the energy of the flux of particles from the plasma incident on the substrate and thus reduce the deposition rate and the nucleation rate. At higher partial pressures, fewer nuclei of critical size are formed on the substrate, and the growth of these larger nuclei is favored. At low partial pressures, the deposition rate remains relatively high and favors the formation of smaller sized grains. The surface of the substrate itself may, in addition, influence film growth and structure. Lattice defects, such as dislocations and point defects, represent nucleation sites of lower energies. High surface roughness can induce a topographic shadowing effect, impeding the growth in the affected areas of the film [75, 76].

PLD has one major difference with respect to RF sputtering: deposition is intermittent rather than continuous and is less stable and more rapid, so it can be expected to produce less ordered film structures. Another typical feature of thin films deposited by PLD, due to the high degree of supersaturation of the vapor plume and the high rate of deposition, is their rather small grain size. Subcritical size nuclei formed on the substrate surface during a laser pulse will tend to dissociate into mobile species during the interval preceding the next pulse, allowing them to renucleate by forming new agglomerates or to be incorporated into existing nuclei with greater than the critical radius.

The ratio between the duration of the pulse and the interval between successive pulses influences both the nucleation and the growth processes. A long pulse will vaporize a greater amount of material from the target, resulting in the formation of a high number of small-sized grains. However, if the pulse length is short, the amount of vaporized material is limited and produces a small number of critically sized nuclei, which grow by diffusion and agglomeration during the intervals between pulses. The pulse length and repetition frequency must therefore be carefully chosen to obtain the required grain size [66].

5. INFLUENCE OF FILM STRUCTURE ON SENSOR RESPONSE

5.1. Radio-Frequency Sputtering

Figure 6.2 shows an array of gas sensors fabricated by depositing a tin dioxide film of a few hundred nanometers thickness on a polycrystalline alumina substrate by reactive RF magnetron sputtering from a 99.9% pure tin dioxide target at 250 °C [77–80]. The active area of each sensor is approximately 1 cm². Alumina is commonly used as the substrate material for thin-film gas sensors because it is an electrical insulator and, therefore, does not interfere with measurements of the film resistance. Film deposition was carried out at a pressure of 50 Pa (3.75 mtorr) with an argon/oxygen ratio of 9:1. A platinum resistance heater and the electrical contacts were deposited by the same method in a pure argon atmosphere. In some cases, the same procedure was used to deposit a 10–20 nm thick layer of platinum to catalyze the reaction with adsorbed gas molecules to increase the sensitivity and the selectivity. Following deposition, a thermal annealing treatment was performed to improve the crystallinity of the oxide film.

The films have a nodular surface structure, in which it is possible to distinguish small granular particles of a few tens of nanometers in diameter

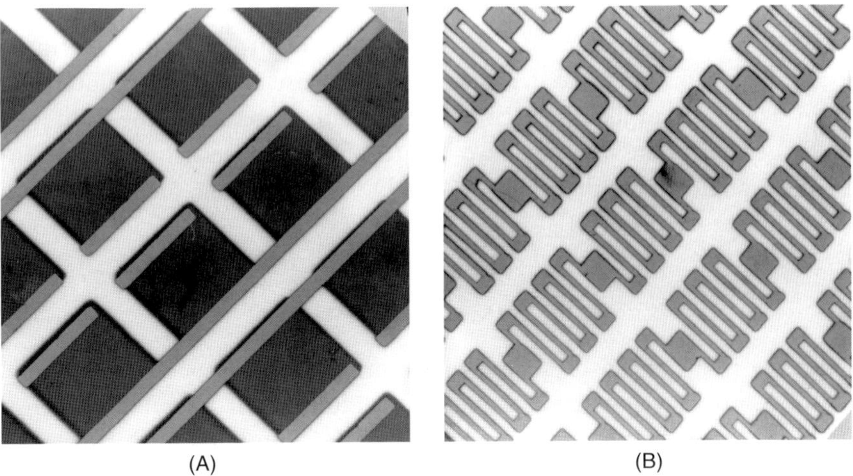

(A)   (B)

Figure 6.2 Thin-film gas sensors fabricated on a polycrystalline alumina substrate by the RF magnetron sputtering technique: (A) nanocrystalline tin dioxide layer with platinum contacts; (B) platinum heater.

Figure 6.3 Surface structure of a tin dioxide film deposited by reactive sputtering at a substrate temperature of 250 °C.

(Fig. 6.3). Their surface morphology replicates that of the underlying alumina substrate, and the resulting surface roughness increases the effective area of the oxide in contact with the air [77]. These films are nanocrystalline with grain diameters ranging from a few nanometers to a few tens of nanometers (Fig. 6.4). The sputtered platinum catalyst has mean grain size of approximately 10 nm [79]. The films deposited at a temperature of 250 °C have a compact columnar structure, in contrast to those formed at room temperature, which have an equiaxed grain structure containing spherical pores of ~5 nm diameter that are especially evident in the first 200 nm or so from the substrate. The deposition temperature appears to influence the grain size less than its influence on the growth structure of the film.

The mean grain size can be estimated from the peak broadening in a glancing angle X-ray diffraction (GARDX) spectra (Fig. 6.5), by applying the Scherrer equation [81]

$$D = K\lambda/(\beta\cos\theta) \tag{5}$$

where D is the mean crystallite diameter, K is the shape factor (~0.9), λ is the X-ray wavelength, β is the line width at half maximum intensity (FWHM), and θ is the Bragg angle. For the RF sputtered films, this lead to values of 3–10 nm [82], compared with that of approximately 10 nm estimated by dark field transmission electron microscopy [21]. This is reasonably close to the optimal grain diameter of approximately 6 nm imposed by the Debye length criterion. It can be concluded that electrical conduction

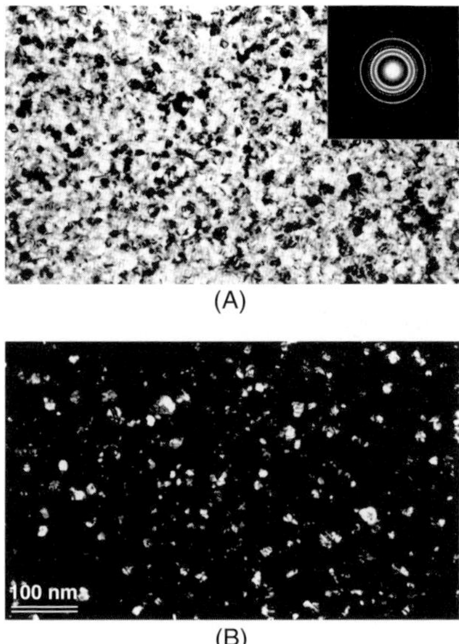

(A)

(B)

Figure 6.4 Nanocrystalline microstructure in RF sputtered tin dioxide: (A) bright-field transmission electron microscopy and selected area electron diffraction pattern; (B) dark-field transmission electron microscopy using the 110 reflection.

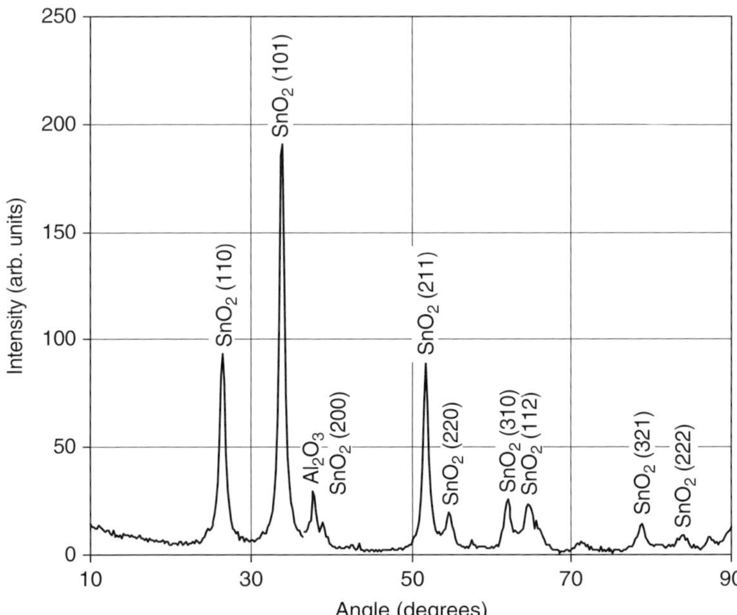

Figure 6.5 Glancing angle X-ray diffraction spectrum obtained from a tin dioxide film deposited by reactive RF sputtering.

in these nanocrystalline semiconductor films can be expected to be dominated mainly by grain boundary effects.

High-resolution electron microscopy studies (Fig. 6.6) indicate that grain boundaries are generally well defined and that planar defects and dislocations are rarely present [78, 79]. Dissociated and undissociated $\{101\}<101>$ dislocations have, however, been observed in nanocrystalline tin dioxide [83] produced by electron beam evaporation. The composition at the cores of dissociated dislocations differs from the exact stoichiometric composition and widely separated pairs of them form oxygen-deficient $\{101\}$ crystallographic shear planes (CSP) that act as sinks for oxygen vacancies. Nonstoichiometric dislocation cores and CSP thus act as traps for free electrons, influencing the electrical conductance of the film. Free carrier densities corresponding to oxygen vacancy concentrations in the range 10^{18}–$10^{19}\,\mathrm{cm}^{-3}$, consistent with a slightly understoichiometric composition, have been determined in nanocrystalline films deposited by the RF sputtering technique by Hall Effect measurements [17]. The free carrier mobility μ_H may vary between 0.1–10 cm^2/V/s depending on the film structure. Values of μ_H for the compact columnar films deposited at 250 °C were 2–3 cm^2/V/s.

The deposition of platinum, either on the surface of the film or between two layers of tin dioxide, results in a metallic layer with a well-defined

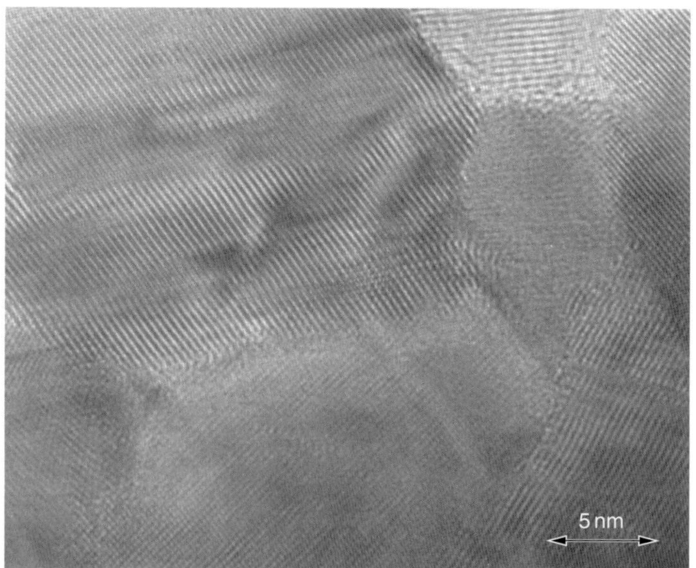

Figure 6.6 High-resolution transmission electron microscopy showing lattice plane contrast and grain boundaries in a nanocrystalline tin dioxide film.

interface with the oxide. The nonohmic contact between the metallic layer and the semiconductor leads to the formation of a Schottky barrier at the interface [59]. The presence of platinum, in or on the surface of the oxide film, influences the sensitivity of the sensor as a result of the spillover effect due to the platinum and the existence of the contact potential between the metal and the semiconductor. The difference between the work function of platinum (5.6 eV) and the electron affinity of tin dioxide (4.5 eV) generates a potential barrier that causes nonlinear behavior of the conductivity. There is, therefore, an increase in the conductivity of the material with respect to the oxide alone, particularly in the case of films deposited at higher temperatures. This is a result of the columnar structure, which provides more continuous contact with the platinum grains compared to the relatively porous structure of the films deposited at room temperature [80].

The sensitivity to carbon monoxide is greatest at temperatures between 300 °C and 350 °C for tin dioxide films without platinum, while below 250 °C it is negligible. The response (R_{air}/R_{CO}), determined by measuring the resistance while increasing the concentration in pure air, is shown in Fig. 6.7 for a sensor with a 12-nm surface layer of platinum. It can be

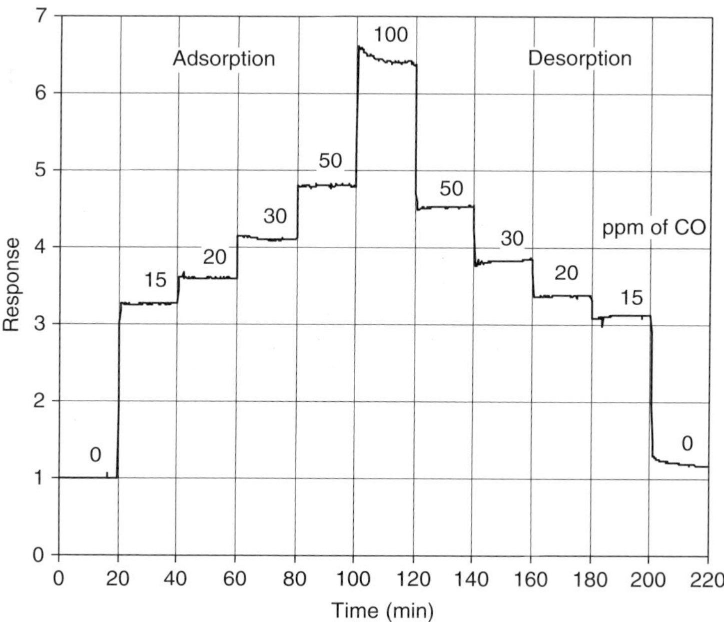

Figure 6.7   Response to carbon monoxide in pure air of a tin oxide sensor with a thin platinum surface layer operated at a temperature of 250 °C.

noted that the response is both rapid and almost completely reversible. The presence of the noble metal increases the response from 2.5 with an oxide film alone to 3.7 with the platinum surface layer, for a carbon monoxide concentration of 20 ppm, and reduces the operating temperature required to attain maximum sensitivity.

For practical applications, it is essential to understand the effects of interference between the various gases present in the air on the overall sensor response. When adsorbed on the surface of the oxide, carbon monoxide acts as an electron donor, whereas nitrogen dioxide acts as an electron acceptor and they, therefore, have opposite effects on the conductance. The effect of a small concentration (0.1 ppm) of nitrogen dioxide in the air on the response of a tin dioxide sensor without a platinum catalyst layer is shown in Fig. 6.8. By comparing the pairs of measurements made at three different operating temperatures, it can be seen that while at 250 °C the effect due to the nitrogen dioxide dominates the response, the interference decreases with increasing operating temperature and is almost negligible at

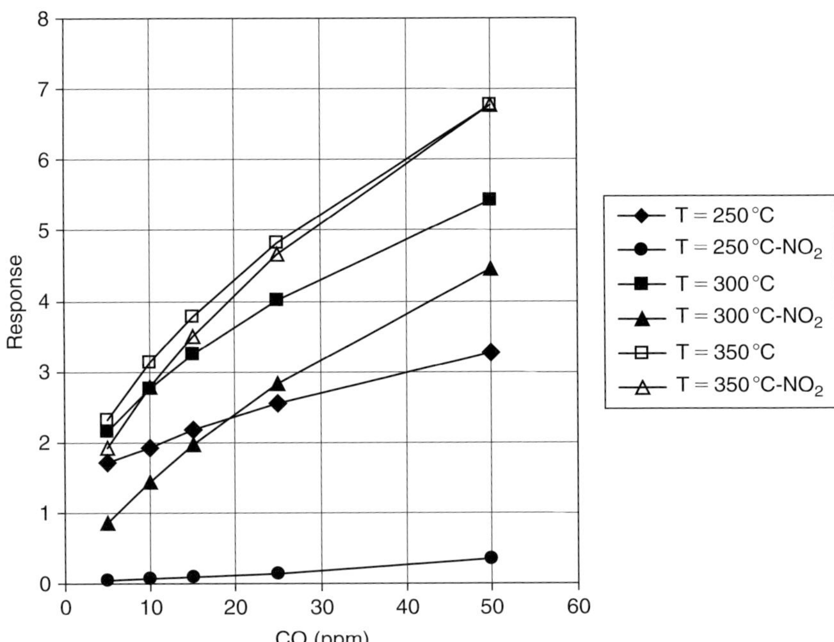

Figure 6.8   Response of a tin dioxide sensor without a platinum catalyst, at three operating temperatures, to different concentrations of carbon monoxide in the presence of 0.1 ppm of nitrogen dioxide.

350 °C. High selectivity can, thus, be achieved by variation of the operating temperature.

The results of measurements made with two different sensors over a range of carbon monoxide concentrations in pure air, at an operating temperature of 300 °C, are shown in Fig. 6.9. Deposition of the tin dioxide layers was carried out at 250 °C (S1) and room temperature (S2). The film deposited at 250 °C has noticeably greater sensitivity than that of the film deposited at room temperature. This is due to the superior crystallinity and the columnar structure in the film deposited at higher temperature. The oxide film deposited at 250 °C also had higher electrical resistance in pure air, i.e., approximately $1^6\,\Omega$ compared with approximately $2^5\,\Omega$ for the one deposited at room temperature [78].

The sensitivity of tin dioxide films deposited at room temperature can be significantly improved by the addition of a thin layer of platinum. This effect is most pronounced when the noble metal layer is deposited between two layers of oxide (Fig. 6.10) so that two metal/oxide interfaces are formed. The presence of the platinum generates a higher concentration of adatoms at the interface with the oxide [84, 85] and thus a higher free carrier density in the semiconductor, which leads to an increase in the sensitivity. The platinum acts to catalyze the rate of reaction between the tin oxide and the gas molecules, without altering the work function. The superior performance of the films with an internal layer of platinum confirms that the reaction is not limited solely to the surface of the film but that substantial diffusion of gas molecules also occurs within the oxide.

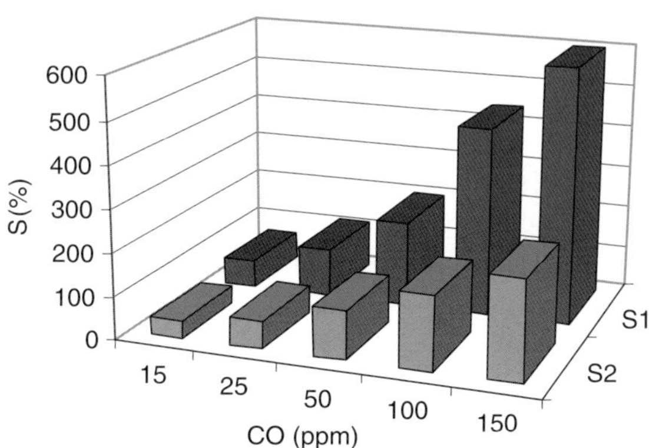

Figure 6.9   Sensitivities of tin dioxide films deposited at 250 °C (S1) and room temperature (S2) to various concentrations of carbon monoxide at an operating temperature of 300 °C.

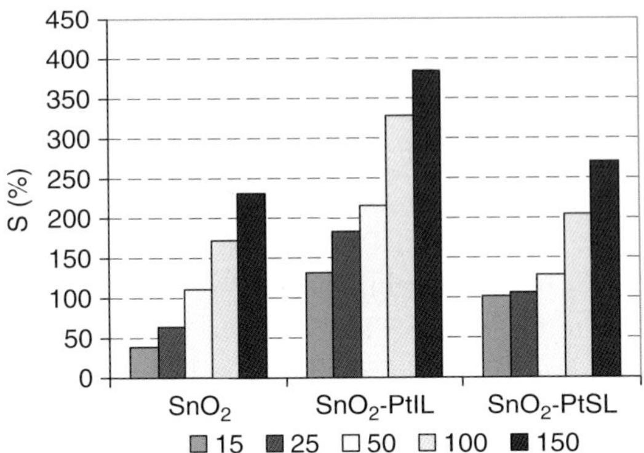

Figure 6.10   Sensitivity to different concentrations of carbon monoxide of a tin dioxide film alone (SnO$_2$), with an internal platinum layer (SnO$_2$-PtIL), and with a surface platinum layer (SnO$_2$-PtSL).

5.2. Pulsed Laser Deposition

The pulsed laser technique has been used to deposit tin dioxide thin films on the same type of alumina substrate using a KrF excimer laser operating at a wavelength of 248 nm [86, 87]. A pulse length of 12 ns was used with a repetition frequency of 39 Hz. The substrate temperature was varied between 25 °C and 600 °C at a constant oxygen pressure of 2 kPa (150 mtorr). In addition, the oxygen pressure was varied between 1 and 200 mtorr at a constant temperature of 300 °C to investigate the effect of pressure during deposition. In-situ doping of the oxide layer was carried out by a co-deposition method by attaching strips of platinum to the tin dioxide target. An annealing treatment was performed after deposition as in the case of the RF sputtered films.

The structure of thin films of tin dioxide produced by the PLD technique has a marked dependence on the substrate temperature during deposition. It shows a gradual transition from an equiaxed granular porous structure to a compact columnar one with increasing temperature [86]. At room temperature, the mean grain diameter is approximately 4 nm, whereas at a temperature of 300 °C, it increases to 12 nm, when individual crystallites begin to agglomerate in clusters of similar orientation and the film starts to take on a more columnar appearance with its porosity aligned perpendicularly to the substrate surface (Fig. 6.11). The columnar structure becomes more pronounced at 450 °C, though residual equiaxed grains persist, until the structure of the film becomes completely columnar at a deposition temperature of 600 °C.

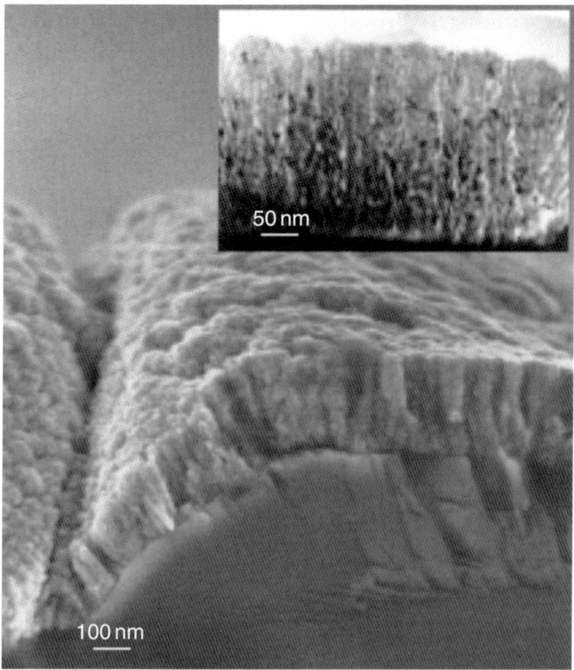

Figure 6.11   Cross-sectional scanning electron microscope image of a tin dioxide film deposited on a polycrystalline alumina substrate, at a temperature of 300 °C, by the pulsed laser method. Inset shows a transmission electron microscope bright-field image of the same film for comparison.

In-situ doping with small amounts of platinum by the co-deposition method resulted in small particles of metallic platinum evenly dispersed in the films. The overall appearance of these films was similar to that of the undoped films deposited at the same substrate temperature. However, the films produced by the PLD technique contained a higher density of dislocations and planar defects compared to the same material deposited by sputtering.

A small partial pressure of oxygen in the vacuum chamber during deposition is required to maintain the correct stoichiometry in the growing film. This plays a fundamental role in the nucleation of tin dioxide while preventing the precipitation of pure tin or monoxide [88]. X-ray diffraction of films deposited under vacuum indicates the presence of both polycrystalline tin dioxide and amorphous tin oxide [86]. The latter disappears only at a higher deposition temperature (600 °C) where oxidation of the film can easily occur. The mean grain size remains constant at approximately

4 nm with increasing partial pressure of oxygen up to 50 mtorr but increases to 8 nm at 200 mtorr (Fig. 6.12). Films produced at low partial pressure of oxygen have compact columnar structure, with columns of 10–15 nm width consisting of small equiaxed or slightly elongated nanocrystals. Films deposited at higher partial pressures of oxygen have granular, porous structure and a larger mean grain size, with individual grains tending to form clusters and the porosity perpendicular to the substrate.

The resistivity decreases approximately from 200 to $10\,\Omega$ cm for films deposited at room temperature and 300 °C and thereafter remains essentially constant with further increase in substrate temperature [89]. The high resistivity of films deposited at room temperature is due to their low density and highly porous structure. Only by using a deposition temperature above 150 °C it is possible to obtain more conductive films. The electrical resistivity reaches a shallow minimum at a deposition temperature of 450 °C. This is possibly because the hybrid structure of films deposited at this temperature tends to be more homogeneous than that of the completely columnar films produced at higher temperature, in which the voids separating the neighboring columns can be expected to reduce the conductivity due to poor electrical contact.

Comparison of the sensitivity to carbon monoxide of films deposited at 300 °C under an oxygen partial pressure of 150 mtorr with that of in-situ doped films containing small amounts of platinum indicated that 2 at% gave the best performance over a range of gas concentrations from 20 to 500 ppm, at operating temperatures from 150 to 350 °C [86]. The presence

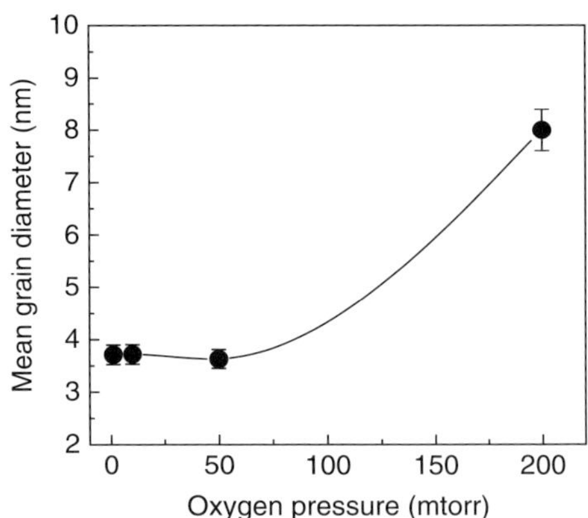

Figure 6.12   Dependence of the mean grain size of films deposited by PLD at 300 °C on the partial pressure of oxygen during deposition.

of the platinum also reduced significantly the temperature at which the maximum sensitivity was obtained from 300 to 200 °C.

The variation of the resistivity in pure air with the partial pressure of oxygen is shown in Fig. 6.13. The films deposited at low partial pressure of oxygen are quite good conductors due to their microstructural characteristics. The extremely small mean grain size and the absence of porosity in these films facilitate electrical conduction in the material. In fact, the grain radius (~2 nm) is smaller than twice the Debye length for tin dioxide (~3 nm).

Electrical transport is, therefore, no longer controlled by the grain boundary contribution but by the material in the interior of the grains because the space charge region occupies the entire grain, and the potential barrier at the interface between adjacent grains is rather low. Reduction of the grain size, therefore, allows transport of electrons from one grain to the next with a negligible influence due to the presence of the boundary. The formation of agglomerates with similar crystallographic orientation constitutes an additional factor that facilitates electrical conduction. As the partial pressure of oxygen is increased, the resistivity increases as a result of the larger grain diameter and the intercolumnar porosity within the film itself.

It can, therefore, be concluded that the grain size is a very important parameter influencing the performance of tin oxide gas sensors. A grain diameter of approximately 10 nm appears to be sufficient to preserve the

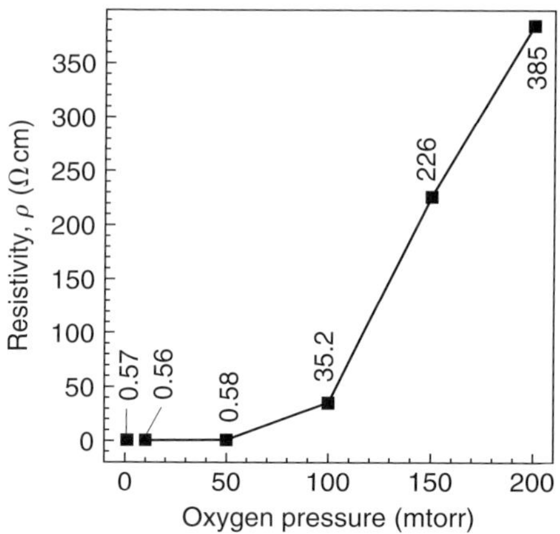

Figure 6.13   Dependence of the resistivity at room temperature of tin oxide films deposited by PLD at 300 °C on the partial pressure of oxygen during deposition.

semiconducting properties of the oxide; however, if the grain size is less than 2L, the material begins to behave more like a conductor.

The presence of water vapor causes an increase in the conductivity of tin oxide due to the adsorption of water molecules on the surface. Adsorption and desorption of $(OH)^-$ species occur at temperatures of approximately 150–160 °C [90], while the optimal temperature for the detection of carbon monoxide is 300–350 °C, at which these $(OH)^-$ species will be already completely desorbed. It is, therefore, important to take into consideration the effect of humidity in the case of detection of NO_2, where the optimal temperature for detection is less than 200 °C.

6. INTEGRATED SOLID-STATE SENSORS

Microelectronic fabrication techniques can be used in combination with thin-film synthesis methods to produce integrated gas microsensors that are fully compatible with CMOS technology. The advantages include cost reduction, device miniaturization, lower power consumption, and the capability to construct multisensor arrays [91]. The use of silicon micromachining methods enables thermally isolated structures to be fabricated to reduce heat dissipation from the metal oxide sensing elements. Thermal isolation can be achieved by depositing the sensing layer on thin free-standing membranes fabricated by micromachining methods. However, it is essential to optimize the membrane dimensions and heater design to ensure a uniform temperature distribution over the sensitive region and to reduce power requirements.

Figure 6.14 shows an integrated sensor array consisting of eight microsensors on a single silicon chip. The sensing elements have dimensions of $0.5 \times 0.5\,mm$, and the entire device is mounted in a standard TO-8 package. Each one of these microsensors can be operated at a different temperature to make it selective to a specific gas. As has already been discussed, metal oxide gas sensors need to be heated during operation to allow the discrimination of different gases, and precise temperature control is, therefore, required to maximize the selectivity. Finite element calculations were used to optimize the design to achieve the correct mechanical and thermal properties for the heater and other components of the device [92]. Silicon nitride was chosen in preference to silica as the membrane material, despite its greater thermal conductivity, because it allows the use of a thinner membrane, resulting in lower power consumption. The tin oxide thin film was deposited by reactive sputtering because this process is compatible with microelectronic fabrication technology.

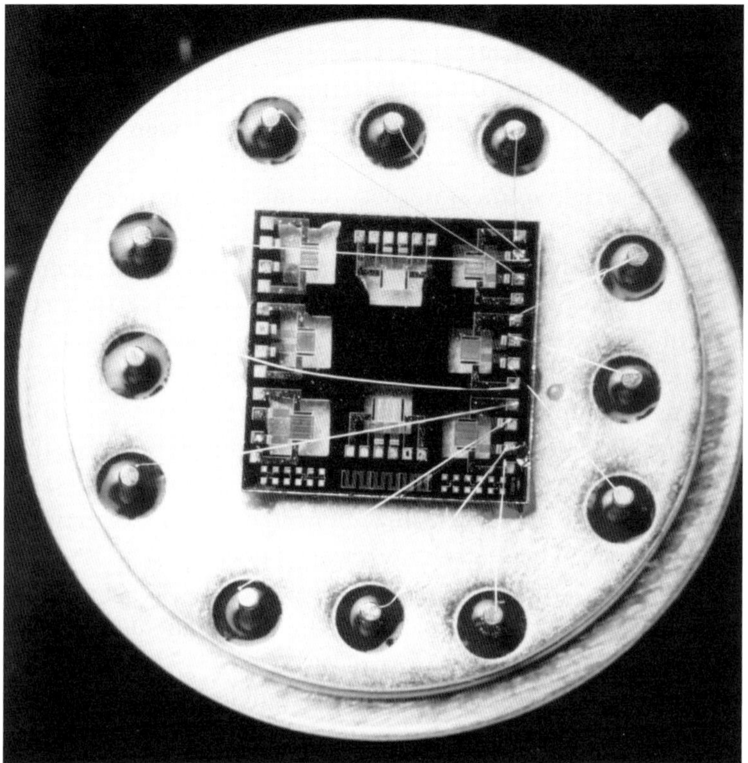

Figure 6.14   A solid-state integrated microsensor array consisting of eight tin dioxide sensing elements on a single silicon chip.

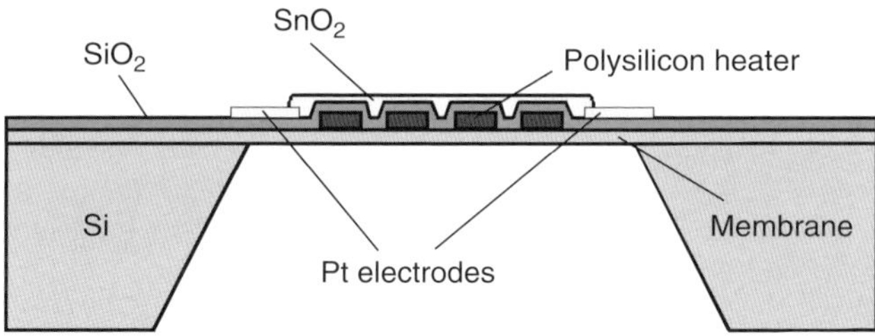

Figure 6.15   Schematic diagram of an individual microsensor element showing details of its construction.

The construction of an individual microsensor is illustrated schematically in Fig. 6.15. It consists of a tin oxide sensing layer with platinum electrodes, a silica layer for electrical insulation, and the polysilicon heater, all of which are thermally and electrically isolated from the substrate by a thin silicon nitride membrane. The entire device is fabricated on a 300-µm thick P-type <100> silicon chip. Details of the various technological processing steps involved in the fabrication of this device can be found elsewhere [93]. The power consumption is less than 50 mW at an operating temperature of 350 °C.

A cross-sectional transmission electron micrograph of a microsensor (Fig. 6.16) indicates that the thickness of the nanocrystalline tin dioxide layer is approximately 300 nm in this case and the mean grain size of the material is approximately 10 nm. Evidence is visible of a columnar type growth structure, with columns approximately 100 nm in width. The intermediate silica layer is 450 nm thick and completely amorphous and, acting as an insulator, provides a smooth surface on which to deposit the oxide film. The polysilicon heater is approximately 600 nm thick and is polycrystalline with a grain diameter of approximately 200–400 nm, although the grains are noticeably elongated in the direction parallel to the substrate. These structures are supported on a 300-nm thick amorphous silicon nitride membrane, produced by low-pressure CVD (**LPCVD**), which has low internal stress and good adherence to the polysilicon layer.

The multisensor array is capable of high sensitivity and fast response times, allowing the detection of low concentrations of aromatic hydrocarbons,

Figure 6.16 Cross-sectional transmission electron micrograph through a microsensor element showing the structure of the nanocrystalline tin dioxide layer, the silica insulating layer, the polysilicon heater, and the amorphous silicon nitride membrane.

carbon monoxide, and nitrogen dioxide. The experimental data shown in Table 6.1 are the results of laboratory tests for the detection of low concentrations of nitrogen dioxide in pure air using sensor elements with tin oxide layers of different thicknesses [94]. The sensitivity increases significantly with increasing film thickness for concentrations of nitrogen dioxide in the 0.5–1.5 ppm range. The times for the sensor to reach equilibrium and for recovery after exposure to the gas were similar and were typically less than 3 min.

The performance of metal oxide gas sensors is greatly influenced by the grain size and the surface morphology of the active layer because these factors determine the effective surface area on which the adsorption of gas molecules can occur. Nanocrystalline films with a porous columnar structure have higher sensitivities and lower optimal operating temperatures than those with larger grain sizes [95]. Films deposited by RF sputtering generally have finer grain sizes and a higher degree of crystallinity than those produced by other techniques such as spraying [96] or chemical vapor deposition [97].

Other configurations of integrated tin oxide gas sensors have been developed that are completely compatible with standard CMOS fabrication methods [98, 99]. Procedures such as silicon etching to create the thermally isolated microhotplate and the deposition of the oxide film by RF sputtering are carried out as postprocessing steps. Devices with a 1-nm platinum layer deposited on a 300 nm thick layer of oxide to increase the selectivity were tested for their response to carbon monoxide concentrations of 200–1000 ppm over an operating temperature range 200–300 °C [100]. The response was fastest at operating temperatures of 250–300 °C and was dependent on the gas concentration; in general, higher concentrations gave the shortest response times.

Table 6.1 Sensitivity to low concentrations of nitrogen dioxide in pure air for tin dioxide films of three different thicknesses at an operating temperature of 250 °C

NO_2 (ppm)	300 nm	400 nm	500 nm
0.5	78	110	180
1.0	151	205	310
1.5	197	275	330

7. THICK-FILM TECHNOLOGY

The original gas sensors were constructed using thick films of tin dioxide. Although thin-film technology has gained in popularity among research workers, due to the ease with which the microstructure can be controlled simply by adjusting the deposition conditions, many commercial gas sensors nevertheless still use thick-film technology based on wet chemistry powder synthesis methods and screen printing technology. Some examples of the application of the older technology in the construction of metal oxide sensors for air-quality monitoring will now be considered.

Nanosized powders of several metal oxides and a mixed oxide were prepared either by a sol–gel process using alkoxide precursors or, in the case of the mixed oxide, by a thermal decomposition of the corresponding hexacyanocomplex [101]. Thick-film sensors were fabricated by screen printing on thin alumina substrates that had a heating element and interdigitated gold electrodes for making the resistance measurements. The thick films made in this way were still extremely porous even after firing at 850 °C, a temperature that was selected to maintain the particle size within the 30–50 nm range.

The performance of sensors fabricated from these synthesized oxide nanoparticles and commercially available nanopowders was tested in extensive field trials [101, 102], in parallel with measurements made conventional air-quality monitoring systems. By using the sensors based on different oxides, it is possible to overcome the problem of interferences in the presence of different gases. The sensors based on tin dioxide and tantalum-doped titanium oxide films were sensitive mainly to carbon monoxide. In contrast, the sensor based on a perovskite-type lanthanum iron oxide, a P-type semiconductor, prepared by thermal decomposition, was sensitive principally to nitrogen dioxide. By comparing the sensor data to measurements made by standard analytical equipment at a conventional monitoring station, calibration curves were generated that allowed the sensor response to be directly related to the concentrations of both these gases.

In the light of the results obtained from these preliminary experiments, a system for air-quality monitoring was developed [103], containing an array of five different sensors based on nanograin films of tin dioxide, tantalum, and niobium-doped titanium dioxide, perovskite-type $LaFeO_3$, and indium oxide (for the detection of ozone). Addition of the metallic ions to titanium dioxide is necessary to maintain the nanocrystalline structure by inhibiting sintering during firing [104]. The indium oxide film, which was made with a commercially available nanopowder, had a somewhat larger particle size than the others, which was in the 200–400 nm range.

In general, the sensor data corresponded quite closely to the evolution with time of the concentration of pollutant gases recorded by the nearby conventional monitoring stations. Figure 6.17 shows the readings obtained from the tin dioxide sensor superimposed on the variation of the level of carbon monoxide measured by the standard instrumentation. The sensor signal was recorded every minute, but values were averaged over an hour to facilitate comparison with the data obtained using the standard analysis methods. The correspondence between the signal from the $LaFeO_3$ mixed oxide sensor and the true concentration of nitrogen dioxide was less good because this device was also sensitive to other reducing gases, such as volatile organic compounds.

The main sources of error in the sensor measurements are due to the influence of ambient temperature and humidity. These effects have to be taken into consideration to compensate the sensor response, which is affected by the temperature of the surroundings independently of the operating temperature of the sensor itself [105]. Adsorption of water molecules on the surface of the oxide leads to the formation of hydroxyl groups, which cause changes in the oxygen affinity and that act as donors. At low gas concentrations and low operating temperatures, reactions with these groups will tend to dominate the sensor response [106]. It is, however, the

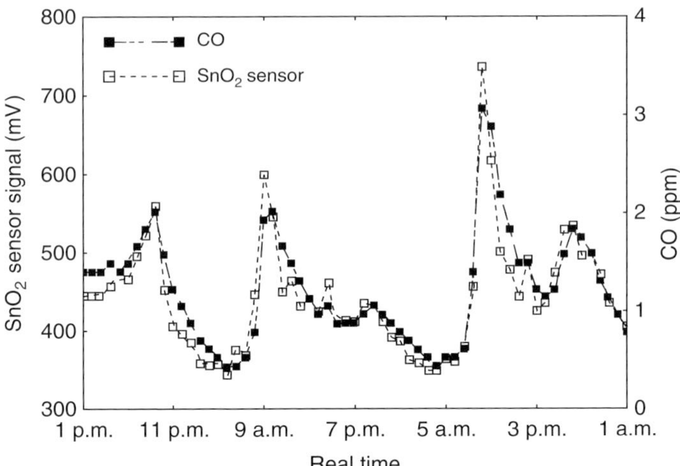

Figure 6.17 Response of a tin dioxide sensor compared with the concentration of carbon monoxide measured in air by standard monitoring equipment.
Reproduced with permission from Carotta MC et al. [103]

partial pressure of the water vapor rather than the relative humidity that determines the sensor response, and the true gas concentration can be derived from an equation of the form [107]:

$$CO_{conc} = \left[\left(G - G_0 - Ap^\alpha \right) / \left(B + Cp \right)^\alpha \right]^{1/\gamma} \tag{6}$$

where CO_{conc} is the concentration of carbon monoxide, p is the water vapor pressure, G is the conductance, G_0 is the conductance in the absence of the gas and water vapor; A, B, C, α, and γ are constants obtained by fitting experimental values for different concentrations of CO and different water vapor partial pressures.

Another thick-film gas sensor developed for air monitoring consists of three thick-film sensors on microhotplates, which are monolithically integrated with digital temperature controllers and the interface circuitry on a single silicon chip [108, 109]. The surfaces of the microhotplates are coated with tin dioxide to create the gas-sensitive layer. A transistor heater was used, rather than a resistive heating element, to reduce the overall power consumption and allows operation at temperatures up to 350 °C. One of the sensors is coated with a pure tin oxide layer, whereas the other two are doped with 0.2 and 3 wt% palladium. The performance of the device was assessed with mixtures of the environmentally relevant gases: carbon monoxide; methane; and nitrogen dioxide. The undoped tin dioxide sensor was selective to nitrogen dioxide, whereas the sensors doped with 0.2 wt% and 3 wt% palladium, respectively, were selective to carbon monoxide and methane.

Extensive testing was carried out [110] to evaluate the calibration accuracy, minimum detection limit, sensitivity, selectivity, long-term stability, cross sensitivity to other gases, and the effects of interference due to humidity. The optimal operation temperatures were 275 °C for the detection of carbon monoxide, 300 °C for the detection of nitrogen dioxide, and 300 °C for the detection of methane. Although the overall performance of the device was generally good, problems were experienced with calibration drift after extended periods of operation. The cross sensitivity to humidity and volatile organic compounds could be compensated, but measurements with binary mixtures of carbon monoxide and methane gave poor results due to the influence of calibration drift. Cross sensitivity to ozone was also a difficulty in measurements of nitrogen dioxide because the sensitivity is higher to ozone than that to nitrogen dioxide. These cross-sensitivity problems could be resolved to some extent by modulation of the operation temperature.

During operation in the field, the complex mixture of gases present in real air can lead to significant systematic and random errors. Standard analytical instruments were, therefore, used to provide a direct calibration of the sensor data because these errors prevented the use of the laboratory calibration. Temperature modulation of the sensor hotplates was not as useful in resolving the problem of analyzing multicomponent gas mixtures as it had been in the laboratory tests. Measurements made at constant temperature gave results in closer agreement with the standard reference methods. Because the conventional analytical instruments provide concentrations that are averaged over 10-min periods compared to readings every second for the sensor system, there were difficulties in correlating the noisy raw data from the sensors to these smoothed averaged data. This problem is even more evident when temperature modulation is used because the rapid variation of the temperature prevents the metal oxide layer from reaching chemical equilibrium, resulting in non-steady state operation.

8. INNOVATIVE METAL OXIDE ARCHITECTURES

As already discussed, the sensitivity of metal oxide gas sensors is dependent on various factors, such as the grain size, porosity, surface morphology, and dopant levels. To some degree, the sensitivity can be increased by tailoring the structure of the metal oxide layer by careful control of the synthesis conditions, particularly when thin-film techniques are used for fabricating the sensor. The interaction of gas molecules with the semiconducting oxide is primarily a surface phenomenon. Further improvements in sensor performance are, therefore, likely to be due to the development of novel film architectures that allow the gas molecules access to much larger surface areas of metal oxide and the integration of these structures in miniaturized integrated devices.

Varghese and Grimes [111] have reviewed some of the currently available routes for the production of metal oxide nanoarchitectures with increased surface areas for gas sensing applications, including mesoporous thin films, nanowires or nanobelts, and nanotubes. Among the techniques considered were mesoporous thin films produced using a sol–gel process, titania nanowires formed using a highly ordered nanoporous alumina membrane as a template, and arrays of titania nanotubes formed by anodization of titanium. The use of nanoporous alumina for humidity sensing and titania nanotubes as hydrogen sensors were considered examples. An excellent response can be achieved, even in the low humidity regime, with a layer of pores of uniform nanodimension in alumina containing adsorbed anionic impurities, while titania nanotubes are highly sensitive to low concentrations of a reducing gas.

Quasi one-dimensional nanobelts of semiconducting oxides were first synthesized by a high-temperature evaporation technique [112]. They are highly pure, monocrystalline, uniformly structured, rectangular in cross section, and generally defect free, with widths of 30–300 nm and lengths of up to several millimeters. This type of structure is ideal for studying dimensionally confined transport in semiconducting metal oxides and offers interesting possibilities for the construction of functional devices. Because of the high specific surface area, surface-related effects, such as catalytic activity and adsorption, are enhanced, and due to the elimination of the grain coalescence and the drift in electrical response typical of polycrystalline materials, they are very promising materials for the development of a new generation of metal oxide sensors with increased stability [113].

The synthesis of metal oxide nanowires can be carried out by thermal decomposition of precursor powders, followed by vapor–solid or vapor–liquid–solid growth. The materials that have been prepared by this method include tin dioxide [114–117], zinc oxide [117], indium oxide [116–119], and tungsten oxide [120]. Figure 6.18 shows a scanning electron micrograph of indium oxide nanobelts that were deposited on an alumina substrate. By varying the temperature of the substrate from 800 to 1100 °C, different crystalline structures could be obtained, ranging from coarse equiaxed grains to nanowires and nanobelts. By seeding the substrate with sputtered particles of pure metallic indium, controlled growth of the nanowires can be achieved along the <100> crystal directions.

Figure 6.18   Indium oxide nanobelts.
Reproduced with permission from Bianchi et al. [118]

Measurements of the electrical characteristics of tin dioxide and indium oxide nanowires at a temperature of 400 °C, in the presence of ozone concentrations of a few hundred ppb, indicated an appreciable sensitivity, whereas the response of zinc oxide nanowires was far poorer [117]. At a temperature of 200 °C, indium oxide nanowires with widths less than 100 nm also showed much better response to nitrogen dioxide concentrations of a few ppm than similar wires of width 500 nm [118]. This behavior can be attributed to their reduced lateral dimension that allows completely depletion of carriers. Three-dimensional orthogonal networks of tungsten oxide nanowires synthesized by thermal evaporation of tungsten powder in oxygen at 1400–1450 °C [120] showed a high sensitivity to nitrogen dioxide, for measurements made at 300 °C over a concentration range 50–1000 ppb, but had only a rather weak response to carbon monoxide at the same temperature. This is, therefore, a promising material for the development of nitrogen dioxide sensors with low cross sensitivity.

Attaching contact electrodes to nanowires is a major technical challenge, and to overcome this difficulty, it has been proposed to use changes in the visible photoluminescence of metal dioxide nanowires due to the presence of nitrogen dioxide, rather than the electrical properties [117, 121]. The adsorbed surface species generate surface states that quench the photoluminescence, and the magnitude of the effect depends on the size and shape of the nanostructure, the crystal surface, and the surface oxygen vacancies. It was possible experimentally to discriminate nitrogen dioxide concentrations in the range 0.1–20 ppm using tin dioxide nanowires at 120 °C. Interferences due to water vapor and reducing gases are negligible, and for zinc oxide, the quenching is maximum at room temperature. This leads to the possibility of applying metal oxide nanowires for the construction of low-power selective optical gas sensors operating at room temperature.

9. SENSOR NETWORKS FOR AIR MONITORING

The integration of metal oxide sensors in CMOS-compatible components allows the construction of inexpensive, portable air-quality monitoring systems based on gas-sensor arrays. A prototype of such a system with high precision and a wide dynamic range has been designed [122] using an array of metal oxide thin-film sensors on micromachined substrates, an integrated front-end with an analog-to-digital converter and temperature control unit, and a data processing system to allow information extraction from acquired data. The sensors are fabricated by spin coating a silicon wafer that contains hundreds of miniature devices, with a 100 nm layer of tin dioxide nanopowder prepared by the sol–gel technique. The gas concentration is directly proportional to the variation in resistance of the sensor. Tests of the system carried out with various concentrations of several volatile organic compounds demonstrated its effectiveness and precision.

Such sensor systems could act as nodes in sensor networks for real-time air monitoring with fast response capability and high spatial resolution. The use of solid-state sensors and appropriate signal processing techniques, in combination with data transmission via the existing telecommunications infrastructure, would enable the creation of intelligent sensor networks able to provide highly localized information on air pollution. A system for monitoring nitrogen dioxide was developed based on a commercial tin dioxide gas sensor [123], and its performance was evaluated by comparing the sensor readings with the concentrations measured at a nearby fixed monitoring station. The network connectivity can also be used for automatic calibration of the sensors, as well as for data collection. Nitrogen dioxide concentration can be determined with reasonable accuracy in this way provided that adequate compensation for temperature and humidity is applied. Among various factors that influence the sensor response, such as interferences from other gases and humidity, baseline drift is the major source of error. This means that the system has to be recalibrated as often as necessary to maintain its accuracy.

Installation of a sensor network in a densely populated urban area can allow the spatial and temporal variation of air pollutants to be determined and correlation of this data with other types of information, such as health effects. Solid-state sensors can generate a data point every second for each sensor, creating problems in processing and analyzing such large quantities of data. Distributed wireless sensor networks can be used together with grid computing technology to resolve the data handling challenge to acquire real-time data on air pollution in a cost-effective manner [124]. This makes it feasible to use sensor networks integrated with telecommunications and informatics systems to monitor, model, and predict the evolution of air pollutants by means of fixed roadside and mobile vehicle-mounted sensors [125, 126].

References

[1] Yamazoe N. Chemical sensors R&D in Japan. Sens Actuators B 1992;6:9–15.
[2] Ikohura K, Watson J. The stannic oxide gas sensor – principles and applications. Boca Raton (FL): CRC Press; 1994.

[3] Schierbaum KD. Engineering of oxide surfaces and metal/oxide interfaces for chemical sensors: recent trends. Sens Actuators B 1995;24–25:239–47.

[4] Schierbaum KD, Geiger G, Weimar U, Göpel W. Specific palladium and platinum doping for SnO_2-based thin film sensor arrays. Sens Actuators B 1993;13–14:143–7.

[5] Bârsan N, Schweizer-Berberich M, Göpel W. Fundamental and practical aspects in the design of nanoscaled SnO_2 gas sensor: a status report. Fresenius J Anal Chem 1999;365: 287–304.

[6] Ogawa H, Abe A, Nishikawa M, Hayakawa S. Electrical properties of tin oxide ultrafine particle films. J Electrochem Soc 1981;128:2020–5.

[7] Schweizer-Berberich M, Zheng JG, Weimar U, Bârsan N, Pentia E, Tomescu A. The effect of Pt and Pd surface doping on the response of nanocrystalline tin dioxide gas sensors to CO. Sens Actuators B 1996;31:71–5.

[8] Sberveglieri G, Groppelli S, Nelli P, Camanzi A. Bismuth-doped tin oxide thin-film gas sensors. Sens Actuators B 1991;3:183–9.

[9] Sberveglieri G, Groppelli S, Nelli P. Highly sensitive and selective NO_x and NO_2 sensor based on Cd-doped SnO_2 thin films. Sens Actuators B 1991;4:457–61.

[10] He YS, Campbell JC, Murphy RC, Arendt MF, Swinnea JS. Electrical and optical characterisation of $Sb:SnO_2$. J Mater Res 1993;8:3131–4.

[11] Bruno L, Pijolat C, Lalauze R. Tin dioxide thin film gas sensor prepared by chemical vapour deposition: Influence of grain size and thickness on the electrical properties. Sens Actuators B 1994;18–19:195–9.

[12] Davazoglou D. Optical properties of SnO_2 thin films grown by atmospheric pressure chemical vapour deposition oxidizing $SnCl_4$. Thin Solid Films 1997;302:204–13.

[13] Kanamori M, Suzuki K, Ohya Y, Takahashi Y. Analysis of the change in the carrier concentration of SnO_2 thin film gas sensor. Jpn J Appl Phys 1994;33:6680–3.

[14] Hassan AK, Gould RD, Keeling AG, Williams EW. Electronic conduction processes in Pt-doped tin oxide thin films prepared by RF magnetron sputtering. J Mater Sci: Mater Electron 1994;5:310–4.

[15] Kissine VV, Sysoev V, Voroshilov SA. Individual and collective effects of oxygen and ethanol on the conductance of SnO_2 thin films. Appl Phys Lett 2000;76:2391–3.

[16] Chang SC. Thin-film semiconductor NO_x sensor. IEEE Trans Electron Devices 1979;ED-26:1875–80.

[17] Gutiérrez FJ, Arés L, Robla JI, Getino JM, Horrillo MC, Sayago I, et al. Hall coefficient measurements for SnO_2 doped sensors, as a function of temperature and atmosphere. Sens Actuators B 1993;15–16:98–104.

[18] Barbi GB, Santos Blanco J. Structure of tin oxide layers and operating temperature as factors determining the sensitivity performances to NO_x. Sens Actuators B 1993; 15–16:372–8.

[19] Vlachos DS, Papadopoulos CA, Avaritsiotis JN. Dependence of sensitivity of SnO_x thin-film gas sensors on vacancy defects. J Appl Phys 1996;80:6050–4.

[20] Capehart TW, Chang SC. The interaction of tin oxide films with O_2, H_2, NO, and H_2S. J Vac Sci Technol 1981;18:393–7.

[21] Rickerby DG, Horrillo MC. Crystallite size distributions and lattice defects in r.f. sputtered nanograin TiO_2 and SnO_2 films. J Nanostruct Mater 1998;10:357–63.

[22] Cricenti A, Generosi R, Scarselli MA, Perfetti P, Siciliano P, Serra A, et al. $Pt:SnO_2$ thin films for gas sensor characterized by atomic force microscopy and x-ray photoemission spectromicroscopy. J Vac Sci Technol B 1996;14:1527–30.

[23] Cirilli F, Kaciulis S, Mattogno G, Galdikas A, Mironas A, Senuliene D, et al. Influence of Cu overlayer on the properties of SnO_2-based gas sensors. Thin Solid Films 1998;315:310–5.

[24] Sberveglieri G, Faglia G, Groppelli S, Nelli P, Perego C. Oxygen gas sensing properties of undoped and Li-doped SnO_2 thin films. Sens Actuators B 1993;13–14: 117–20.

[25] Kaciulis S, Mattogno G, Galdikas A, Mironas A, Setkus A. Influence of surface oxygen on chemoresistance of tin oxide film. J Vac Sci Technol A 1996;14:3164–8.

[26] Chung YS, Hubenko A, Meyering L, Schade M, Zimmer J. Morphology and phase of tin oxide thin films during their growth from metallic tin. J Vac Sci Technol A 1997;15:1108–12.

[27] Williams G, Coles GSV. Gas sensing properties of nanocrystalline metal oxide powders produced by a laser evaporation technique. J Mater Chem 1998;8:1657–64.

[28] Hu WS, Liu ZG, Wu ZC, Feng D. Comparative study of laser ablation techniques for fabricating nanocrystalline SnO_2 thin films for sensors. Mater Lett 1996;28:369–72.

[29] Hu WS, Liu XG, Zheng JG, Hu XB, Guo XL, Göpel W. Preparation of nanocrystalline SnO_2 thin films used in chemisorption sensors by pulsed laser reactive ablation. J Mater Sci: Mater Electron 1997;8:155–8.

[30] Koinkar VN, Ogale SB. Pulsed excimer laser processing of optical thin films. Thin Solid Films 1991;206:259–63.

[31] Lal R, Grover R, Vispute RD, Viswanathan R, Godbole VP, Ogale SB. Sensor activity in pulsed laser deposited and ion implanted tin oxide thin films. Thin Solid Films 1991;206:88–93.

[32] Philips HM, Li Y, Bi Z, Zhang B. Reactive pulsed laser deposition and laser induced crystallization of SnO_2 transparent conducting thin films. Appl Phys A 1996;63:347–51.

[33] Cabot A, Arbiol J, Morante JR, Weimar U, Bârsan N, Göpel W. Analysis of the noble metal catalytic additives introduced by impregnation of as obtained SnO_2 sol-gel nanocrystals for gas sensors. Sens Actuators B 2000;70:87–100.

[34] Wu NL, Wang SY, Rusakova IA. Inhibition of crystallite growth in the sol-gel synthesis of nanocrystalline metal oxides. Science 1999;285:1375–7.

[35] Wu NL, Wu LF, Yang YC, Huang SJ. Spontaneous solution-sol-gel process for preparing tin oxide monolith. J Mater Res 1996;11:813–20.

[36] Cao X, Cao L, Yao W, Ye X. Influences of dopants on the electronic structure of SnO_2 thin films. Thin Solid Films 1998;317:443–5.

[37] Martinelli G, Carotta MC. Sensitivity to reducing gas as a function of energy barrier in SnO_2 thick-film gas sensor. Sens Actuators B 1992;7:717–20.

[38] Kappler J, Bârsan N, Weimar U, Diéguez A, Alay JL, Romano-Rodriguez A, et al. Correlation between XPS, Raman and TEM measurements and the gas sensitivity of Pt and Pd doped SnO_2 based gas sensors. Fresenius J Anal Chem 1998;361:110–4.

[39] Matsuo Y, Hashimoto K, Sugie Y, Nakamura T, Tsunoda J. Micromachining of tin oxide film by electrochemical reduction process. J Electrochem Soc 1998;145:3067–9.

[40] Rumyantseva M, Labeau M, Delabouglise G, Ryabova L, Kutsenok I, Gaskov A. Copper and nickel doping effect on interaction of SnO_2 films with H_2S. J Mater Chem 1997;7:1785–90.

[41] Brinzari V, Korotcenkov G, Golovanov G. Factors influencing the gas sensing characteristics of tin dioxide films deposited by spray pyrolysis: understanding and possibilities for control. Thin Solid Films 2001;391:167–75.

[42] Yagi I, Hagiwara Y, Murakami K, Kaneko S. Growth of highly oriented SnO_2 thin films on glass substrate from tetra-n-butyltin by the spray pyrolysis technique. J Mater Res 1993;8:1481–3.

[43] Vallet-Regì M, Ragel V, Román J, Martínez JL, Labeau M, González-Calbet JM. Texture evolution of SnO2 synthesized by pyrolysis of an aerosol. J Mater Res 1993;8:138–44.

[44] Sberveglieri G. Classical and novel techniques for the preparation of SnO_2 thin-film gas sensors. Sens Actuators B 1992;6:239–47.

[45] Yamazoe N. New approaches for improving semiconductor gas sensors. Sens Actuators B 1991;5:7–19.

[46] Siegel RW. Nanostructured materials – mind over matter. Nanostruct Mater 1993;3:1–18.

[47] Hahn H. Gas phase synthesis of nanocrystalline materials. Nanostruct Mater 1997;9:3–12.

[48] Gleiter H. Materials with ultrafine structure: retrospectives and perspectives. Nanostruct Mater 1992;1:1–19.

[49] Suryanarayana C. Nanocrystalline materials. Int Mater Rev 1995;40:41–64.

[50] Göpel W. Solid-state chemical sensors: atomistic models and research trends. Sens Actuators 1989;16:167–75.

[51] Horrillo MC, Gutiérrez FJ, Arés L, Robla JI, Sayago I, Getino J, Agapito JA. Hall effect measurements to calculate the conduction control in semiconductor films of SnO_2. Sens Actuators A 1994;41–42:619–21.

[52] Mosely PT. Materials selection for semiconductor gas sensors. Sens Actuators B 1992;6:149–56.

[53] Ogawa H, Nishikawa M, Abe A. Hall measurement studies and an electrical conduction model of tin oxide ultrafine particle films. J Appl Phys 1982;53:4448–55.

[54] Williams G, Coles GSV. The gas-sensing potential of nanocrystalline tin dioxide produced by a laser ablation technique. MRS Bull 1999;24(6):25–9.

[55] Göpel W, Schierbaum KD. SnO_2 sensors: current status and future prospects. Sens Actuators B 1995;26–27:1–12.

[56] Xu C, Tamaki J, Miura N, Yamazoe N. Grain size effects on gas sensitivity of porous SnO_2-based elements. Sens Actuators B 1991;3:147–55.

[57] Barbi GB, Santos JP, Serrini P, Gibson PN, Horrillo MC, Manes L. Ultrafine grain-size tin-oxide films for carbon monoxide monitoring in urban environments. Sens Actuators B 1995;24–25:559–63.

[58] Morrison SR. In: Hagenmuller P, editor. Proceedings of the second international meeting on chemical sensors. Bordeaux (France): Imprimerie Biscaye; 1986. p. 39–48.

[59] Madou MJ, Morrison SR. Chemical sensing and solid state devices. New York: Academic Press; 1989.

[60] Aaronson AJ. The thin film book of basics. 5th ed. New York: Materials Research Corporation; 1988.

[61] Petrov NN. Ejection of electrons from metals by ions. Soviet Phys Solid State 1960;2:1182–8.

[62] Rol PK, Onderdelinden D, Kistemaker J. In: Adam H, editor. Transactions of the third international vacuum congress. Vol. 1. Oxford: Pergamon Press; 1966. p. 75–82.

[63] Woodyard JR, Cooper CB. Mass spectrometric study of neutral particles sputtered from Cu by 0–100 eV Ar ions. J Appl Phys 1964;35:1107–17.

[64] Wehner GK, Laegreid N. Sputtering yields of metals for Ar^+ and Ne^+ ions with energies from 50 to 600 eV. J Appl Phys 1961;32:365–9.

[65] Horrillo MC. Estudio y Realización de Sensores para CO basados en la Modulación de la Conductividad Eléctrica del Semiconductor SnO_2. PhD Thesis, Universidad Complutense de Madrid; 1992.

[66] Chrisey DB, Hubler GK. Pulsed laser deposition of thin films. New York: John Wiley; 1994.

[67] Cheung J, Horowitz J. Pulsed laser deposition history and laser-target interactions. MRS Bull 1992;17:30–6.

[68] Hubler GK. Pulsed laser deposition. MRS Bull 1992;17:26–7.

[69] Metev S, Meteva K. Nucleation and growth of laser-plasma deposited thin films. Appl Surf Sci 1989;43:402–8.

[70] Ozegowski M, Meteva K, Metev S, Sepold G. Pulsed laser deposition of multicomponent metal and oxide films. Appl Surf Sci 1999;138–139:68–74.

[71] Thornton JA. High rate thick film growth. Ann Rev Mater Sci 1977;7:239–60.

[72] Sanders JV. Structure of evaporated metal films. In: Anderson JR, editor. Chemisorption and reactions on metallic films. Vol. 1. New York: Academic Press 1971; p. 1–38.

[73] Movchan BA, Demschishin AV. Study of the structure and properties of thick vacuum condensates of nickel, titanium, tungsten, aluminum oxide, and zirconium dioxide. Phys Met Metallogr 1969;28:83–90.

[74] Norton MG, Carter CB. On the optimization of the laser ablation process for the deposition of $YBa_2Cu_3O_{7-\delta}$ thin films. Physica 1990;C172:47–56.

[75] Dew SK, Smy T, Tait RN, Brett MJ. Modeling bias sputter planarization of metal films using SIMBAD. J Vac Sci Technol A 1991;9:519–23.

[76] Dew SK, Smy T, Brett MJ. Simulation of elevated temperature aluminum metallization using SIMBAD. IEEE Trans Electron Devices 1992;39:1599–606.

[77] Rickerby DG, Horrillo MC, Santos JP, Serrini P. Microstructural characterization of nanograin tin oxide gas sensors. Nanostruct Mater 1997;9:43–52.

[78] Horrillo MC, Serventi A, Rickerby DG, Gutiérrez J. Influence of tin oxide microstructure on the sensitivity to reductor gases. Sens Actuators B 1999;58: 474–7.

[79] Serventi AM, Rickerby DG, Horrillo MC, Saint-Jacques RG, Gutiérrez J. Transmission electron microscopy investigation of SnO_2 thin films for sensor devices. Nanostruct Mater 1999;11:813–9.

[80] Serventi AM, Rickerby DG, Horrillo MC, Saint-Jacques RG. Transmission electron microscopy investigation of the effect of deposition conditions and a platinum layer in gas-sensitive r.f.-sputtered SnO_2 films. Thin Solid Films 2003;445:38–47.

[81] Patterson AL. The Scherrer formula for x-ray particle size determination. Phys Rev 1939;56:978–82.

[82] Rickerby DG, Horrillo MC. In: Gardener JW, Persaud KC, editors. Olfaction and electronic noses 2000. London: Institute of Physics; 2000. p. 49–54.

[83] Zheng JG, Pan X, Schweizer M, Weimar U, Göpel W, Rühle M. Dislocations in nanocrystalline SnO_2 thin films. Philos Mag Lett 1996;73:93–100.

[84] Gland JL, Sexton BA, Fisher GB. Oxygen interactions with the Pt(111) surface. Surf Sci 1980;95:587–602.

[85] Yamazoe N, Tamaki J, Miura N. Role of hetero-junctions in oxide semiconductor gas sensors. J Mater Sci Eng B 1996;41:178–81.

[86] El Khakani MA, Dolbec R, Serventi A, Horrillo MC, Trudeau M, Saint-Jacques RG, et al. Pulsed laser deposition of nanostructured tin oxide films for gas sensing applications. Sens Actuators B 2001;77:383–8.

[87] Serventi AM, Dolbec R, El Khakani MA, Saint-Jacques RG, Rickerby DG. High-resolution transmission electron microscopy investigation of the nanostructure of undoped and Pt-doped nanocrystalline pulsed laser deposited SnO_2 thin films. J Phys Chem Solids 2003;64:2097–103.

[88] Godbole VP, Vispute RD, Chaudhari SM, Kanetkar SM, Ogale SB. Dependence of the properties of laser deposited tin oxide films on process variables. J Mater Res 1990;5:372–7.

[89] Dolbec R, El Khakani MA, Serventi AM, Trudeau M, Saint-Jacques RG. Microstructure and physical properties of nanostructured tin oxide thin films grown by means of pulsed laser deposition. Thin Solid Films 2002;419:230–6.

[90] Horrillo MC, Serrini P, Santos J, Manes L. Influence of the deposition conditions of SnO_2 thin films by reactive sputtering on the sensitivity to urban pollutants. Sens Actuators B 1997;45:193–8.

[91] Götz A, Gràcia I, Cané C, Lora-Tamayo E, Horrillo MC, Getino J, et al. A micromachined solid state integrated gas sensor for the detection of aromatic hydrocarbons. Sens Actuators B 1997;44:483–7.

[92] Götz A, Gràcia I, Cané C, Lora-Tamayo E. Thermal and mechanical aspects for designing micromachined low-power gas sensors. J Micromech Microeng 1997;7:247–9.

[93] Horrillo MC, Sayago I, Arés L, Rodrigo J, Gutiérrez J, Götz A, et al. Detection of low NO_2 concentrations with low power micromachined tin oxide gas sensors. Sens Actuators B 1999;58:325–9.

[94] Rickerby DG, Waechter N, Horrillo MC, Gutiérrez J, Gracià I, Cané C. Structural and dimensional control in micromachined integrated solid state gas sensors. Sens Actuators B 2000;69:314–9.

[95] Zhu Y, Lu H, Lu Y, Pan X. Characterization of SnO_2 films deposited by d.c. gas discharge activating reaction evaporation onto amorphous and crystalline substrates. Thin Solid Films 1993;244:82–6.

[96] Melsheimer J, Tesche B. Electron microscopy studies of sprayed thin tin dioxide films. Thin Solid Films 1986;138:71–8.

[97] Sundaram KB, Bhagavat GK. X-ray and electron diffraction studies of chemically vapour-deposited tin oxide films. Thin Solid Films 1981;78:35–40.

[98] Tang Z, Fung SKH, Wong DTW, Chan PCH, Sin JKO, Cheung PW. An integrated gas sensor based on tin oxide thin-film and improved micro-hotplate. Sens Actuators B 1998;46:174–9.

[99] Sheng L-Y, Tang Z, Wu J, Chan PCH, Sin JKO. A low-power CMOS compatible integrated gas sensor using maskless tin oxide sputtering. Sens Actuators B 1998;49:81–7.

[100] Sharma RK, Tang Z, Chan PCH, Sin JKO, Hsing I-M. Compatibility of CO gas sensitive SnO_2/Pt thin film with silicon integrated circuit processing. Sens Actuators B 2000;64: 49–53.

[101] Carotta MC, Martinelli G, Crema L, Gallana M, Merli M, Ghiotti G, et al. Array of thick film sensors for atmospheric pollutant monitoring. Sens Actuators B 2000; 68:1–8.

[102] Traversa E, Sadaoka Y, Carotta MC, Martinelli G. Environmental monitoring field tests using screen-printed thick-film sensors based on semiconducting oxides. Sens Actuators B 2000;65:181–5.

[103] Carotta MC, Martinelli G, Crema L, Malagù C, Merli M, Ghiotti G, et al. Nano-structured thick-film gas sensors for atmospheric pollutant monitoring: quantitative analysis on field tests. Sens Actuators B 2001;76:336–42.

[104] Carotta MC, Ferroni M, Gnani D, Guidi V, Merli M, Martinelli G, et al. Nanostructured pure and Nb-doped TiO_2 as thick film gas sensor for environmental monitoring. Sens Actuators B 1999;58:310–7.

[105] Carotta MC, Benetti M, Ferrari E, Giberti A, Malagù C, Nagliati M, et al. Basic interpretation of thick film gas sensors for atmospheric application. Sens Actuators B 2007;126:672–7.

[106] Bârsan N, Weimar U. Understanding the fundamental principles of metal oxide based gas sensors the example of CO sensing with SnO_2 sensors in the presence of humidity. J Phys Condens Mater 2003;15:R813–39.

[107] Carotta MC, Ferrari E, Giberti A, Malagù C, Nagliati M, Gherardi S, et al. Semiconductor gas sensors for environmental monitoring. Adv Sci Technol 2006;45:1818–27.

[108] Graf M, Frey U, Reichel P, Taschini S, Bârsan N, Weimar U, et al. Monitoring of environmentally relevant gases by a digital monolithic metal-oxide microsensor array. Sensors, Proc IEEE 2004;24–27:776–9.

[109] Graf M, Gurlo A, Bârsan N, Weimar U, Hierlemann A. Microfabricated gas sensor systems with nanocrystalline metal-oxide sensitive films. J Nanopart Res 2006;8: 834–9.

[110] Reichel P. Development of a chemical gas sensor system. PhD Thesis, University of Tübingen; 2005.

[111] Varghese OK, Grimes CA. Metal oxide nanoarchitectures for environmental sensing. J Nanosci Nanotech 2003;3:277–93.

[112] Pan ZW, Dai ZR, Wang ZL. Nanobelts of semiconducting oxides. Science 2001;291: 1947–9.

[113] Comini E, Baratto C, Faglia G, Ferroni M, Vomiero A, Sberveglieri G. Quasi-one dimensional metal oxide semiconductors: preparation, characterization and application as chemical sensors. Prog Mater Sci 2009;54:1–67.

[114] Calestani D, Zha M, Salviati G, Lazzarini L, Zanotti L, Comini E, et al. Nucleation and growth of SnO_2 nanowires. J Cryst Growth 2005;275:2083–7.

[115] Comini E, Bianchi S, Faglia G, Ferroni M, Vomiero A, Sberveglieri G. Functional nanowires of tin oxide. Appl Phys A 2007;89:73–6.

[116] Vomiero A, Ferroni M, Comini E, Faglia G, Sberveglieri G. Preparation of radial and longitudinal nanosized heterostructures of In_2O_3 and SnO_2. Nano Lett 2007;7: 3553–8.

[117] Sberveglieri G, Baratto C, Comini E, Faglia G, Ferroni M, Ponzoni A, et al. Synthesis and characterization of semiconducting nanowires for gas sensing. Sens Actuators B 2007;121:208–13.

[118] Bianchi S, Comini E, Ferroni M, Faglia G, Vomiero A, Sberveglieri G. Indium oxide quasi-monodimensional low temperature gas sensor. Sens Actuators B 2006;118: 204–7.

[119] Vomiero A, Bianchi S, Comini E, Faglia G, Ferroni M, Sberveglieri G. Controlled growth and sensing properties of In_2O_3 nanowires. Cryst Growth Des 2007;7: 2500–4.

[120] Ponzoni A, Comini E, Sberveglieri G, Zhou J, Deng SZ, Xu NS, et al. Ultrasensitive and highly selective gas sensors using three-dimensional tungsten oxide nanowire networks. Appl Phys Lett 2006;88:203101.

[121] Faglia G, Baratto C, Sberveglieri G, Zha M, Zappettini A. Adsorption effects of NO_2 at ppm level on visible photoluminescence response of SnO_2 nanobelts. Appl Phys Lett 2005;86:011923.

[122] Baschirotto A, Capone S, D'Amico A, Di Natale C, Ferragina V, Ferri G, et al. A portable integrated wide-range gas sensing system with smart A/D front-end. Sens Actuators B 2008;130:164–74.

[123] Tsujita W, Yoshino A, Ishida H, Moriizumi T. Gas sensor network for air-pollution monitoring. Sens Actuators B 2005;110:304–11.

[124] Richards M, Ghanem M, Osmond M, Guo Y, Hassard J. Grid-based analysis of air pollution data. Ecol Model 2006;194:274–86.

[125] Ma Y, Richards M, Ghanem M, Guo Y, Hassard J. Air pollution monitoring and mining based on sensor grid in London. Sensors 2008;8:3601–23.

[126] Völgyesi P, Nádas A, Koutsoukos X, Lédeczi Á. In: Kaiser WJ, editor. Proceedings of the 7th international conference on information processing in sensor networks. Washington (DC): IEEE Computer Society; 2008. p. 529–30.

CHAPTER 7

Hydrogen Storage on Carbon Adsorbents: A Review

Tengyan Zhang[*,**], L. T. Fan[**], Walter P. Walawender[**], Maohong Fan[***], Alan E. Bland[*], Tianming Zuo[*] *and* Donald W. Collins[*]

[*] Western Research Institute, Laramie, Wyoming, USA
[**] Department of Chemical Engineering, Kansas State University, Manhattan, Kansas, USA
[***] Department of Chemical and Petroleum Engineering, University of Wyoming, Laramie, Wyoming, USA

Contents

1. INTRODUCTION

In recent years, global energy demand and global warming due to greenhouse gas emissions have spurred intensive research for alternative fuel. Hydrogen (H_2), an ideal energy carrier, is obviously a promising candidate. Its utilization efficiency is high; for instance, it is 2.75 times greater than gasoline for the same weight. Moreover, it is environmentally benign with negligible pollutant emissions [1–7]. H_2 can be produced through the conversion of various resources, including different fossil fuels, such as coal, oil, and natural gas, as well as a variety of biomass. The energy required for this conversion can be generated from such resources themselves or from other energy sources, such as wind, solar, and nuclear. As an example, H_2

DOI: 10.1016/B978-0-08-054820-3.00007-1

can be produced indirectly from coal gasification and reforming processes for which advanced technologies are available; these processes, combined with carbon dioxide (CO_2) separation and sequestration, have the potential to manufacture substantial quantities of H_2 with minimum greenhouse gas emissions [7, 8].

Synthesized H_2 can be deployed to generate electricity from fuel cells; alternatively, it can be combusted for providing energy for space heating, replacing natural gas in industry, and fueling aircraft [9]. For such deployments, H_2 needs to be effectively stored for stationary or mobile application. The former entails the installation of storage facilities in a relatively large area, execution of multistep chemical charging/recharging cycles at low temperatures and under high pressures, and provision of extra capacity to compensate for slow kinetics. In contrast, the latter demands the operation with minimum volume and weight specifications, adequate supply of H_2 for the acceptable driving range, i.e., over 380 km (300 miles) on a full tank, with the charge/recharge near room temperature, and supply of H_2 at sufficiently high rates for fuel-cell locomotion of various vehicles [10, 11]. Nevertheless, barriers to gassing up these vehicles with H_2 are substantial. At present, it is impossible to store H_2 as compactly and simply as the conventional liquid fuel. According to George J. Thomas, formerly with Sandia National Laboratories, "…one of the key enabling technologies on which a future hydrogen economy rests is hydrogen storage. The topic is central to all aspects of hydrogen usage…" [12]. Thus, one of the most challenging technical issues remaining unresolved in developing a H_2-based energy system is the efficient and safe storage of a sufficient quantity of H_2 on board without a concomitant, appreciable increase in the weight or volume to the vehicle.

To meet the requirements for H_2 as a transportation fuel, the US Department of Energy (DOE) has mandated that the target energy densities for the development of onboard H_2-storage systems be 5.5 wt% (1.8 kWh/kg system) and 40-kg H_2/m^3 (1300 kWh/m^3) by year 2015 and 7.5 wt% (2.5 kWh/kg system) and 70-kg H_2/m^3 (2300 kWh/m^3) ultimately [6]. Currently, three major approaches are available for H_2 storage: (1) pressurization of H_2 in high-pressure vessels, (2) liquefaction of H_2 at the cryogenic temperature, and (3) materials-based H_2 storage [6].

For most of the current prototype fuel-cell vehicles, numbering hundreds, H_2 is stored in high-pressure cylinders at a pressure ranging from 5000 psi (~35 MPa) to 10 000 psi (~70 MPa), which can be readily tapped for use [13]. Strong- and light-weight tanks, such as those fabricated from carbon-fiber reinforced composites, render it possible to safely store H_2 at high pressures.

Nevertheless, the density of stored H_2 does not proportionally increase with the increase in the pressure. Even at a pressure of 10 000 psi, the energy content of H_2 is 4.4 MJ/L, significantly less than that for the same volume of gasoline, which is 31.6 MJ/L. Available high–pressure tanks can contain merely about 3.5–4.5 wt% of H_2, which is far from attaining the aforementioned target energy density.

Naturally, the second approach, i.e., liquefaction of H_2 at the cryogenic temperature, can increase the energy density of H_2 to yield a system with 14 wt% and 51-kg H_2/m^3, corresponding to the energy content of 8.4 MJ/L, which approximate the aforementioned target energy densities. Transporting or handling liquefied H_2, however, tends to be exceedingly costly. In fact, the cost of liquefied H_2 as a transportation fuel is nearly twice that of gaseous H_2, thereby hindering its deployment. In addition, the extremely low boiling point of H_2 necessitates special precautions for safe handling. Furthermore, any substantial improvement can only be achieved at the expense of severe operating conditions and/or high fuel consumption [14]. Consequently, much effort is needed to reduce the cost of liquefaction for it to be economically viable.

The most promising method for H_2 storage appears to be through materials–based storage technologies, which are inherently safe and should be more energy efficient than pressurization or liquefaction. Materials for such technologies include carbon-based materials, metal-organic frameworks, metal hydrides, conducting polymers, and clathrates [6]. Among them, it is highly likely that carbon-based materials, or carbon adsorbents, have the greatest potential. They are light, inexpensive, and ideal from the standpoint of the gravimetric requirements for deploying fuel cells of automobiles. Moreover, the adsorbent-manufacturing industry has substantial expertise in the production of carbon adsorbents capable of effectively controlling their micro- and nano-structures for large-scale production. Intensive investigations have been performed on H_2 storage on carbon adsorbents, including activated carbons, carbon nanofibers (CNFs), single–walled carbon nanotubes (SWCNTs), and multiwalled carbon nanotubes (MWCNTs).

2. FUNDAMENTALS OF ADSORPTION

Adsorption is a process, which increases in the density of a gas or solute (adsorbate) in the vicinity of the surface of a substrate (adsorbent) due to molecular interactions between the adsorbate and the adsorbent. Gases can undergo supercritical and subcritical adsorption. The former takes place above the gas critical temperature, thereby exhibiting no gas–liquid phase transition. The latter is observed below the gas critical temperature, and its isotherm is characterized by a steep increase in the adsorbed density close to the saturation pressure associated with the formation of a liquid layer on the surface of the adsorbate. Because of the extremely low critical temperature of H_2, i.e., 33.18 K, only supercritical adsorption needs to be taken into account for H_2 storage [15].

2.1. Volumetric and Gravimetric Densities

The H_2-storage system is characterized by its volumetric and gravimetric densities. The amount of H_2 in the adsorbent-filled container includes the H_2 adsorbed by the adsorbent and the bulk H_2 in the nonadsorbing volume of the container. To facilitate the delineation of the role of the adsorbent on H_2 storage, a thermodynamically significant notion of the "excess amount" has been proposed; it is defined as "…the excess material present in the pores over that which would be present under the normal density at the equilibrium pressure" [16, 17]. A plot of the excess amount versus pressure often exhibits a maximum; beyond this maximum, the bulk gas density increases more quickly with increasing pressure than does the adsorbed density. It is highly plausible that H_2-storage capacity can be elevated in a given container by removing the carbon adsorbent [17].

Naturally, the excess number of adsorbate molecules, N_{ex}, is defined as the difference between the total amount of adsorbate molecules, N_a, within the volume, V_{void}, of a measurement cell containing the adsorbent at a given temperature and pressure, and the number of molecules N_g that would have been found in the void volume in the absence of adsorbate–adsorbent interactions; thus,

$$N_{ex} = N_a - N_g. \tag{1}$$

Obviously,

$$N_g = \rho_g V_{void} \tag{2}$$

where ρ_g is the bulk gas density (obtained from the equation of state of the adsorbate) at the same temperature and pressure, and V_{void} is the sum of the pore volume of the adsorbent, V_{pore}, and the volume of the section of the cell containing no adsorbent, V_{empty}, i.e., $V_{void} = V_{empty} + V_{pore}$. Combining Eqs (1) and (2) gives rise to

$$N_{ex} = N_a - \rho_g V_{void}. \tag{3}$$

Dividing N_{ex} by the mass of the adsorbent, M, in the cell results in the excess adsorbed density, n_{ex}, i.e.,

$$n_{ex} = \frac{N_{ex}}{M}. \tag{4}$$

Note that N_{ex}/M is not equal to the absolute adsorbed density, n_a, since it also includes molecules in the volume, V_{empty}. In terms of the local density, Eq (4) can be rewritten as

$$n_{ex} = \frac{1}{M}\int_{V_{void}} \left(\rho(\bar{r}) - \rho_g\right) dr. \tag{5}$$

Note that Eq (3) can be expressed in terms of the absolute adsorption, without any reference to the empty volume of the measurement cell, as

$$n_{ex} = n_a - \rho_g v_{pore}, \tag{6}$$

where

$$n_a = \frac{N_a - \rho_g V_{empty}}{M}. \tag{7}$$

The quantity of H_2 (in moles) stored in an adsorption–based storage unit of total volume V_{sys} can be estimated from the following expression:

$$N_{stored} = N_a + \rho_g V_{empty} = \left(n_{ex} + \rho_g v_{pore}\right)M + \rho_g V_{empty}. \tag{8}$$

Thus, the stored volumetric density is

$$\frac{N_{stored}}{V_{Sys}} = \frac{\left(n_{ex} + \rho_g v_{pore}\right)M}{V_{Sys}} + \frac{\rho_g V_{empty}}{V_{Sys}}, \tag{9}$$

and the stored gravimetric density is

$$x_{stored} = \frac{N_{stored}}{M_{Sys} + N_{stored} \cdot m_{H_2}} \tag{10}$$

In fact, the gravimetric density of the system is often estimated by multiplying the excess density, n_{ex}, with the molar mass of the adsorbate (H_2): V_{Sys} and M_{Sys} are known only after the system is designed. Nevertheless, such estimation assumes that $M_{Sys} = M$ and also neglects the contribution of the bulk phase in the pore volume to the total amount of gas in the storage unit as well as the contribution of the stored H_2 to the total mass of the filled storage system.

2.2. Physisorption

The binding energy for physisorption via the van der Waals bonds ranges from 0.04 to 0.12 eV with the largest value corresponding to the regime of low H_2 coverage, where the H_2–H_2 repulsion is insignificant [18, 19]. For conventional physisorption, the gas-adsorption performance of a porous solid is maximized when the pore size is less than a few molecular diameters [20]. As such, the potential fields from the walls of the so-called micropores overlap to produce a stronger interaction than would be possible for adsorption on a semi-infinite plane. If the escaping tendency of the gas is much less than the adsorption potential, the entire micropore may be filled with a condensed adsorbate phase. For the case of H_2, with a kinetic diameter of approximately 2.9 Å, pores would have to be significantly smaller than 40 Å to begin to condense H_2 by a nanocapillary filling mechanism.

Any of the suitable or appropriate models for the excess-adsorption isotherms of microporous adsorbents should contain a minimum number of parameters with physical interpretation. Such models currently available include those based on viral expansion, Langmuir isotherm, self-consistent Ono–Kondo approach, and Dubinin pore-filling approach. The details of these models are available [15, 21].

2.3. Chemisorption

In addition to physisorption, H_2 can be adsorbed onto the carbon adsorbents through chemical adsorption, i.e., chemisorption. The binding energy for chemisorption is between 2 and 3 eV, much stronger than the van der Waals interactions [22, 23]. The usual ab initio methods, e.g., density functional theory (DFT), are well established to study the nature and energetics of chemical bonding. The DFT methods include the local density approximations and the generalized gradient approximations. The former tends to overestimate the binding energies; the latter usually reproduces experimental values to within 0.1 or 0.2 eV, thereby providing more accurate information [24, 25]. In fact, chemisorption of H_2 in graphitic nanocarbons is inapplicable for reversible H_2 storage. It is difficult to control, and the adsorbed H_2 can only be released at high temperatures, e.g., 400 K [26]. Thus, the focus of the research has been on physisorption, which gives rise to a much closer energy range for practical H_2 storage than chemisorption.

3. CARBON ADSORBENTS

Carbon-based adsorbents, including activated carbons, CNFs, and carbon nanotubes (CNTs), have been widely recognized as profoundly effective adsorbents. The attributes of ideal carbon adsorbents for H_2 storage should be such that they have uniform and small micropores in high density, minimal macroporosity, and high thermal conductivity. The first attribute is required for increasing heat of adsorption, which might render possible adsorption at the ambient temperature, while the first two attributes collectively insure that the internal volume of the adsorbent is not wasted. The third attribute is required for an enhanced heat flux, which tends to be relatively large for CNTs. Besides the adsorption capacity, the viability of a H_2-storage device hinges heavily on its capability for releasing the adsorbed H_2 [12]. Note that the properties of carbon-based adsorbents are strongly dependent on those of precursors and on processing conditions, none of which is sufficiently well understood at the molecular level.

3.1. Activated Carbons

The term, activated carbons, defines a group of materials with highly developed internal surface areas and porosities. Activated carbons constitute the only variety of carbon adsorbents that are manufactured in abundance and thus inexpensive.

3.1.1. Structure

Carbonized materials, including activated carbons and chars, have structures and properties similar to those of graphite [27, 28]. The density of graphite is $2.267\,g/cm^3$, and its atoms are bonded together in hexagonal two-dimensional networks. The layers of these atoms are held loosely together by van der Waals force. They are principally in the form of A-B-A-B, i.e., the hexagonal form; a small fraction (<10%) in natural graphite stacks in the form of A-B-C-A-B-C, i.e., the rhombohedral form. Note that the latter can convert irreversibly to the former at approximately 2400 K; see Fig. 7.1.

The structure of activated carbons is less perfectly ordered than that of graphite, generally representing an intermediate between the organic precursor and the single graphite crystal [30]. Activated carbons are either graphitic or nongraphitic depending on the degree of crystallographic ordering. The former includes all varieties of carbonized materials in which elemental carbon is in the allotropic form of graphite, irrespective of the presence of structural defects. The latter includes all varieties of carbonized materials in which elemental carbon has two-dimensional long-range-ordered carbon atoms in planar hexagonal networks; nevertheless, in the third dimension, apart from more or less parallel stacking, no crystallographic order can be detected [28].

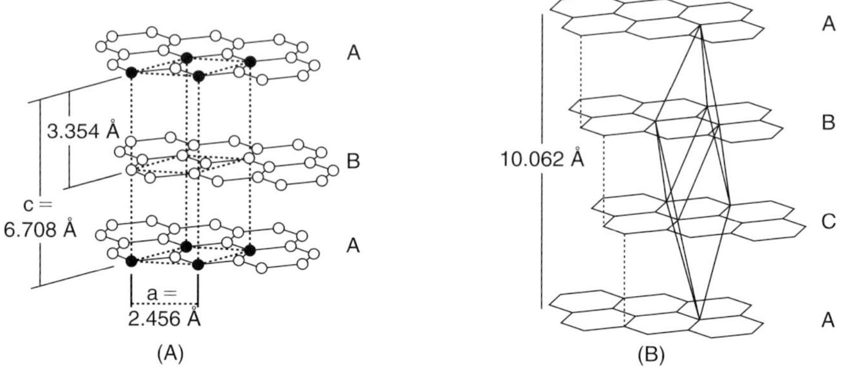

Figure 7.1   Two structures of graphite: (A) hexagonal unit cell and (B) rhombohedral unit cell [29].

The nongraphitic carbons can be further divided into the graphitizable and nongraphitizable carbons. Graphitizable carbons can be converted into graphitic carbons by heat treatment; the degree of graphitization depends on the temperature of heat treatment and the time allowed to anneal the structure [30]. Such carbons are essentially cokes resulting from the carbonization of pitch materials derived from coal and petroleum [28]. On the other hand, a nongraphitizable carbon cannot be transformed into a graphitic carbon solely by heat treatment under inert conditions even up to a temperature of 3500 K [30]. The majority of such carbons are generated from pyrolysing cellulose and lignin within wood and nut shells or from nonfusing coals, e.g., anthracite and subbituminous coals. These materials lose molecules of water (H_2O), CO_2, and organics of low molecular weights on pyrolysis, with only the cross-linking and basic back-bone structure remaining. The voids created by the loss of small molecules, including H_2 and CO_2, at higher temperatures (500 ~ 1000 °C) constitute the pores, 0.5–2.0 nm in diameter and high surface areas of about or greater than 1000 m^2/g, within the carbons [27, 28, 31].

A distinct difference in the structures of graphitizable and nongraphitizable carbons has been identified by Byne and Marsh [30]. According to them, all carbonized materials, including activated carbons, have structures comprising imperfect sections of graphitic lamellae. It is plausible that these lamellae can be sliced into sections, 1.0 nm–500 μm in thickness. The smaller sections of 1.0 ~ 10 nm thick can be crumpled and bonded together to create a three-dimensional network. Such a structure, with defects and irregularities, approximates that of porous activated carbons. Because of the crumpled, defective nature of the constituent lamellar-based molecules (LCMs), their packing density is low, approximately $1.0\,g/cm^3$, and it is the spaces or voids between the LCM that constitute porosity. The graphitizable cokes, however, comprise LCMs that are larger, contain fewer defects, and are less crumpled than those in the porous carbons. Accordingly, they stack parallel to each other over relatively larger dimensions of micrometers, thus increasing their densities, which approach $2.0\,g/cm^3$ (the density of graphite is $2.2\,g/cm^3$); they have little or no pores with dimensions comparable to those of molecules, i.e., nanometers [30]. Figure 7.2 shows a schematic representation of the structures of nongraphitizable and graphitizable carbons [27].

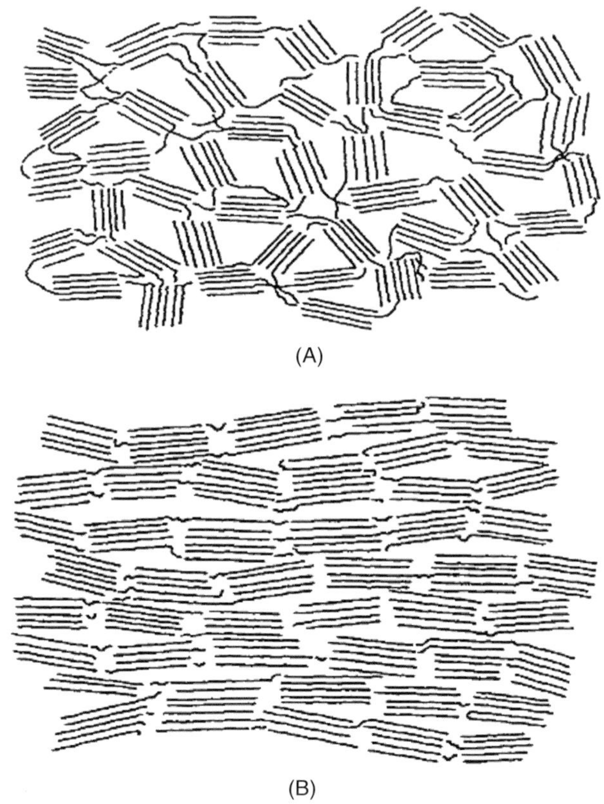

(A)

(B)

Figure 7.2   Schematic representation of activated-carbon structures: (A) nongraphitizing and (B) graphitizing [27].

The structure of an activated–carbon particle is formed by pores wide ranging in size. According to the classification of the International Union of Pure and Applied Chemistry (IUPAC), the pores of activated carbons can be divided into the following categories based on their diameters; see Fig. 7.3. The diameters of "macropores" for admission and diffusion are greater than 50 nm, those of "mesopores" range between 2.0 and 50 nm, and those of "micropores" are less than 2.0 nm. Pores having diameters of the same order of magnitude as the diameters of molecules, i.e., <0.8 nm, are called "submicropores."

The macropores serve as transport conduits enabling the molecules of the adsorbate to reach smaller pores situated in the interior of a carbon particle. Thus, although the macropores do not contribute appreciably to the total surface area, they significantly affect the admission rate of adsorbates into the carbon. The mesopores, branching off from the macropores, serve as secondary passages to the micropores for the adsorbate. In mesopores, capillary condensation may take place with the formation of a meniscus of the liquid adsorbate. This phenomenon usually produces the hysteresis loop on adsorption isotherms [27]. The mesopore volume is usually between 0.02 and 0.1 cc/g. In most activated carbons, the surface area contributed

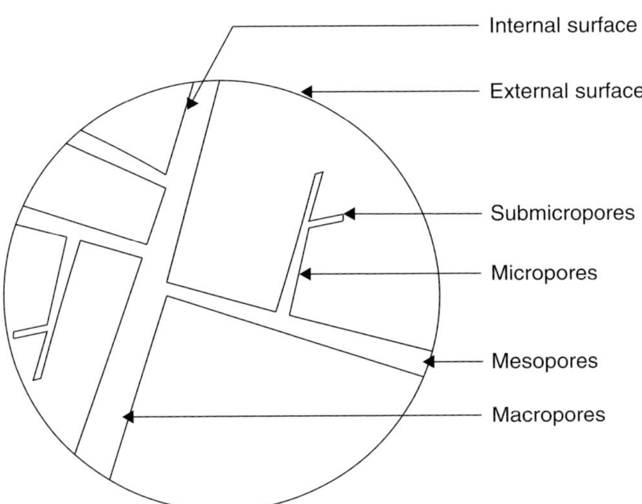

Figure 7.3 Schematic representation of the various types of pores in a particle of carbonaceous adsorbent with different diameters: macropores, >50 nm; mesopores, between 2 and 50 nm; micropores, between 0.8 and 2 nm; and submicropores, <0.8 nm [32].

by mesopores is relatively low but not insignificant. The micropores contribute most to the total surface area of activated carbons, thereby providing a high adsorptive capacity for molecules of small dimensions; the majority of the adsorption takes place within them. Their small size prevents capillary condensation. They are occupied by the adsorbate at low relative pressures.

3.1.2. Preparation

Activated carbons have been prepared from a number of precursors, such as lignite, coconut shell, wood, and grains, through one-stage or two-stage processes [33–45]. In the one-stage process, activated carbons are generated directly from raw materials via physical or chemical activation. Physical activation is carried out most frequently by burning off some of the raw carbon in an environment of oxidizing gas to create micropores. The usual choices of oxidizing gas are steam, CO_2, air, or their mixtures. Activation normally takes place at temperatures between 700 and 1000 °C in steam and CO_2 and lower temperatures in air. In chemical activation, a carbon precursor is impregnated with a dehydrating agent before heated between 500 and 800 °C. The most widely used dehydrating agents include zinc chloride, potassium sulfide, potassium thiocyanate, phosphoric acid, sulfuric acid, hydroxides of the alkali metals, magnesium chloride, and calcium chloride. The resultant activated carbons from the one-stage process have small surface areas and high mesoporosities.

The two-stage process comprises carbonization followed by physical or chemical activation. Carbonization is performed by pyrolyzing raw materials at a temperature less than 700 °C to produce char. The char obtained is physically activated through oxidation in the presence of an oxidizing gas or chemically activated through pyrolysis in an inert gas after mixing with the dehydrating agent. The resultant activated carbons usually have high surface areas and high microporosities.

3.1.3. Hydrogen Adsorption

It is entirely plausible that microporous activated carbons will be effective for H_2 storage in light of their high porosities and low cost. Nevertheless, in one of the first investigations [46], the maximum excess H_2 amounting to merely 20.2-g H_2/kg activated carbon was reported at 25 atm and 76 K, corresponding to a gravimetric storage density of approximately 2.0 wt%, which is far from reaching the target H_2 storage density set by DOE [6].

Intensive efforts have been made to increase the H_2-adsorption capacity of microporous activated carbons. One attempt is through metal modification of activated carbons: H_2 is well known to be inordinately adsorptive to adsorbents doped with a transition metal, i.e., a metal in group VIII of the Periodic Table [47–52]. Upon contact with a transition metal in elemental form, H_2 molecules are preferentially adsorbed onto the metal's surface; some of these H_2 molecules dissociate into H atoms, which are more efficient at filling the available active sites on the activated carbons than H_2 molecules [47, 48, 50]. The effect of this phenomenon, known as H_2 spillover, on catalytic reactions has been extensively documented, although the chemical form of H_2 that spills over is still uncertain [52]. According to Schwarz [50], the transition metals are themselves specifically active chemically. Moreover, they do not affect the hydroxyl, aldehyde, carbonyl, or other active groups on the surface of activated carbons. Such active groups possibly play a significant role in the adsorption of the dissociated H. Apparently, the minute amount of a transition metal can appreciably enhance the adsorbability of H_2 by activated carbons. Nevertheless, the metal-impregnated activated carbons adsorb only approximately 4.8 wt% H_2 at a temperature of 87 K and a pressure of 59 atm [51]. A simple calculation by adopting the parameters specified by Schwarz [51] has revealed that 4.8 wt%, or 16.5 kg/m^3, of H_2 could be stored in the tank at the same pressure and temperature without the activated carbons [53].

X-ray diffraction of the nanocrystalline graphite resulting from mechanical ball milling under an H_2 atmosphere has indicated that the graphite inter-layer spacing expands accompanied by the increasing disappearance of long-range order with milling time [54]. Up to 7.4 wt% of H_2 adsorption has been reported after milling for 80 h. Neutron-diffraction measurements have revealed that both molecular H_2 and covalently bonded H_2 are present [54]. Although the carbon adsorbents fabricated via intensive milling appear quite promising for H_2 storage, the adsorbed H_2 is extremely difficult to be desorbed from them.

Inclusion of activated carbons in a storage tank can increase the overall H_2 energy storage density under certain pressure and temperature. For any given activated carbons, this increase is maximized at low temperatures and high pressures, corresponding to the situation where the adsorbent surface is populated, but the density in the gas phase above the adsorbent is not appreciable [14]. Although the overall storage capacity of a tank is maximized at the high-pressure and low-temperature conditions, the effectiveness of the adsorbent is minimized under these conditions [51, 55]. The magnitude of the maximum gain depends on the surface area and microporous volume. Ideal adsorbents should be such that they are highly microporous with relatively high density, i.e., low in macroporous volume. In a recent study, various carbon adsorbents, including activated carbons, carbon black, carbon aerogels, and carbon molecular sieves, were tested to determine if they could augment the capacities of compressed H_2 gas storage systems at the ambient temperature (298 K), acetone and dry-ice temperature (190 K), and liquid-nitrogen temperature (80 K) [17]. The study has revealed that at pressures typical of vehicular compressed H_2-storage systems (approximately 200 bars), only one of 10 carbon sorbents tested can marginally augment the capacity of the storage vessel at 190 and 298 K; such augmentation is nill at 80 K [17]. Relatively little has been published on adsorbents of the activated-carbon type, which are capable of increasing the storage capacities of a high-pressure tank at 298 K [56, 57]. In fact, the H_2-adsorption capacity of activated carbons is low at room temperature and moderate pressure. For instance, according to Jin et al. [58], H_2 adsorption is less than 1 wt% at 100 bars and 298 K, despite the surface area of 2800 m^2/g. Although H_2 is mainly adsorbed by physical interaction, H_2 adsorption can be promoted by incorporating surface groups and other reactive sites into the activated carbons. Such measures, however, tend to operate mostly via chemisorption and are ineffective in enhancing the reversible storage of H_2.

As mentioned earlier, activated carbons inevitably contain macropores, mesopores, and micropores. In certain types of activated carbons, approximately 50% of the total pore volume comprises macropores, which hardly contribute to H_2 storage. The poor H_2-adsorption capacity of activated carbons can be largely attributed to the nonuniformity of their pores. The development of the specifically designed carbon adsorbents with uniform pores, therefore, has been extensively pursued for H_2 storage [53].

3.2. Carbon Nanofibers

In an attempt to elevate H_2 storage densities beyond those attainable by activated carbons, various carbon nanostructures have been investigated. Among them are CNFs.

3.2.1. Structure

CNFs, or graphite carbon nanofibers, are usually nongraphitic. They consist of graphene planes arranged in platelet stacks, either in parallel or in angled arrangements, resulting in a conical fishbone (herringbone) structure. Three distinct structures exist: tubular (90°), platelet (~ 0°), and herringbone (45°), where the angle in parentheses indicates the direction of the nanofiber axis relative to the vector normal to the graphene sheets. The spacing between

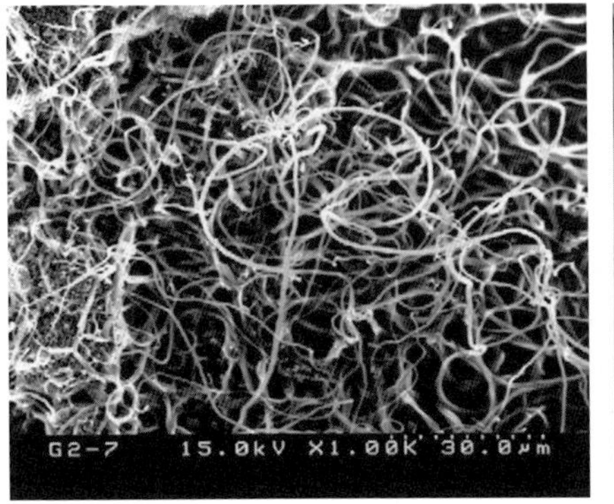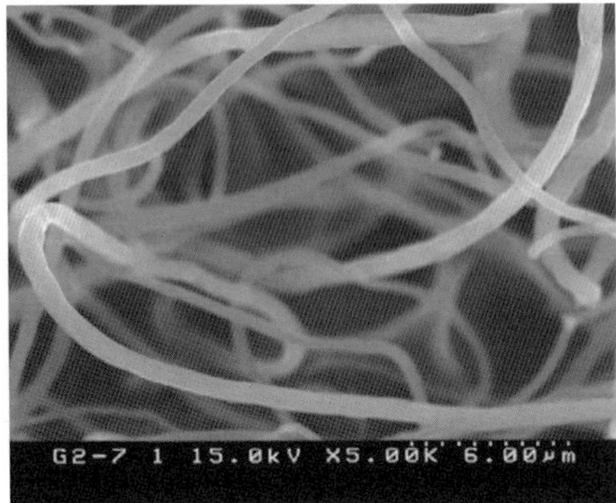

Figure 7.4   SEM photographs of CNFs [60].

graphite layers in each case is the same as that found in conventional graphitic carbon, i.e., about 3.4 Å. From the physical point of view, CNFs vary from 5 to 100 µ in length and from 5 to 100 nm in diameter. One distinct feature of CNFs is the abundance of open edges, thereby favoring H_2 sorption [59]. Figure 7.4 shows the SEM photographs of CNFs prepared by decomposing a mixture of C_2H_4 and H_2 with a ratio of 1 to 4 over $Ni_{0.5}Cu_{0.5}$ alloy powder [60].

3.2.2. Preparation

CNFs have been catalytically synthesized. Specifically, they can be produced by decomposing mixtures of hydrocarbon and H_2 on selected metal and alloy catalysts at temperatures ranging from 400 to 800 °C [61]. Figure 7.5 shows a schematic diagram of catalytically grown CNFs [62]. When a hydrocarbon is adsorbed on a metal surface and conditions exist favoring the scission of a carbon–carbon bond in the molecules, the resulting atomic species may dissolve in the catalyst particle, diffuse to the rear faces, and ultimately precipitate at the interface to form CNFs. The degree of crystalline perfection of the deposited fiber depends on three main factors. They are the chemical nature of the catalyst particle, composition of the reactant gas, and reaction temperature. Surface science studies have revealed that certain faces favor precipitation of carbon in the form of graphite, whereas less ordered carbon will be deposited from other faces [63–65]. By judicious choice of the catalyst, the ratio of the hydrocarbon to H_2 in the reactant mixture, and reaction conditions, it is possible to tailor the morphological characteristics, the degree of crystallinity, and the orientation of the precipitated graphite crystallites relative to the fiber axis [62].

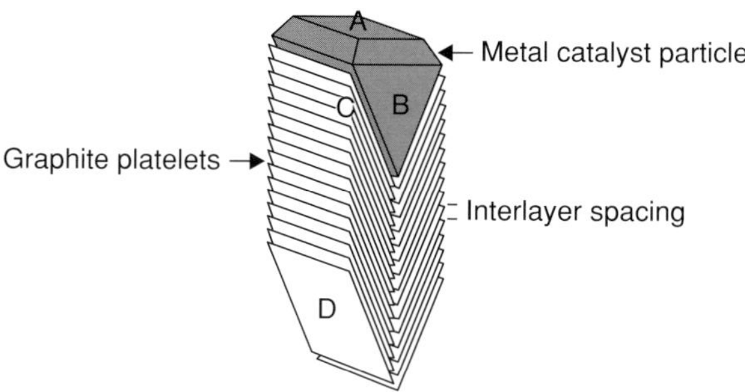

Figure 7.5   Schematic diagram of catalytically grown CNFs: (A) metal surface, (B) metal catalyst particle, (C) interface, and (D) deposited CNFs [62].

3.2.3. Hydrogen Adsorption

The H_2 adsorption on CNFs has been studied extensively at temperatures ranging from 77 to 300 K. The H_2-storage capacities exceeding 60 wt%, i.e., 2-g H_2/g carbon, were achieved for certain CNFs at the ambient temperature under the pressure between 50 and 120 bars [66]. Nevertheless, these results have never been replicated by others. For instance, H_2 adsorption of only approximately 0.08 wt% was attained with similarly fabricated CNFs under similar experimental conditions [67]. Another maximum H_2-storage capacity reported was 1.52 wt% at the ambient temperature under the pressure of 125 atm [68].

To rationalize the results of Chambers and his collaborators [69], they have stipulated that the presence of water vapor would have significantly deteriorated the CNFs, thereby resulting in the expansion of their graphitic layers, which, in turn, gives rise to the CNFs' high H_2-adsorption capacities; unfortunately, they were unable to probe in situ such an expansion. X-ray data of a pristine sample and the same sample following both adsorption and desorption steps showed the lattice expansion of only 0.007 nm, which is considerably less than 0.03 nm observed for H_2 adsorbed in potassium-intercalated graphite [70].

To enhance the H_2-storage capacities of CNFs, they were grown on a Pd-based catalyst system. Pd has a high potential to dissociate an H_2 molecule into H atoms to be adsorbed at the Pb–C interface [71]. The H_2 uptake experiments were performed volumetrically in a Sievert–type installation by varying the mole ratios of Pd to C from 0.05 to 0.9. The quantity of desorbed H_2 was between 0.04 and 0.33 wt% under the pressure ranging from 1 to 100 bars. This study has indicated that a direct correlation exists between the ratio of Pd and C and the quantity of H_2 absorbed. Unfortunately, a saturation value of approximately 1.5 wt% reached at a high ratio of about 1:1 of Pd to C is far from meeting the energy-density criterion set by DOE [6].

3.3. Single-Walled Carbon Nanotubes

CNTs are allotropes of carbon, representing highly ordered, hexagonal-packed nanostructures that can have a length-to-diameter ratio of up to 28 000 000:1 [72]. CNTs comprise SWCNTs and MWCNTs; the former is covered in what follows.

3.3.1. Structure

An individual SWCNT can be visualized as a single graphene sheet rolled up in an elongated and seamless tube with a diameter in the order of 1 nm and a length in hundreds of micrometers. The pattern of the grapheme-sheet wrapping is represented by a pair of indices (n,m) termed the chiral vector. The integers, n and m, denote the numbers of unit vectors along two directions in the honeycomb crystal lattice of graphene. If $m = 0$, the nanotubes are called "zigzag." If $n = m$, they are called "armchair." Otherwise, they are called "chiral"; see Fig. 7.6 [7]. During synthesis, SWCNTs tend to self-organize and form thick nanoropes, consisting of bundles of hundreds of units, via van der Waals interactions, generally in hexagonl arrangement [73].

3.3.2. Preparation

SWCNTs can be synthesized through an electric arc technique. It involves vaporizing carbon in an electric arc discharge or laser vaporization in the presence of cobalt (Co) or nickel (Ni) catalyst. SWCNTs were first

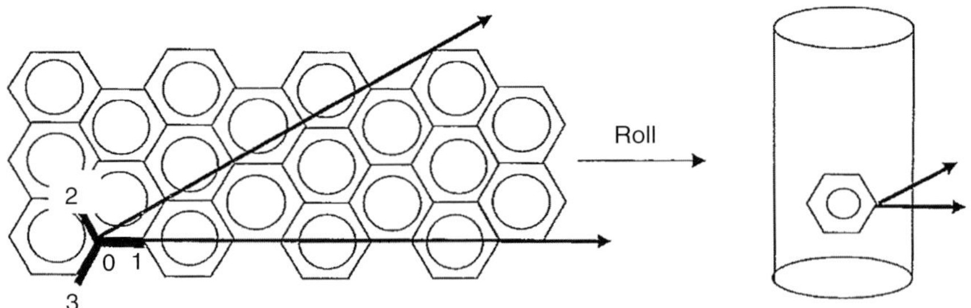

Figure 7.6 Structural relationship between graphene sheet and SWCNT: arrows point to two alternative directions of rolling; circles inside rings reflect on aromatic property of carbon rings and delocalization of resonant electrons inside them [7].

prepared by coevaporating a Co catalyst and graphite in an electric arc. The resultant fibers typically consisted of 7–14 bundled, highly impure SWCNTs with 10–15 Å in diameter [74, 75].

SWCNTs can be produced at a much higher yield at approximately 20–50 wt% by laser vaporization [76]. Crystalline ropes containing hundreds of individual SWCNTs, microns in length, were easily obtained [73]. The soot generated via laser vaporization also contained metal–nanoparticle catalysts and various carbon components. As a result, various purification techniques involving filed flow filtration, acid washing, and annealing steps have been developed to prepare SWCNTs of much higher quality [77–82]. In fact, SWCNTs with purity greater than 98 wt% can be fabricated if micron-sized graphite particles are not generated in a vaporization regime during synthesis [83].

3.3.3. Hydrogen Adsorption

The adsorption processes on bundled SWCNTs proceed at four different types of adsorption sites: the internal sites within the individual SWCNTs, the interstices among the nanotubes within the bundle, the grooves between pairs of SWCNTs at their surfaces, and the external surface sites [15]. It is expected that the cylindrical geometry will deepen adsorbate–adsorbent interaction potential wells inside SWCNTs with small diameters. Nevertheless, the H_2 entering into the interior of SWCNTs is not facilitated; instead, H_2 is preferentially adsorbed on the exterior wall at the interstitial sites in a bundle of SWCNTs due to the overlap of the molecular force fields of an individual SWCNT. In fact, the small pore volume of SWCNTs associated with such sites and the small diameter of each SWCNT reduce the overall density of adsorbed H_2 [15]. Thus, the theoretical maximum amount of adsorbed H_2 is merely 3.0 wt% for SWCNTs with a specific surface area of $1315 \, m^2/g$ at 77 K [10].

SWCNTs were first proposed as H_2-storage media by Dillon et al. [84]. In their study, the soot contained only approximately 0.1–0.2 wt% of SWCNT bundles with 7–14 nanotubes of approximately 12 Å in diameter. Physisorption was identified as the predominant mechanism for H_2 adsorption; the heat of adsorption was estimated to be 19.6 kJ/mol. The study has demonstrated an enhanced interaction between H_2 and SWCNTs compared to planar graphite; for the latter, the heat of adsorption has been reported to be only approximately 4 kJ/mol [18].

The first study of H_2 adsorption on purified laser-generated SWCNTs attained H_2 adsorption of 8 wt% at approximately 40 atm and 80 K [85]; nevertheless, this study failed to elevate high adsorption capacities at 300 K and pressures below 1 atm. It was determined that H_2 was physisorbed on the exposed surfaces of the tubes [85]. In another study, the quantum rotation of H_2 adsorbed on laser-generated SWCNTs was observed via inelastic neutron scattering, demonstrating that H_2 was physisorbed at 25 K under 11 MPa [86].

High H_2-storage capacities on a total sample weight basis were achieved on SWCNTs with a large mean pore diameter of about 1.85 nm produced by a semicontinuous arc-discharge method; the yield from this method tends to be high [87]. The purity of the nanotube soot was estimated to be approximately 50–60%. A sample, first soaked in HCl and subsequently heat-treated in vacuum, was shown to adsorb 4.2 wt% of H_2 at room temperature and 10 MPa; approximately 80% of the adsorbed H_2 could also be released at room temperature. Note that high-capacity adsorption at room temperature was first demonstrated for arc-generated SWCNTs and not for laser-produced nanotubes. This difference might be attributable to a much smaller number of ends or defects in the laser-produced tubes and/or to an enhancement in their stability during the procedures for opening or cutting. A newly developed high-power ultrasonic cutting procedure incorporating a $TiAl_{0.1}V_{0.04}$ alloy has been applied to purified, laser-generated SWCNT materials, thus rendering high-capacity H_2 adsorption possible under ambient conditions [88]. The maximum adsorption capacity of the resultant SWCNTs is approximately 7 wt%, and H_2 adsorption occurs in two separate sites. Approximately 2.5 wt% of the H_2 evolves at 300 K, while the remainder desorbs at 475–850 K. In addition to cutting, the procedure generates an alloy of composition $TiAl_{0.1}V_{0.04}$ due to decomposition of the ultrasonic probe. The presence of the alloy might stimulate H_2 adsorption and desorption. Several control experiments have revealed that the observed uptake is not solely due to the presence of the alloy; some of the experimental results indicate that partial electron transfer is responsible for the stability of the adsorbed H_2 [88].

Electrochemical H_2 storage has also been demonstrated for carbon SWCNTs. Electrodes were fabricated by mixing arc-generated SWCNT soot, containing a few percent of SWCNTs whose diameters range from 0.7 to 1.2 nm, with either copper or gold as a compacting powder in a ratio of 1:4. The kinetics of the SWCNT electrode was found to be relatively poor. The H_2-storage capacity at low discharge currents, however, was as high as 110 mAh/g, corresponding to approximately 0.39 wt% of H_2 [89]. On the assumption that the stored H_2 is totally contained in the small fraction of SWCNTs in the sample, these results indicate that electrochemical H_2 storage on highly pure SWCNTs should not be ignored. In a more recent study [90], SWCNT-composite electrodes, fabricated by mixing SWCNTs, Ni powders, and an organic polytetrafluoroethylene in a ratio of 40:50:10, demonstrated charge/discharge capacities of 160 mAh/g. Unfortunately, no information was provided on either the synthesis or the purity of the SWCNT material.

One approach to enhance H_2 storage resorts to a supported metallic catalyst to promote the dissociation of H_2 molecules into H atoms, which can bind directly with the carbon nanostructures. For instance, the enhancement by a factor of 1.6 was observed for the palladium–catalyst-doped SWCNTs [91].

Nondissociative enhancement has been achieved by modifying the nanostructure to intensify average adsorbate–adsorbent interaction [92, 93]. In some of the approaches, SWCNTs are doped with a metal through the formation of Kubas complexes. This leads to enhanced nondissociatively binding of H_2 molecules to transition metal complexes; such binding is much stronger than van der Waals forces [93]. Calculations based on ab initio quantum chemistry have indicated that a titanium atom embedded SWCNT can store about 7 wt% of H_2 at high titanium coverage under ambient conditions [94]. Such enhancement hinges on the existence of a synthesis path leading to a meta–stable unclustered titanium-covered nano-structure separated from the Ti-clustered configuration by a large energy barrier. It has been speculated that metals, such as Sc, Cr, Pd, and Pt, also enhance H_2 adsorption [92].

3.4. Multiwalled Carbon Nanotubes

MWCNTs constitute another major class of CNTs. They are macromo-lecular arrangements of two or more interpenetrating SWCNTs.

3.4.1. Structure

MWCNTs consist of layers of nested concentric cylinders of graphite with a hollow center; see Fig. 7.7. The spacing between any pair of adjacent concentric cylinders is similar to the interplanar spacing in graphite, and the number of shells varies from 2 up to about 50. MWCNTs have inner and outer diameters that are typically 2–10 and 15–30 nm, respectively, and are generally microns in length. Large bundles of MWCNTs with diameters up to 200 µm and very long individual tubes, whose length exceeds 2 mm [95], have been observed.

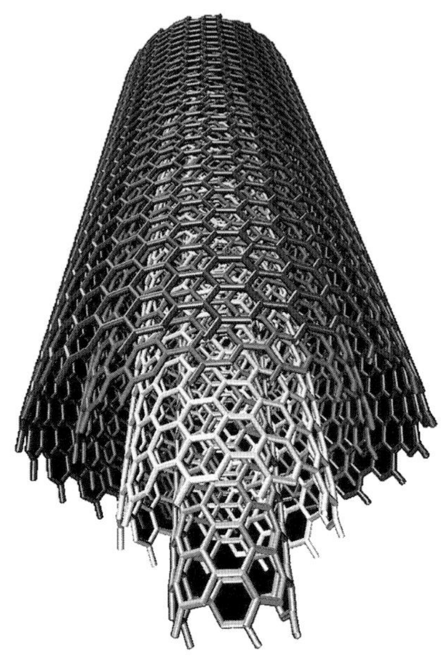

Figure 7.7   Structure of MWCNTs [96].

3.4.2. Preparation

Similar to SWCNTs, MWCNTs can be produced via arc-generation; in the generation process, the carbon contained in the negative electrode sublimates due to the high discharge temperature [97–99]. In fact, MWCNTs were observed in the carbon soot of graphite electrodes during an arc discharge with a current of merely 100 amps, which was intended to produce fullerenes [97]. Arc-generated MWCNTs have been purified with air oxidation. To exhaustively remove the nanocrystalline graphite particles present as impurities, however, entails oxidizing 99% of the total material [100]. It is extremely difficult, if not impossible, to completely remove the impurities from arc-generated MWCNTs. For instance, the nanocrystalline graphite again could not be removed even when oxidized with permanganate [101].

MWCNTs have also been produced by the catalytic decomposition of hydrocarbons, such as benzene pyrolysis [102] and acetylene decomposition with metal catalysts [101, 103]. The crystallinity and stability of the tube caps may be compromised when MWCNTs are produced by the catalytic decomposition of hydrocarbons [101]. CNTs produced by the catalytic decomposition of acetylene over a Co-incorporated zeolite were purified with hydrofluoric acid to remove the catalyst support followed by a permanganate or air oxidation to remove amorphous carbon impurities [101].

3.4.3. Hydrogen Adsorption

H_2-storage capacities were reported to be appreciable for alkali-metal-doped MWCNTs formed by the catalytic decomposition of CH_4 [104]. Li and K were incorporated through solid-state reactions with the metal carbonates or nitrates. H_2 desorption and adsorption were measured by thermogravimetric analysis (TGA) and temperature-programmed desorption. The H_2 uptake was 20 wt% for Li-doped MWCNTs at 653 K and 14 wt% for K-doped MWCNTs at room temperature; the latter was reported to combust upon exposure to air. The H_2 adsorption was believed to proceed by a dissociative mechanism. An infrared spectrum indicated the presence of both Li–H and C–H species [104]; nevertheless, the results have not been confirmed. For instance, in an investigation, Li-doped MWCNTs formed under identical conditions only exhibited a weight increase of 12 wt% when exposed to "wet H_2" and only 2.5 wt% when exposed to "dry H_2" [105]. An infrared spectrum of $LiOH:H_2O$ was strikingly similar to the spectrum acquired by Chen et al. [104]. When K-doped nanotubes were exposed to "dry H_2," a weight increase of only 1.8% was observed [105]. In another investigation, the observation of large H_2 adsorption on Li-doped MWCNTs, similar to those observed by Chen et al. [104], was attributed to the presence of water as an impurity in the TGA atmosphere. In fact, H_2 adsorption by the Li-doped MWCNTs was not evident [106].

Wu et al. [107] studied H_2 adsorption on MWCNTs synthesized by the catalytic decomposition of CO or CH_4 on powdered La_2O_3 catalysts. The CO-generated tubes comprised concentric cylinders, while the CH_4-generated tubes contained graphite layers that were tilted with respect to the tube axis, thus forming cones. In both cases, the catalyst was removed by stirring in dilute nitric acid. The purified nanotubes were then annealed to 1100 °C in vacuum to increase their crystallinity. TGA in flowing H_2 revealed that the CO-generated tubes were capable of adsorbing only a small quantity of H_2 (0.25 wt%) when the sample was cooled at a temperature between 200 °C and room temperature. Another study indicated that MWCNTs could also be charged with H_2 by electrochemical methods [89]. Materials deployed were synthesized by an arc process and contained 10–40 wt% of MWCNTs with diameters ranging from 2 to 15 nm. Stable electrodes were formed by pressing the MWCNT material with palladium powder in a 1:4 ratio. The equilibrium curve of the MWCNT/palladium electrode indicated the occurrence of two separate electrochemical reactions, one of which was not observed for pure palladium electrodes. Although the overall capacity of the palladium/MWCNT electrode was less than that anticipated for pure palladium, the study showed that electrochemical H_2 storage in MWCNTs might be possible. One of the most recent studies of H_2 adsorption on MWCNTs, grown on a $Fe:Co:CaCO_3$ catalytic system and purified by acid cleaning and air oxidation, gave rise to H_2 uptake value of 0.1–0.2 wt% [71].

To summarize, CNTs have been intensively investigated as H_2-storage media; however, their viability remains uncertain. It appears that the CNTs fabricated to date are far from meeting the requirements of high-capacity, room-temperature H_2 storage for automotive application. Furthermore, the cost of scaling-up the process for manufacturing CNTs is formidable; it is reported that "…making a 20-gal hydrogen fuel tank that uses carbon nanotubes could cost \$5.5 million; …" [108]. Economically, therefore, disordered activated carbons might be a more desirable choice than CNTs.

4. CONCLUDING REMARKS

The onboard storage of H_2 is obviously one of the most, if not the most, critical issues in developing an H_2-based energy system. As such, it has lately been a major focus of academic and industrial researches. The storage of H_2 in carbon adsorbents, especially in carbon nanostructures, is still at its infancy and thus is far from mature for commercial application. Many of complexities involved in its large-scale production remain unresolved. They include scale-up, system design, optimization of operating-conditions, and cost estimation.

Adsorption is exothermic, while desorption is endothermic. Hence, the uptake of H_2 into a storage vessel during charging will heat up the bed, thereby reducing its capacity, and the bed will cool on discharging, thus decreasing the deliverable quantity of H_2. Moreover, any impurities in the H_2 gas will accumulate in the storage vessel over repeated charge/discharge cycles; this also will reduce the capacity for H_2. Naturally, it is expected that further extensive research and development need to be undertaken in the near or foreseeable future for the ultimate deployment of carbon adsorbents in the practical onboard storage systems for H_2.

ACKNOWLEDGMENTS

This is contribution No. 10-073-B, Department of Chemical Engineering, Kansas Agricultural Experiment Station, Kansas State University (Manhattan, KS), from which the second and third authors received financial support.

References

[1] Veziroglu TN, Barbir F. Hydrogen: the wonder fuel. Int J Hydrogen Energy 1992; 17:391–401.

[2] Choi B, Nam G, Choi D, Lee B, Kim S, Lee C, et al. Adsorption and regeneration dynamic characteristic of methane and hydrogen binary system. Korean J Chem Eng 1994;21:821–8.

[3] Das LM. Onboard hydrogen storage systems for automotive application. Int J Hydrogen Energy 1996;21:789–800.

[4] Veziroglu TN. Hydrogen energy system as a permanent solution to global energy environmental problems. Chem Ind 1999;53:383–93.

[5] Walker G. Hydrogen storage technologies. In: Walker G, editor. Solid-state hydrogen storage: materials and chemistry (Woodhead Publishing in Materials). New York: CRC Press; 2008. p. 3–17.

[6] DOE. Targets for onboard hydrogen storage systems for light-duty vehicles, http://www1.eere.energy.gov/hydrogenandfuelcells/storage/current_technology.html, Sept. 2009.

[7] Varin RA, Czujko T, Wronski ZS. Nanomaterials for solid state hydrogen storage (fuel cells and hydrogen energy). New York: Springer; 2009.

[8] Hashimoto K, Habazaki H, Yamasaki M, Meguro S, Sasaki T, Katagiri H, et al. Advanced materials for global carbon dioxide recycling. Mater Sci Eng A 2001; 304–306:88–96.

[9] Bockris JO'M. The origin of ideas on a hydrogen economy and its solution to the decay of the environment. Proceedings – Electrochemical Society. 2000–20 (Global Climate Change). 2001. p. 1–24.

[10] Sandi G. Hydrogen storage and its limitations. Electrochem Soc Interface 2004;13:40–5.

[11] Gutowski M, Autrey T. Hydrogen gets on board. Chem World 2006;3:44–8.

[12] Jacoby M., Filling up with hydrogen. Chem Eng News 2005;83:42–7.

[13] Satyapal S, Petrovic J, Thomas G. Gassing up with hydrogen. Sci Am 2007;296:80–7.

[14] Chahine R, Bénard P. Adsorption storage of gaseous hydrogen at cryogenic temperature. In: Kittel P, editor. Advance in cryogenic engineering, vol. 43. New York: Plenum Press; 1998. p. 1257–64.

[15] Benard P, Chahine R. Carbon nanostructures for hydrogen storage. In: Walker G, editor. Solid-state hydrogen storage: materials and chemistry (Woodhead Publishing in Materials). New York: CRC Press; 2008. p. 261–87.

[16] Coolidge AS. Adsorption at high pressures. I. J Am Chem Soc 1934;56:554–61.

[17] Hynek S, Fuller W, Bentley J. Hydrogen storage by carbon sorption. Int J Hydrogen Energy 1997;22:601–10.

[18] Pace EL, Siebert AR. Heat of adsorption of parahydrogen and orthodeuterium on Graphon. J Phys Chem 1959;63:1398–400.

[19] Okamoto Y, Miyamoto Y. Ab initio Investigation of Physisorption of Molecular Hydrogen on Planar and Curved Graphens. J Phys Chem B 2001;105:3470.

[20] Gregg SJ, Sing KSW. Adsorption, surface and porosity. London: Academic; 1982.

[21] Ruthven DM. Principles of adsorption and adsorption processes. New York: John Wiley & Sons; 1984.

[22] Lee SM, Lee YH. Hydrogen storage in single-walled carbon nanotubes. Appl Phys Lett 2000;76:2877.

[23] Gulseren O, Yildirim T, Ciraci S. Tunable adsorption on carbon nanotubes. Phys Rev Lett 2001;87:116–802.

[24] Kong XJ, Chan CT, Ho KM, Ye YY. Cohesive Properties of Crystalline Solids by the General Gradient Approximation. Phys Rev B: Condens Matter 1990;42:9357–64.

[25] Perdew JP, Burke K, Ernzerhof M. Generalized gradient approximation made simple. Phys Rev Lett 1996;77:3865.

[26] Lee SM, An KH, Lee YH, Seifert G, Frauenheim T. A hydrogen storage mechanism in single-walled carbon nanotubes. J Am Chem Soc 2001;123:5059–63.

[27] Smisek M, Cerny S. Activated carbon: manufacture, properties, and applications. Amsterdam: Elsevier; 1970.

[28] Marsh H. Structure in carbons. In: Figueiredo JL, Moulijn JA, editors. Carbon and coal gasifi cation. Dordrecht: Martinus Nijhoff Publishes; 1986. p. 27–57.

[29] Edwards IAS. Structure in carbon and carbon forms. In: Marsh H, editor. Introduction to carbon science. Oxford: Butterworth-Heinemann; 1989. p. 2–36.

[30] Byrne JF, Marsh H. Introduction Overviews. In: Patrick JW, editor. Porosity in carbons: characterization and applications. London: Edwards Arnold; 1995. p. 1–48.

[31] Capelle A. Adsorption of Cationic Copolymers from Dilute Aqueous Solutions on Powdered Activated Carbon. In Capelle A, De Vooys F, editors. Activated Carbon. A Fascinating Material: Some Thoughts on Activated Carbon. Amersfoort, The Netherlands: Norit, 1983. p.191–204.

[32] Jüntgen H. New applications for carbonaceous adsorbents. Carbon 1977;15:273–83.

[33] Spiro CL, McKee DW, Kosky PG, Lamby EJ. Catalytic CO_2-gasification of graphite versus coal char. Fuel 1983;62:180–4.

[34] Laine J, Calafat A, Labady M. Preparation and characterization of activated carbons from coconut shell impregnated with phosphoric acid. Carbon 1989;27:191–5.

[35] Hall RC, Holmes RJ. The preparation and properties of some activated carbons modified by treatment with phosgene or chlorine. Carbon 1992;30:173–6.

[36] Solum MS, Pugmire RJ, Jagtoyen M, Debyshire F. Evolution of carbon structure in chemically activated wood. Carbon 1995;33:1247–54.

[37] Toles C, Rimmer S, Hower JC. Production of activated carbons from a Washington lignite using phosphoric acid activation. Carbon 1996;34:1419–26.

[38] Venkatraman A, Walawender WP, Fan LT. Production and characterization of activated carbon from cereal grains. Fuel chemistry division, ACS 1996;41:260–4.

[39] Wu C-C, Walawender WP, Fan LT. Chemical agents for production of activated carbons from extrusion cooked grain products. Extended abstracts and program, 23rd biennial conference on carbon. Penn State University; 1997. p. 116–7.

[40] Diao Y, Walawender WP, Fan LT. Production of activated carbons from wheat using phosphoric acid activation. Adv Environ Res 1999;3:333–42.

[41] Diao Y, Walawender WP, Fan LT. Activated carbons prepared from phosphoric acid activation of grain sorghum. Bioresour Technol 2002;81:45–52.

[42] Zhang T. Preparation and characterization of carbon molecular sieves and activated carbon. Ph.D. Dissertation, Kansas State University; 2004.

[43] Zhang T, Walawender WP, Fan LT, Fan M, Daugaard D, Brown RC. Preparation of activated carbon from forest and agricultural residues through CO_2 activation. Chem Eng J 2004;105:53–9.

[44] Zhang T, Walawender WP, Fan LT. Preparation of carbon molecular sieves by carbon deposition from methane. Bioresour Technol 2005;96:1929–35.

[45] Zhang T, Walawender WP, Fan LT. Enhancing the microporosities of activated carbons. Sep Sci Technol 2005;44:247–9.

[46] Kidnay AJ, Hiza MJ. High pressure adsorption isotherms of neon, hydrogen, and helium at 76 K. Adv Cryog Eng 1967;12:730–40.

[47] Sinfelt JH, Yates DJC. Catalytic hydrogenolysis of ethane over the noble metals group VIII. J Catal 1967;8:82–90.

[48] Kubicka H. The specific activity of technetium, rhenium, ruthenium, platinum, and palladium in catalytic reactions of benzene with hydrogen. J Catal 1968;12:223–37.

[49] Sermon PA, Bond GC. Hydrogen spillover. Cat Rev 1973;8:211–39.

[50] Schwarz JA. Method and apparatus for cold storage of hydrogen. EU Patent 0230384, 1987.

[51] Schwarz JA, Amankwah KAG. The effect of impurities on hydrogen storage capacity on activated carbons at refrigeration temperature. World hydrogen energy conference, 8th. Honolulu and Waikoloa, Hawaii, U.S.A., 22–27 July 1990. p. 973–81.

[52] Kasaini H, Goto M, Furusaki S. Selective separation of Pd(II), Rh(III), and Ru(III) ions from a mixed chloride solution using activated carbon pellets. Sep Sci Technol 2000;35:1307–27.

[53] Dillon AC, Heben MJ. Hydrogen storage using carbon adsorbents: past, present and future. Appl Phys A 2001;72:133–42.

[54] Orimo S, Majer G, Fukunaga T, Zuttel A, Schlapbach L, Fujii H. Hydrogen in the mechanically prepared nanostructured graphite. Appl Phys Lett 1999;75:3093–95.

[55] Carpetis C, Peschka W. A study on hydtrogen storage by use of cryoadsorbents. Int J Hydrogen Energy 1980;5:539–54.

[56] Hynek S, Fuller W, Bentley J, McCullough J. Hydrogen storage by carbon sorption. Hydrogen energy progress: world hydrogen energy conference. Coral Gables, FL; 1994. p. 985–1000.

[57] Chahine R, Bose TK. Low-pressure adsorption storage of hydrogen. Int J Hydrogen Energy 1993;19:161–4.

[58] Jin H, Lee YS, Hong I. Hydrogen adsorption characteristics of activated carbon. Catal Today 2007;120:399–406.

[59] Schur DV, Tarasov BP, Yu. Zaginaichenko S, Pishuk VK, Veziroglu TN, Shul'ga YM, et al. The prospects for using carbon nanomaterials as hydrogen storage systems. Int J Hydrogen Energy 2002;27:1063–69.

[60] Xu W-C, Takahashi K, Matsuo Y, Hattori Y, Kumagai M, Ishiyama S, et al. Investigation of hydrogen storage capacity of various carbon materials. Int J Hydrogen Energy 2007;32:2504–12.

[61] Chambers A, Rodriguez NM, Baker RTK. Catalytic engineering of carbon nanostructures. Langmuir 1995;11:3862–6.

[62] Baker RTK. http://www.wtec.org/loyola/nano/us_r_n_d/09_03.htm; Jan 1998.

[63] Goodman DW, Kelley RD, Madey TE, White JM. Kinetics of carbon deposition from CO on Cu(110) and Ni(100)1980. J Vac Sci Technol 1980;17:143.

[64] Chen JP, Yang RT. Chemisorption of hydrogen on different planes of graphite-a semi-empirical molecular orbital calculation. Surf Sci 1989;216:481–8.

[65] Nakamura J, Hirano H, Xie M, Matsuo I, Yamada T, Tanaka K. Formation of a hybrid surface of carbide and graphite layers on Ni(100) but no hybrid surface on Ni(111). Surf Sci 1989;222: L809–17.

[66] Chambers A, Park C, Baker RTK, Rodriguez NM. Hydrogen storage in graphite nanofibers. J Phys Chem B 1998;102:4253–6.

[67] Ahn CC, Ye Y, Ratnakumar BV, Whitham C, Bowman Jr RC, Fultz B. Hydrogen desorption and adsorption measurements on graphite nanofibers. Appl Phys Lett 1998;73:3378–80.

[68] Ströbel R, Jörissen L, Schliermann T, Trapp V, Schütz W, Bohmhammel K, et al. Hydrogen adsorption on carbon materials. J Power Sci 1999;84:221–4.

[69] Park C, Anderson PE, Chambers A, Tan CD, Hidalgo R, Roderiguez NM. Further studies of the interaction of hydrogen with graphite nanofibers. J Phys Chem B 1999;103:10572–81.

[70] Doll GL, Eklund PC, Senatore G. Elastic neutron scattering studies of H_2 and D_2 physisorbed stage 2 graphite-potassium. In: Dresselhaus MS, editor. NATO ASI Series, v. 148, Intercalation in layered materials. New York: Plenum Press; 1986. p. 309–11.

[71] Biris AR, Lupu D, Dervishi E, Li Z, Saini V, Saini D, et al. Hydrogen storage in carbon-based nanostructured materials. Part Sci Technol 2008;26:297–305.

[72] Pant KK, Gupta RB. Hydrogen Storage in Carbon Materials. In: Gupta RB, editor, Hydrogen fuel: production, transport, and storage. New York: CRC Press; 2009. p. 409–36.

[73] Thess A, Lee R, Nikolaev P, Dai H, Pitit P, Robert J, et al. Crystalline ropes of metallic carbon nanotubes. Science 1996;273:483–7.

[74] Iijima S, Ichihashi T. Single-shell carbon nanotubes of 1-nm diameter. Nature 1993;363:603–5.

[75] Bethune DS, Kiang C-H, de Vries MS, Gorman G, Savoy R, Vasquez J, et al. Cobalt-catalysed growth of carbon nanotubes with single-atomic-layer walls. Nature 1993;363:605–7.

[76] Guo T, Nikolaev P, Thess A, Colbert DT, Smalley RE. Catalytic growth of single-walled nanotubes by laser vaporization. Chem Phys Lett 1995;243:49–54.

[77] Tohji K, Goto T, Takahashi H, Shinoda Y, Shimizu N, Jeyadevan B, et al. Purifying single-walled nanotubes. Nature 1996;383:679.

[78] Bandow S, Rao AM, Williams KA, Thess A, Smalley RE, Eklund PC. Purification of single-wall carbon nanotubes by microfiltration. J Phys Chem B 1997;101:8839–42.

[79] Dujardin E, Ebbesen TW, Krishnan A, Treacy MMJ. Purification of single-shell nanotubes. Adv Mater 1998;10:611–3.

[80] Rinzler AG, Lui J, Dai H, Nikolaev P, Huffman CB, Roderiguez-Macias FJ, et al. Large-scale purification of single-wall carbon nanotubes: process, product, and characterization. Appl Phys A 1998;67:29–37.

[81] Shelimov KB, Esenaliev RO, Rinzler AG, Huffman CB, Smalley RE. Purification of single-wall carbon nanotubes by ultrasonically assisted filtration. Chem Phys Lett 1998;282:429–34.

[82] Dillon AC, Gennett T, Jones KM, Alleman JL, Parilla PA, Heben MJ. A simple and complete purification of single-walled carbon nanotube materials. Adv Mater 1999;11:1354–8.

[83] Dillon AC, Parilla PA, Alleman JL, Perkins JD, Heben MJ. Controlling single-wall nanotube diameters with variation in laser pulse power. Chem Phys Lett 2000;316:13–8.

[84] Dillon AC, Jones KM, Bekkedahl TA, Kiang CH, Bethune DS, Heben MJ. Storage of hydrogen in single-walled carbon nanotubes. Nature 1997;386:377–9.

[85] Ye Y, Ahn CC, Witham C, Fultz B, Liu J, Rinzler AG, et al. Hydrogen adsorption and cohesive energy of single-walled carbon nanotube. Appl Phys Lett 1999;74:2307–9.

[86] Brown CM, Yildirim T, Neumann DA, Heben MJ, Gennett T, Dillon AC, et al. Quantum rotation of hydrogen in single-wall carbon nanotubes. Chem Phys Lett 2000;329:311–6.

[87] Liu C, Fan YY, Liu M, Cong HT, Cheng HM, Dresselhaus MS. Hydrogen storage in single-walled carbon nanotubes at room temperature. Science 1999;286:1127–9.

[88] Dillon AC, Gennett T, Parilla PA, Alleman JL, Jones KM, Heben MJ. Nanotubes and related materials. Mater Res Soc Symposium Proceedings, vol. 633. 2001. P. A5.2.1–6.

[89] Nutzenadel C, Zuttel A, Chartouni D, Schlapbach L. Electrochemical storage of hydrogen in nanotube materials. Electrochem Solid-State Lett 1999;2:30–2.

[90] Lee SM, Park KS, Choi YC, Park YS, Bok JM, Bae DJ, et al. Hydrogen adsorption and storage in carbon nanotubes. Synth Met 2000;113:209–16.

[91] Lachawiec Jr AJ, Qi G, Yang RT. Hydrogen storage in nanostructured carbons by spillover: bridge-building enhancement. Langmuir 2005;21:11418–24.

[92] Dag S, Ciraci S. Coverage and strain dependent magnetization of titanium-coated carbon nanotubes. Phys Rev B: Condens Matter 2005;71:165414-1–6.

[93] Kim Y-H, Zhao Y, Williamson A, Heben MJ, Zhang SB. Nondissociative adsorption of H_2 molecules in light-element-doped fullerenes. Phys Rev Lett 2006;96:1–4.

[94] Yildirim T, Ciraci S. Titanium-decorated carbon nanotubes as a potential high-capacity hydrogen storage medium. Phys Rev Lett 2005;94:1–4.

[95] Wang XK, Lin XW, Dravid VP, Ketterson JB, Chang RPH. Stable glow discharge for synthesis of carbon nanotubes. Appl Phys Lett 1995;66:2430–2.

[96] Rochefort A. http://www.nanotech-now.com/nanotube-buckyball-sites.htm. 2009.

[97] Iijima S. Helical microtubules of grahpitic carbon. Nature 1991;354:56–8.

[98] Ebbesen TW, Ajayan PM. Large-scale synthesis of carbon nanotubes. Nature 1992; 358:220–2.

[99] Ishigami M, Cumings J, Zettl A, Chen S. A simple method for the continuous production of carbon nanotubes. Chem Phys Lett 2000;319:457–9.

[100] Ebbesen TW, Ajayan PM, Hiura H, Tanigaki K. Purification of nanotubes. Nature 1994;367:519.

[101] Colomer J-F, Piedigrosso P, Wilems I, Journet C, Bernier P, Tendeloo GV, et al. The production and structure of pyrolytic carbon nanotubes (PCNTs). J Chem Soc Faraday Trans 1998;94:3753–8.

[102] Endo M, Takeuchi K, Igarashi S, Kobori K, Shiraishi M, Kroto HW. The production and structure of pyrolytic carbon nanotubes (PCNTs). J Phys Chem Solids 1993; 54:1841–8.

[103] Ivanov V, Nagy JB, Lambin P, Lucas A, Zhang XB, Zhang XF, et al. The study of carbon nanotubules produced by catalytic method. Chem Phys Lett 1994;223:329–35.

[104] Chen P, Wu X, Lin J, Tan KL. High H_2 uptake by alkali-doped carbon nanotubes under ambient pressure and moderate temperature. Science 1999;285:91–3.

[105] Fan Y-Y, Liao B, Liu M, Wei Y-L, Lu M-Q, Cheng H-M. Hydrogen uptake in vapor-grown carbon nanofibers. Carbon 1999;37:1649–52.

[106] Pinkerton FE, Wicke BG, Olk CH, Tibbetts GG, Meisner GP, Meyer MS, et al. Thermogravimetric measurement of hydrogen absorption in alkali-modified carbon materials. J Phys Chem B 2000;104:9460–7.

[107] Wu XB, Chen P, Lin J, Tan KL. Hydrogen uptake by carbon nanotubes. Int J Hydrogen Energy 2000;25:261–5.

[108] Feather fibers fluff up hydrogen storage capacity. CEP. August 2009; p.14–5.

CHAPTER 8

Treatment of Nanodiamonds in Supercritical Water

Vlamidir I. Anikeev *and* **Anna Yermakova**
Boreskov Institute of Catalysis SB RAS, Novosibirsk, Russia

Contents

1. INTRODUCTION

Carbon nanostructures have been studied quite intensively in recent years. Among various synthesized carbon nanomaterials, of particular interest are the so-called ultradispersed nanodiamonds (UDD) obtained by detonation of solid carbon-containing explosives with a negative oxygen balance [1–4]. Current and potential application areas of UDD are very wide [3]. UDD can be used in metal–diamond galvanic coatings, in polishing pastes and suspensions, for lubricating oils and lubricant coolants, as addition to polymers. Due to the high adsorption capacities, UDD is considered promising for medical and biological applications.

Although this method is highly efficient, the resulting product – detonation carbon (soot) or diamond blend – contains different structures and forms of carbon, with the diamond phase constituting approximately 35–45 wt%. Moreover, depending on the production process, detonation carbon may include the sorbed impurities represented by metals, their oxides, and their carbides [3–5].

To recover the diamond phase and remove water-insoluble impurities, the initial diamond blend is usually treated with liquid or gaseous oxidants [3, 6, 7]. Mixtures of sulfuric and nitric acids and sulfurous and chromic anhydrides are used as the liquid oxidants. The gaseous thermo-oxidants are represented by oxygen or ozone. In practice, the most used technique consists in treating the diamond blend with concentrated nitric acid under thermobaric conditions – in an autoclave at elevated pressure and temperature. This method allows one to oxidize nanodiamond carbon and remove metals, their oxides, and some other impurities. To remove or change the composition of surface functional groups (NH, $C=O$, $C=N$, NO_3, SO_3H), the starting mix is treated also with organic solvents, alcohols, or ion-exchange resins [3]. Thus, purified detonation carbon may contain up to 90–97% of different nanodiamond species and 3–10% of nanodiamond carbon and other impurities.

However, chemical purification with the use of strong acids may lead, on the one hand, to the oxidation of the diamond phase and contamination of the diamond surface with different functional groups, and on the other hand, to a large release of aggressive wastes. Thus, irrespective of a satisfactory removal of nondiamond phases and impurities provided by the acid purification, it is necessary to suggest and test alternative methods for treatment of diamond-containing mix and UDD recovery. These methods may lead to the formation of new properties of ultradispersed diamonds, which can underlie the development of novel materials and substances with specified properties. New methods of treatment can be used individually, combined with the conventional techniques and supplement them. Note that the ability to decrease detrimental and aggressive wastes should be taken into account in the development of new methods.

For this purpose, authors of the present work carried out a study aimed at the treatment of the detonation carbon mix produced at the Federal Scientific and Production Center "ALTAI," Biysk, Russia, both in supercritical water (SCW) and with the addition of hydrogen peroxide. The goal of this work was to study the morphological and structural changes of carbon species and to choose the composition of supercritical fluid and the process conditions. Note that determination of conditions and parameters providing the maximum purification of UDD, as well as a thorough investigation of its surface properties, phase composition, and microstructure, was beyond the scope of our work because these aspects have received much attention in the literature [3, 4, 8–11].

2. THERMODYNAMICS OF SOLID GRAPHITE AND DIAMOND CONVERSION IN SCW

A relatively large number of publications focused on the studies of stability of various nanocarbon phases, including the diamond and graphite ones, are available in the literature [12–16]. Nanocarbon phase diagrams were studied as a function, besides temperature and pressure, of the third parameter — size of cluster, which may contain from several hundreds to tens of thousands of carbon atoms [12, 17]. As a rule, the nanocarbon phase transitions proceed at high temperatures and pressures or under exposure to electron beam [18].

Quantum mechanical calculation of the carbon–cluster formation energy, including potential surface energy, which controls the dynamics of the graphite- to diamond-phase transition, proved relative stability of various nanocarbon phases. It was shown that stability of carbon phases increased in raw fullerene $\longrightarrow$ diamond $\longrightarrow$ graphite with increasing cluster size [13, 14].

However, for studying chemical reactions of different carbon phases in SCW under relatively soft conditions, it seems reasonable to estimate the differences in the conversion of diamond and graphite in SCW in the presence of oxygen. For this purpose, let us use classic thermodynamics for the equilibrium calculation and introduce in the Gibbs free energy equation an item (summand) representing the contribution of the surface energy of a cluster comprised n atoms of ith carbon phase:

$$G_i(T, P, n) = dE_i n^{-1/3} + G_i(T, P) \tag{1}$$

where G_i is the Gibbs energy of ith carbon phase and dE_i is the surface energy of a cluster comprised of n atoms of ith carbon phase. Gibbs energy is assumed to be 70 kcal/mol for diamond and 40 kcal/mol for graphite [12].

Calculation showed that at temperature 390 °C and pressure 285 atm (corresponding to experimental values), both graphite and diamond phases convert completely in SCW to yield H_2, CO, CO_2, and CH_4 (in the presence of oxygen).

As graphite with low n number is oxidized in SCW with deficient oxygen, the H_2 yield is considerably higher and the CO yield is considerably lower than respective values in the oxidation of diamond.

At temperature and pressure similar to experimental values and in oxygen deficiency, the oxidation of graphite in SCW yields a larger amount of H_2 and CO_2, but lower amount of CO and CH_4, than the oxidation of diamond.

As the temperature of carbon (both diamond and graphite species) reaction with SCW increases from 220 to 400 °C, the CO concentration in the reaction products increases considerably, whereas the CO_2 and CH_4 concentrations decrease slightly.

Thus, the calculation of equilibrium conversion of diamond and graphite in SCW showed no considerable differences related to individual carbon phases or number of carbon atoms in the cluster. Therefore, it is reasonable to expect that experiments on the oxidation of these carbon species in SCW will show only different kinetic characteristics.

3. EXPERIMENTAL PROCEDURE

The experiments were performed in an autoclave reactor of approximately $70 \, cm^3$ capacity equipped with electric heater and magnetic stirrer. When detonation carbon was treated only in SCW, the initial sample of the mix was placed with water into an autoclave and heated to the temperature of experiment, 390 °C (± 5), at the reactor pressure 285 atm (± 5). The temperature and pressure were selected slightly higher of the critical parameters of pure water ($P_{cr} \approx 22.1 \, MPa$, $T_{cr} \approx 374 \, °C$). After establishing the temperature and pressure, the process lasted from 2 to 4 h.

Phase composition of the solid phase was studied before and after the treatment of the mix in SCW using the high-resolution transmission electron microscopy (HRTEM), scanning electron microscopy, and X-ray diffraction analysis. The composition of gaseous reaction products was analyzed chromatographically, with the measurement of their volume.

A quantitative ratio of diamond and graphite phases in all the samples, including the initial samples, was determined with an X'TRA diffractometer (Switzerland) using the $CuK\alpha$- radiation ($\lambda = 1.5418 \, Å$). Virtually, each diffraction micrographs shows the considerably broadened peaks from graphite-like and diamond phases, as well as the diffraction peaks from impure crystal phases (Fe_3O_4 and others).

A quantitative analysis of the relative content of graphite-like and diamond phases was based on the ratio of intensities of the largest peaks from these phases: the graphite peak (002) and the diamond peak (111). The calibration plot showing the I_D/I_G ratio of integral intensities of the diamond (I_D) and graphite (I_G) peaks (111) as a function of weight percent ratio of

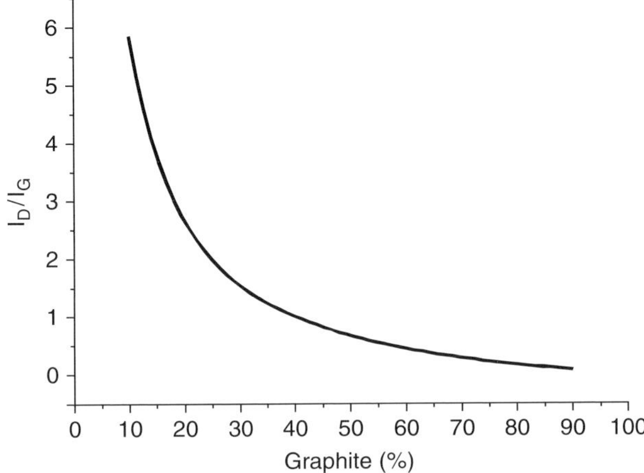

Figure 8.1   The ratio of integral intensities of the diamond (I_D) and graphite (I_G) peaks (111) in relation to the weight percent ratio of these phases.

these phases was obtained from theoretical calculation of intensities made with the use of PowderCell for Widows (PCW) program (Fig. 8.1).

4. RESULTS AND DISCUSSION

4.1. Initial Sample

Initial sample of the diamond blend, detonation carbon (sample #1), was a carbon material represented by a highly dispersed black powder. From the HRTEM data, initial sample of the mix consists of a carbon material with quite uniform density, which includes several types of nanostructures. Figure 8.2A shows the diamond crystals (marked by squares) embedded in the carbon material as flakes. The sample contains also a large amount of amorphous carbon, for example, in the micrograph region marked by a circle. A high magnification of the initial sample in Figs 8.2B and C shows the presence of onion-like carbon species with inclusion of diamond nanoparticles and diamond nanocrystals not covered with the graphite-like layers (marked by squares in Fig. 8.2C).

4.2. Results of the X-ray Diffraction Analysis

The diffraction micrograph of the initial sample (sample #1) in Fig. 8.3 shows the broadened peaks from graphite-like carbon phase and peaks (111) and (220) assigned to the diamond phases. From the XRD data, mean size of diamond crystals is $30 \pm 5\,\text{Å}$, and a quantitative ratio makes

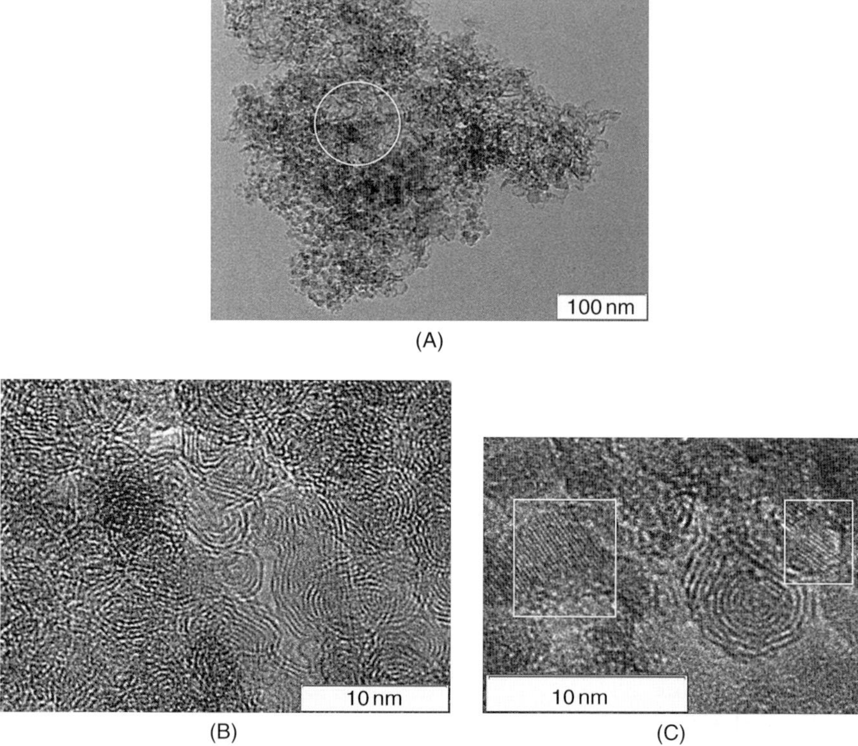

(A)

(B)   (C)

Figure 8.2   (A–C) HRTEM images of the initial sample of detonation carbon (sample #1).

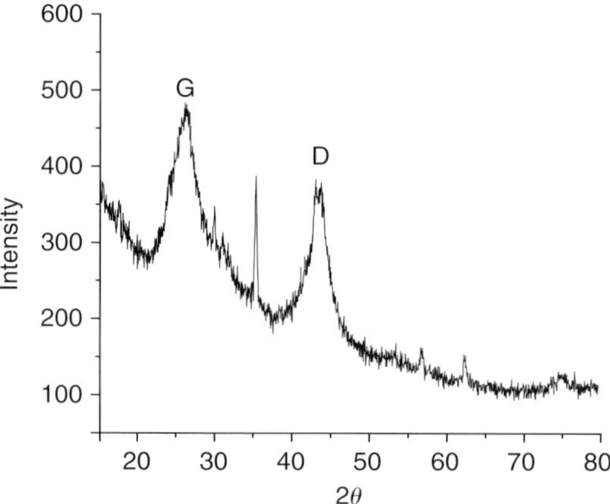

Figure 8.3   Diffraction micrograph of the initial sample of detonation carbon (sample #1).

up 42% for diamond and 58% for graphite-like phase ($I_D/I_G = 0.72$). The micrograph of initial sample shows also sharp diffraction peaks from impurity phases, with the peak corresponding to crystal iron being the most pronounced. The peak from graphite-like phase partially splits, which may evidence a morphological inhomogeneity of this phase in the initial sample of diamond-containing mix.

4.3. Pretreatment of the Diamond Blend

Detonation carbon pretreated in hydrochloric acid was used as the second initial sample of the diamond blend (sample #2) to be treated in SCW. In this case, the mix was boiled in the presence of hydrochloric acid in a glass flask on oil bath for approximately 5 h at 128–132 °C. After boiling, solid phase was filtered and repeatedly washed with water to bring pH to 7. After the secondary filtration, the sample was dried at 50–60 °C for a day. The treated powder did not change its color, whereas the waste acid became yellow and green.

The UV spectroscopy data (spectrophotometer Shimadzu 2501-PC) showed the absorption in the region of 11000–12000 cm^{-1}, which is typical of aqua complexes of bivalent iron. Hence, treatment of the diamond blend in acid leads to partial or complete removal of iron, which is reliably confirmed by the XRD data (Fig. 8.4). The XRD pattern of the sample contains the broadened peaks from diamond and graphite-like phases; impurity phases are not observed. Mean size of the diamond crystals and a quantitative ratio of diamond and graphite-like phases remained the same as in sample #1.

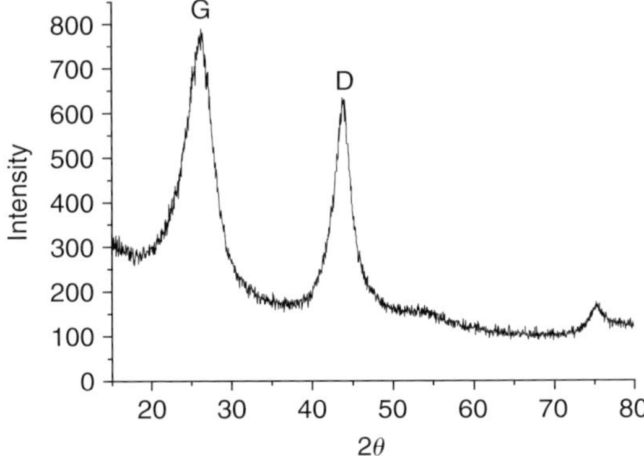

Figure 8.4   Diffraction micrograph of the treated initial sample of detonation carbon (sample #2).

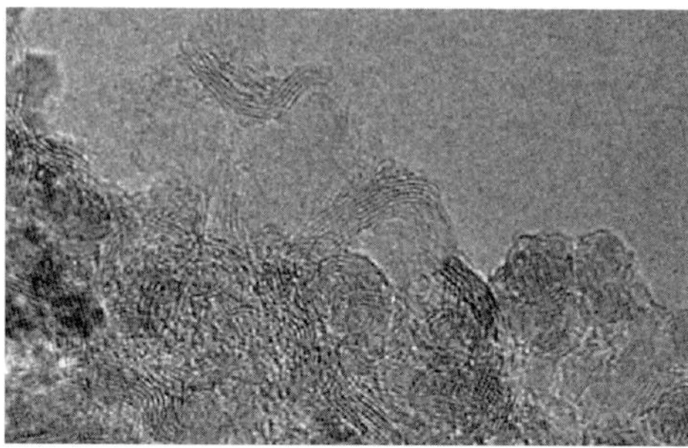

Figure 8.5   Image of sample #2 of detonation carbon treated in hydrochloric acid.

The HRTEM analysis of sample #2 of detonation carbon after processing in hydrochloric showed (see Fig. 8.5) that onion-like carbon structures are partially destroyed and transform into the so-called graphene sheets.

4.4. Treatment of Detonation Carbon in SCW

A sample of the diamond blend with water was placed into an autoclave reactor and heated under stirring to the temperature of experiment, 390 °C, and pressure 285 atm. After establishing the temperature and pressure corresponding to water transition to a supercritical state, the process ran from 2 to 6 h. After cooling the reactor, pressure and gas volume of the reaction products were measured. The measurement showed a 2 to 2.5-fold increase of the gas volume of the products compared with the initial free volume of the reactor, which indicates a strong oxidation of the carbon-containing phase. Analysis of gaseous product shown that CO_2 constituted approximately 70–80 vol.% of the resultant gases; in addition, the reaction products contained CO and CH_4 (up to 20%) and trace amounts of H_2. A weight loss of the solid diamond blend ranged from 15 to 35%, depending on the process time, which also evidences the oxidation of carbon component of the diamond blend to occur in the process with SCW. However, this gives no information about carbon structures subjected to preferential oxidation.

4.5. Results of the X-ray Diffraction Analysis

The analysis of the diffraction spectra of SCW-treated samples #1 shows that the I_D/I_G ratio of integral intensities of the peaks for this treatment depends on the process time and increases up to 0.8–0.9. All these spectra of treated samples also contain the diffraction peaks from impurities, with the maximum peak corresponding to the crystal phase of Fe_3O_4. The peaks appear because water in supercritical state, due to its dissociation, exhibits the acidic properties, which leads to oxidation of the reactor walls made of stainless steel and to ingress of metal oxides into the subject of investigation residing in the reactor.

4.6. Results of the HRTEM Analysis

A comparison of the images of initial sample #1 (Fig. 8.2A–C) with the images obtained after its treatment in SCW (Fig. 8.6) demonstrates essential distinctions: first, a considerably decreased amount of amorphous carbon phase in the treated sample; second, an increased concentration of the cubic modification of carbon.

Sample #2 was subjected to SCW treatment by the procedure similar to that described above for sample #1. The HRTEM images show that the treated sample consist of three parts: partially retained onion-like carbon species, homogeneous carbon structures with embedment of crystal diamond (Fig. 8.7A), and carbon (graphite) fibers 20–60 nm in length (Fig. 8.7B). The formation of so different carbon nanostructures in the

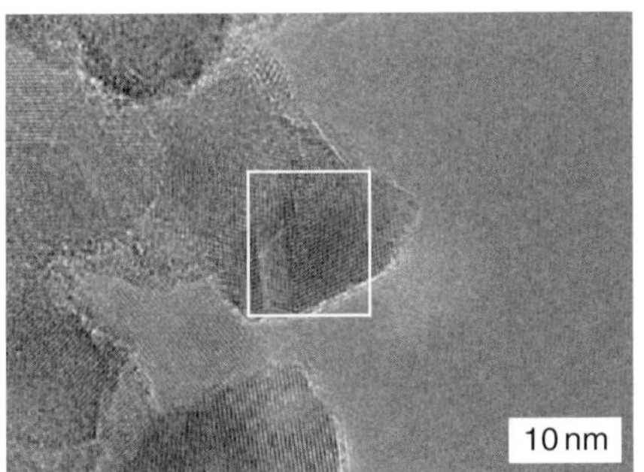

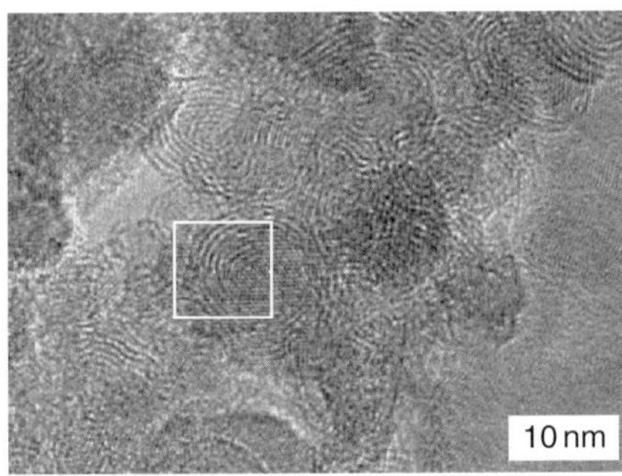

Figure 8.6   Images of detonation carbon after SCW treatment of sample #1.

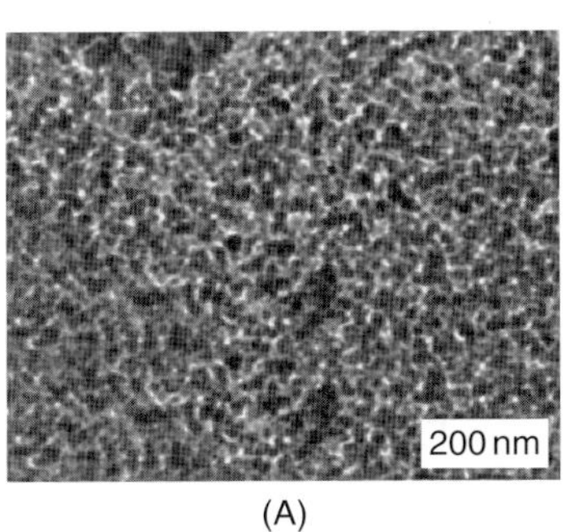

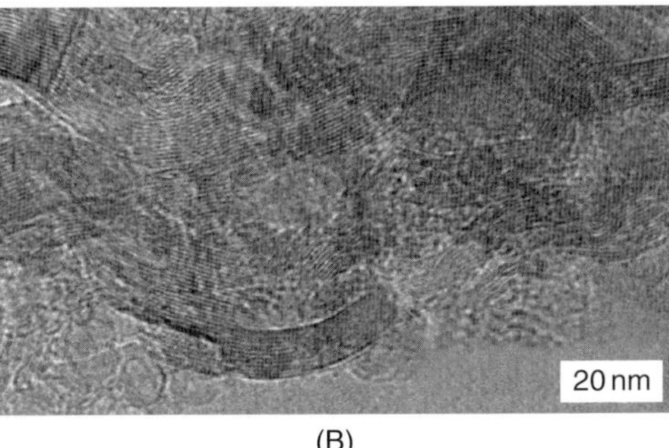

(A)   (B)

Figure 8.7   HRTEM images of detonation carbon after SCW treatment of sample #2.

samples of SCW treated material may have several reasons. First, reactor walls may affect the structure of condensed carbon upon contacting with SCW. Second, a part of carbon material may settle to the reactor bottom due to weak stirring, which leads to insufficient contact with SCW.

4.7. Treatment of Detonation Carbon in SCW in the Presence of Hydrogen Peroxide Decomposition Products

The detonation carbon with water (sample #1 or #2) was placed into an autoclave reactor and heated up to 300 °C, then a specified amount of 30% solution of hydrogen peroxide was injected under pressure by syringe pump to provide the molar ratio of oxygen/carbon, 0.25–1. After establishing the stationary temperature 390 °C ($\pm$5) and pressure 285 atm ($\pm$5), the process ran from 1 to 4 h. After the treatment of sample #2 in SCW in the presence of oxygen that formed at hydrogen peroxide decomposition, detonation carbon changed its color from black to gray. According to the measurement performed after reactor cooling, the volume of evolved gas, including the unreacted oxygen, exceeded more than 8-fold the initial free volume of the reactor.

During the treatment of sample #2, chromatographic analysis of gaseous reaction products showed the reaction mixture to contain, along with oxygen, more than 85% of CO_2, H_2, CO, and CH_4. The injected hydrogen peroxide decomposed completely to yield oxygen and water just after its introduction into the reactor. Thus, the oxidation of solid carbon involved also the oxygen dissolved in SCW. The amount of carbon that passed from solid to gas phase can be calculated from the volume of nascent carbon-containing gases.

4.8. Results of the X-ray Diffraction Analysis

The I_D/I_G ratios of integral intensities for the samples of detonation carbon treated in SCW in the presence of oxygen show a pronounced change toward increasing the fraction of diamond phase and attain the value of 1.86 for samples #2 during 4 h of the process, which corresponds to increasing the fraction of diamond phase up to 75 wt.%. Treatment of samples #1 under the same conditions gives the I_D/I_G ratio not exceeding 1.0. Thus, although the initial samples #1 and #2 contain virtually equal amounts of diamond and graphite phases, pretreatment of sample #2 with acid, according to [3], may result in loosening and etching of the carbon matrix along the structural defects, which provides a more considerable removal of carbon at SCW treatment of sample #2.

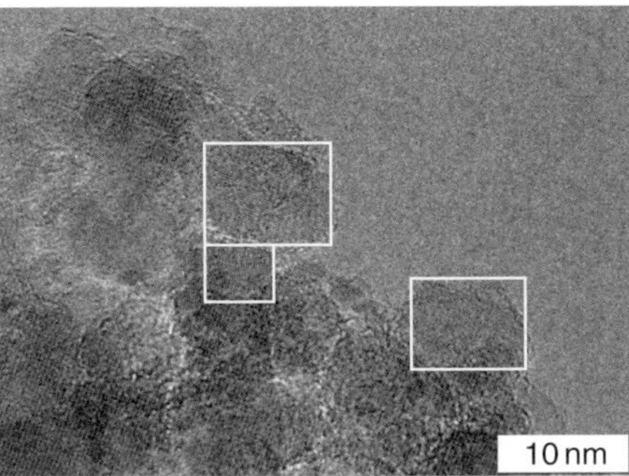

Figure 8.8   HRTEM image of detonation carbon after SCW treatment of sample #2 in the presence of hydrogen peroxide decomposition products.

4.9. Results of the HRTEM Analysis

The analysis of HRTEM images of samples #2 of detonation carbon after their treatment in SCW in the presence of hydrogen peroxide decomposition products shows, first, a high degree of purification (Fig. 8.8); second, nearly a complete absence of amorphous carbon phase and onion–like carbon; third, constancy of the carbon nucleus size, which indicates that the diamond phase is not oxidized.

Note that a disadvantage of the above experiments on treatment of the detonation carbon in SCW performed in a stainless–steel autoclave reactor consists in contamination of a sample with metals (their oxides) from the reactor walls due to their corrosion in the presence of SCW. This problem can be solved in future by using different materials and alloys.

5. CONCLUSIONS

The study showed that the interaction of detonation carbon with SCW is accompanied by the formation of large amounts of gaseous products, mainly CO and CO_2, which is evidence of a deep oxidation of the carbon phase. The transformation of nondiamond carbon phase is most pronounced when the detonation carbon is treated in SCW that contains oxygen formed at decomposition of hydrogen peroxide. The SCW treatment of detonation carbon black was shown to increase considerably the weight fraction of diamond phase. It was found that oxidation of the diamond nucleus does not occur when detonation carbon is treated in SCW.

References

[1] Liamkin AI, Petrov EA, Ershov AP, Sakovich GV, Staver AM, Titov VM. Production of diamonds from explosive substances. Dokl Acad Nauk 1988;302(3):611–13.

[2] Volkov KV, Danilenko VV, Elin VI. Diamond synthesis from the carbon of explosives detonation products. Combustion Explosion Shock Waves 1990;26(3):123–5.

[3] Yu. Dolmatov V. Ultradisperse diamonds produced by detonation synthesis: properties and application. Russian Chem Rev 2001;70(7):687–708.

[4] Shenderova OA, Zhirnov VV, Brenner DW. Carbon nanostructures. Crit Rev Solid State Mater Sci 2002;27(3/4):227–356.

[5] Bogatyreva GP, Voloshin MN. Characterization and properties of diamond powders obtained by explosion. Supersolid Mater 1998;4:82–7.

[6] Putyatin AA, Nikolskaya IV, Kalashnikov N Ya. Chemical methods of diamond recovery from synthesis products. Supersolid Mater 1982;2:20–8.

[7] RF Patent 2109683. A method for recovery of synthetic ultradisperse diamonds. 1998.

[8] Aleksensky AE, Baidakova MV, Vul A Ya, Siklitsky VI. The structure of diamond nano-cluster. Phys Solid State 1999;41(4):740–3.

[9] Kurdyumov AV, Ostrovskaya NF, Zelyavsky VB, Borimchuk NI, Yarosh VV. Structural features of nanodisperse diamonds obtained by dynamic synthesis. Supersolid Mater 1998;4:23–9.

[10] Baidakova MV, Vul A Ya, Siklitsky VI, Faleev NN. Fractal structure of ultradisperse diamond clusters. Phys Solid State 1998;40(4):776–80.

[11] Aleksensky AE, Baidakova MV, Vul A Ya, Davydov VI, Pevtsova Yu A. Diamond-graphite phase transition in ultradisperse diamond clusters. Phys Solid State 1997;39(6):1125–34.

[12] Ree FH, Winter NW, Glosli JN, Viecelli JA. Kinetics and thermodynamic behavior of carbon clusters under high pressure and high temperature. Phys B 1999;265:223–9.

[13] Barnard AS, Russo SP, Snook IK. Thermal stability of graphene edge structure and graphene nanoflakes. J Chem Phys 2003;118:5094–97.

[14] Barnard AS, Snook IK. Phase stability of nanocarbon in one dimension: nanotubes versus diamond nanowires. J Chem Phys 2004;120(8):3817–21.

[15] Nuth III JA. Small-particle physics and interstellar diamonds. Nature 1987;329:589.

[16] Badziag P, Verwoerd WS, Ellis WP, Greiner NR. Nanometer-sized diamonds are more stable then graphite. Nature 1990;343:244–5.

[17] Bundy FP, Bassett WA, Weathers MS, Hemley RJ, Mao HK, Goncharov AF. The pressure-temperature phase and transformation diagram for carbon; updated through 1994. Carbon 1996;34:141–53.

[18] Banhart F, Ajayan PM. Carbon onions as nanoscopic pressure cells for diamond formation. Nature 1996;382:433–35.

CHAPTER 9

Spectrophotometric Flow-Injection System Using Multiwalled Carbon Nanotubes as Solid Preconcentrator for Copper Monitoring in Water Samples

Giovana de Fátima Lima[*], Polyana Maria de Jesus Souza[*], Mariana Gava Segatelli[**], Pedro Orival Luccas[*] *and*
César Ricardo Teixeira Tarley[*]

[*] Departamento de Ciências Exatas, Universidade Federal de Alfenas, Unifal-MG, Alfenas, Minas Gerais, Brazil

[**] Instituto de Química, Departamento de Química Inorgânica, Universidade Estadual de Campinas, Unicamp, Campinas, São Paulo, Brazil

Contents

1. INTRODUCTION

Copper is one of a relatively small group of metallic elements, which are essential for human, plants, and animal enzymes. Copper ions are introduced into human and animal bodies by food, water, air, etc. On the other hand, rain, snow, fertilizer, and irrigation water are the most common routes for copper absorption by plants [1]. Also, copper enters the marine environment primarily from riverine transport and aerosols. Anthropogenic and geothermal sources can also be included [2]. Depending on its concentration, copper may be a hazard to aquatic organisms [3]. Thus, although copper is essential to life, its determination in water samples is warranted by the narrow window of the concentration between essential and toxic [3]. In this way, copper levels in natural environments continue to be of interest, and many efforts to determine copper levels have been focused [4, 5]. In this context, one of the most difficult and complicated analytical tasks is the development of accurate and sensitive quantitative analysis of several kinds of samples especially at trace levels. At the same time, the demand for very simple and much cheaper analysis has been evident [6]. In this way, despite the importance of the highly sensitive atomic spectrometric techniques, such as inductively coupled plasma mass spectrometry or graphite furnace atomic absorption spectrometry is significant for metal determination at trace levels, flame atomic absorption spectrometry (FAAS), owing to simplicity and fairly low initial cost and running cost, still remains as a reliable method for metal determination. However, this technique suffers from lack of sensitivity for metal determination at trace levels, thus requiring usually preconcentration procedures prior determination [7]. UV–vis molecular spectrophotometry is another technique that also can be widely used for very simple and cheap analysis. Moreover, the existence of different chromogenic agents offers a widespread application of UV–vis molecular spectrophotometry for metal determination, enabling enhancements in the selectivity and attains lower detection limits [8]. In addition, several approaches can be found in literature for improving the sensitivity and selectivity of this technique, including formation of mixed aggregates with surfactants [9], measurements with long path length [10], and preconcentration procedures (liquid–liquid extraction, solid-phase extraction, and cloud point extraction) [11–13]. The preconcentration procedures using solid sorbent have gained special importance in analysis of complex matrices owing to some advantageous features, such as higher preconcentration factor (PF), simplicity, easy coupling between flow injection analysis (FIA), better repeatability, high sample throughput, and easy regeneration of solid phase [14]. Basically, a solid sorbent can be used as chelate-forming functional group or as support, and the properties of both components determine the features and the application of the respective material. Hence, the choice of the effective sorbent regarding selectivity and efficiency of the preconcentration in the analytical method is made by taking into account the nature of functional group and the physicochemical properties of the sorbent, such as mechanical and chemical stability, surface area, porous volumes, and kinetic characteristics [15]. The literature presents a variety of sorbents, and the most prominent among them are C_{18} silica, activated carbon, and polymeric resins [16]. Nevertheless, most of them have shown some drawbacks, for which the limited breakthrough volumes and the narrow pH stability range for C_{18} silica and poor selectivity of some polymeric resins can be pointed out [16]. Therefore, currently, there is interest in the development of new solid–phase extractors with high sorption capacity, selectivity, and PF. There are numerous attempts in this direction, and recently, nanomaterials have shown advantages over traditional sorbents, mainly owing to high chemical activity once their surface atoms are unsaturated and, therefore, can bind strongly with other atoms; large surface area; and high chemical stability. As examples, nano-particles of titanium dioxide and alumina have been used as solid sorbents for the enrichment of metal ions following their determination by FAAS and inductively coupled plasma optical emission spectrophotometry (ICP-OES), respectively [17, 18]. Nanomaterials based on only saturated carbon atoms also known as carbon nanotubes, in both single-walled carbon nanotubes and multiwalled carbon nanotubes (MWCNTs), are becoming a potential sorbent for metal ions due to their unique electronic, mechanical, and chemical properties [19]. Literature survey showed that the MWCNTs were first used as sorbent in off-line preconcentration mode for cadmium, manganese, and nickel metals using ICP-OES as quantification technique [20]. Recently, we published two works that were based on sorbent flow preconcentration using MWCNT for cadmium and lead, using, respectively, thermospray flame furnace atomic absorption spectrometry and FAAS for metals determination [21, 22]. These works have emphasized the powerful performance of MWCNT when associated to an element-selective technique. However, there is no investigation with regard to selective performance of MWCNT for metal ions using flow injection preconcentration coupled to UV–vis molecular spectrophotometry, a technique naturally less selective and cheaper than atomic absorption spectrometry. Thus, the aim of this work was to develop a reliable flow injection preconcentration system coupled to UV–vis molecular spectrophotometry for copper determination using MWCNT as sorbent. Copper determination was accomplished after preconcentration followed by elution and complexation with diethyldithiocarbamate (DDTC), a complexant more broadly applicable to heavy metals [23]. The feasibility of the proposed method was assessed by its application for copper determination in water samples (mineral water, tap water, river water, and synthetic seawater) and reference material. Overall, chemometric tools based on factorial design and response surface methodology (RSM) have been used for the methodology optimization.

2. EXPERIMENTAL

2.1. Apparatus

A Femto UV–vis spectrophotometer (São Paulo, Brazil) model 482 equipped with flow cell of 1 cm optical length was used for FIA measurements. Data acquisition was carried out from an interfaced (Advantech) PCL 711S mode, and a computational program was developed in an EXCEL® spreadsheet, using macros of Visual Basic®. A Shimadzu UV–vis molecular absorption spectrophotometer (Tokyo, Japan) model AA-6800 with glass cell of 1 cm optical length was used for spectrum scan in the wavelength range from 400 to 700 nm. An Ismatec Model IPC peristaltic pump (Ismatec IPC-08, Glattzbrugg, Switzerland) furnished with Tygon® tubes was used to propel all sample and reagent solutions. The preconcentration/elution steps were selected by using a home-made injector commutator made of Teflon® (polytetrafluoroethylene). Adjustment of sample pH was done by using a Handylab 1 Schott pHmeter (Stafford, UK). A microwave oven (Milestone, Sorisole, Italy) was used for decomposition of reference material (beech leaves). The morphological characteristic of carbon nanotubes was evaluated using a JEOL JMT-300 scanning electron microscope (Tokyo, Japan) with an electron acceleration voltage of 20 kV. The samples were previously coated with a thin Au/Pd layer in a Bal-Tec MED 020 equipment.

2.2. Reagents and Solutions

The solutions were prepared with analytical grade chemical reagents, as well as with water obtained from a Milli-Q purification system (Millipore, Bedford, Massachusetts, USA). In order to prevent metal contamination from laboratory glassware, it was kept overnight in a 10% (v/v) HNO_3 solution.

Copper standard solutions were prepared daily by appropriate dilution of $1000\,mg\ L^{-1}$ copper solution (Biotec, Darmstadt, Germany).

DDTC solution [0.50% (w/v)] was prepared by solving $1.00\,g$ of sodium DDTC in $25\,mL$ of ethanol (Cetus, São Paulo, Brazil) and $175\,mL$ of hot water and further stored in a freezer until use.

CTAB solution [6.0% (w/v)] was prepared by solving $6.00\,g$ of cetyltrimethylammonium bromide in $50\,mL$ of ethanol (Cetus, São Paulo, Brazil) and $150\,mL$ of water.

Acetate buffer at $2.0\,mol\ L^{-1}$ solution without further purification was prepared by dilution of $57.36\,mL$ of acetic acid (Vetec, Rio de Janeiro, Brazil) in volumetric flask of $500\,mL$ with water. The pH adjustment was carried out with sodium hydroxide at $1.0\,mol\ L^{-1}$ (Vetec, Rio de Janeiro, Brazil).

HNO_3 solution at $1.0\,mol\ L^{-1}$ concentration used as eluent was prepared from concentrated nitric acid (Merck, Darmstadt, Germany).

MWCNTs were supplied by CNT Co., Ltd. (Yeonsu-Gu, Incheon, Korea) with >93% purity, diameter between 10 and $40\,nm$ and length of 5–$20\,\mu m$. Prior to use, MWCNTs were submitted to acid treatment to create carbonyl and carboxyl groups onto MWCNT surface which are potentially involved on metal sorption. For relevant details, the reader is referred to the article by Barbosa et al. [22] and Tarley et al. [21]. The structural changes in the MWCNT morphology induced during acid treatment can be seen in Fig. 9.1. The MWCNT diameter was open-ended, thus favoring a better mass transport of analyte toward MWCNT surface. After this treatment, $30\,mg$ of MWCNT was packed into the minicolumn ($6.0 \times 1.0\,cm$ i.d.) made of polyvinyl chlorine containing glass wool at both ends to prevent losses of sorbent during preconcentration/elution steps.

2.3. Sample Preparation

The methodology developed was applied for copper determination in mineral water, tap water, river water, and synthetic seawater. Mineral water samples, acquired from local supermarkets, were spiked with a known amount of copper followed by pH adjusting with buffer solution. Tap water sample was collected from Unifal-MG campus and analyzed immediately after pH adjustment. River water samples were collected in polypropylene flasks from a lake near Alfenas city and immediately acidified with HNO_3 solution until a pH sample at 2.0 was attained. Afterwards, the samples were filtered through $0.45\,\mu m$ cellulose acetate membranes under vacuum, and the pH was adjusted with buffer solution. Synthetic seawater, which was prepared according to literature [22], also was analyzed without filtering procedure.

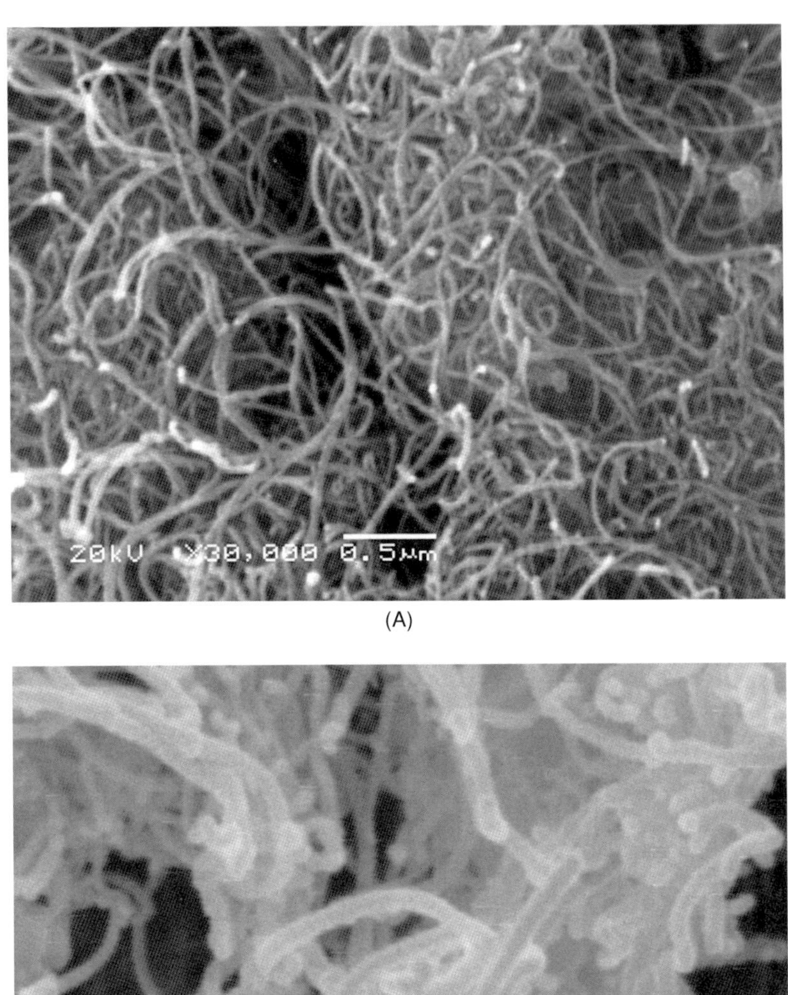

(A)

(B)

Figure 9.1   Scanning electron micrographs of raw MWCNT (amplified 30,000 times) (A) and MWCNT oxidized with concentrated HNO_3 (amplified 50,000 times) (B).

Beyond these samples, the feasibility of the methodology also was evaluated by analysis of reference material (beech leaves CRM No 100). Approximately 300 mg of the sample was placed into Teflon® flasks followed by addition of 10 mL concentrated HNO_3 and 4 mL 30% (v/v) H_2O_2. The heating program of the microwave oven was accomplished using two steps. A potency of 400 W during 5 min was used with further increase to 700 W, which was maintained during 5 min. After that, digested samples were heated on a hot plate to near dryness, then cooled at room temperature, and dissolved in buffer solution. Any contamination source was checked by using blank solutions.

2.4. Sorbent Flow Preconcentration System

The diagram of the flow preconcentration system is displayed in Fig. 9.2. At the preconcentration step position (Fig. 9.2A), the sample buffered at pH 5.2 was percolated through a minicolumn of MWCNT at 5.1 mL min^{-1} flow rate during 2 min and 47 s. After this stage, by switching the central part of the injector commutator (Fig. 9.2B), a stream of 1.0 mol L^{-1} HNO_3

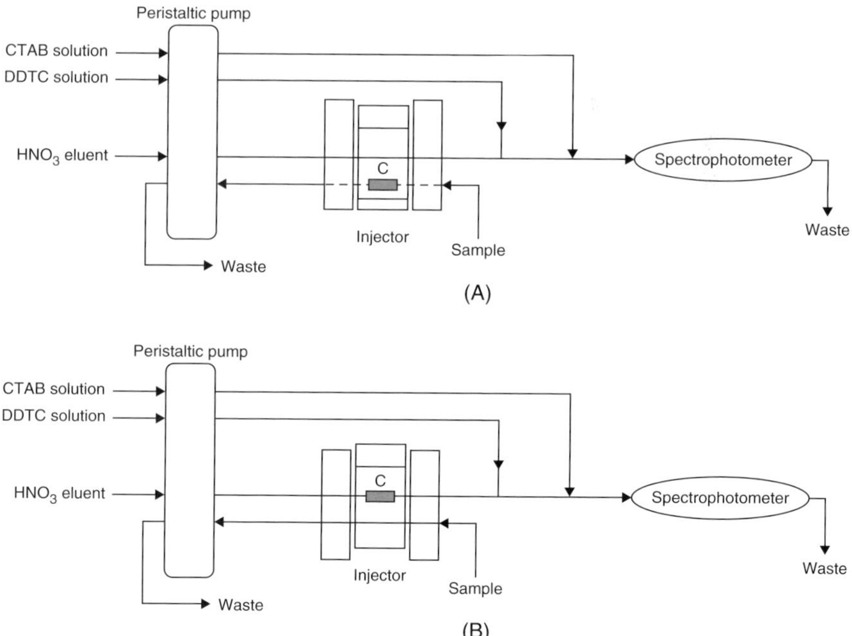

Figure 9.2 Schematic diagram of the sorbent flow preconcentration system of copper ions onto MWCNT with spectrophotometry determination: (A) preconcentration stage and (B) elution stage. C, minicolumn packed with MWCNT. For more details, see text.

displaces the copper ions at $1.0\,mL\ min^{-1}$ flow rate. Afterwards, copper ions eluated react with 0.15% (w/v) DDTC solution at $3.0\,mL\ min^{-1}$ flow rate forming colloidal slurry that is further solubilized by 1.0% (w/v) CTAB surfactant at $0.6\,mL\ min^{-1}$ flow rate. The $Cu(DDTC)_2$ complex is driven to the spectrophotometer where the absorbance measurements are continuously made at 452 nm. All absorbance signals were taken as peak height.

2.5. Optimization Procedure

The optimization procedure was performed using 2^{6-2} fractional factorial design followed by Doehlert design for three factors [24], to attain the best sensitivity and to reduce the number of essays. All essays in duplicate were carried out by percolating 10 mL of $254\,\mu g\ L^{-1}$ copper solution through minicolumn of MWCNT. The total of six factors was investigated, and their respective levels are shown in Table 9.1. Doehlert design was then used to establish the final optimization of those significant factors, using the RSM. It is important to point out that other factors that exert paramount importance in a sorbent flow preconcentration system, such mass of sorbent, type, and flow rate of the eluent, were fixed according to previous study [21]. Thus, 30 mg of MWCNT and HNO_3 $1.0\,mol\ L^{-1}$ as eluent at $1.0\,mL\ min^{-1}$ flow rate were adopted in this study. In addition, the complexant flow rate and buffer concentration were also fixed at $3\,mL\ min^{-1}$ and $0.1\,mol\ L^{-1}$. The analyses of experimental essays were processed using the STATISTICA software package (StatSoft, Tulsa, USA).

Table 9.1   Factors and experimental domain used in the 2^{6-2} fractional factorial design for copper preconcentration onto MWCNT

Factors	Low level (−)	High level (+)
pH	3.8	5.5
Surfactant flow rate $(mL\ min^{-1})$	1.0	3.0
Sampling flow rate $(mL\ min^{-1})$	5.1	7.5
Complexant concentration (% w/v)	0.05	0.1
Surfactant coil	Without coil	With coil $(250\,\mu L)$
Surfactant concentration (% w/v)	1.0	2.5

3. RESULTS AND DISCUSSION

3.1. Absorption Spectra

The DDTC complexes have been studied thoroughly, and detailed information on their behavior is readily accessible in the literature, and one of them is that DDTC form chelates with metals even at high acidities [23]. This is a favorable aspect once copper elution from MWCNT is commonly carried out with strong mineral acids, mainly nitric acid solution. In order to confirm the best sample pH of the reaction between DDTC and copper ions and to achieve the maximum wavelength of absorption, a scan spectrum was carried out from 400 to 700 nm, against a reagent blank. The absorption spectra were taken using 50 mg L^{-1} copper solution previously buffered at different pH range 2.5–3.5–5.5–7.0 and 9.0 containing CTAB at 1.0% (w/v) concentration. The absorbance values of $Cu(DDTC)_2$ complexes at different pH values did not show significant differences, and a maximum absorbance was attained at 452 nm.

3.2. Optimization Procedure Based on Fractional Factorial and Doehlert Designs

The screening of factors involved in the sorbent flow preconcentration system was made from a 2^{6-2} fractional factorial design which enables investigation of the main estimated effect of each factor using only 16 essays (Table 9.2), against 64 essays in regard to a full factorial design. The fractional factorial design was built by using the two generator matrices $I = pH \cdot SFR \cdot C$ and $I = pH \cdot SFR \cdot SR \cdot SC$. Thus, the signals of surfactant coil (C) column were achieved by multiplying the signals of pH and surfactant flow rate (SFR), while the signals of surfactant concentration (SC) were obtained by multiplying the

Table 9.2 Experimental matrix of the 2^{6-2} fractional factorial design and responses (height peak) for copper preconcentration onto MWCNT

Runs	pH	SFR	SR	CC	C	SC	Height peak
1	−	−	−	−	+	−	0.4158/0.4226
2	+	−	−	−	−	+	0.58360.5487
3	−	+	−	−	−	+	0.32460.3167
4	+	+	−	−	+	−	0.32040.3073

Continued

Table 9.2   (*Continued*)

Runs	pH	SFR	SR	CC	C	SC	Height peak
5	−	−	+	−	−	+	0.3146/0.2964
6	+	−	+	−	+	−	0.3063/0.2871
7	−	+	+	−	+	−	0.2343/0.2498
8	+	+	+	−	−	+	0.3002/0.3050
9	−	−	−	+	−	−	0.5921/0.6303
10	+	−	−	+	+	+	0.5013/0.5281
11	−	+	−	+	+	+	0.3040/0.3129
12	+	+	−	+	−	−	0.3359/0.3302
13	−	−	+	+	+	+	0.2563/0.2660
14	+	−	+	+	−	−	0.4464/0.4548
15	−	+	+	+	−	−	0.2767/0.3079
16	+	+	+	+	+	+	0.3409/0.3365

Note: SFR = surfactant flow rate; SR = sampling flow rate; CC = complexant concentration; C = surfactant coil; SC = surfactant concentration.
Generator matrices were $I = pH \cdot SFR \cdot C$ and $I = pH \cdot SFR \cdot SR \cdot SC$.

signals of pH, SFR, and sampling flow rate (SR). The main estimated effects of each factor in the system studied were calculated taking into account the height peak as experimental response. Analysis of variance (ANOVA) was successfully used for establishing the significance of estimated effects, which was represented from Pareto Chart (Fig. 9.3). This chart leads to an interpretation with a 95% confidence interval of those significant factors, whereas the horizontal bars higher than vertical line demonstrate statistically significant factors. Analyzing the Pareto Chart, it can be observed that the majority of factors were statistically significant [SFR, SR, C, sample pH, and CC (complexant complexation)], except the SC. So, the low level [1.0% (w/v)] of this latter factor was adopted in this study. In addition, it can be seen from Pareto Chart that some interaction effects were also significant, but they were not considered once their effects present serious confounding pattern, which stem from random experimental errors taking in consideration that a 2^{6-2} fractional factorial design (1/4 of a full factorial design) was used.

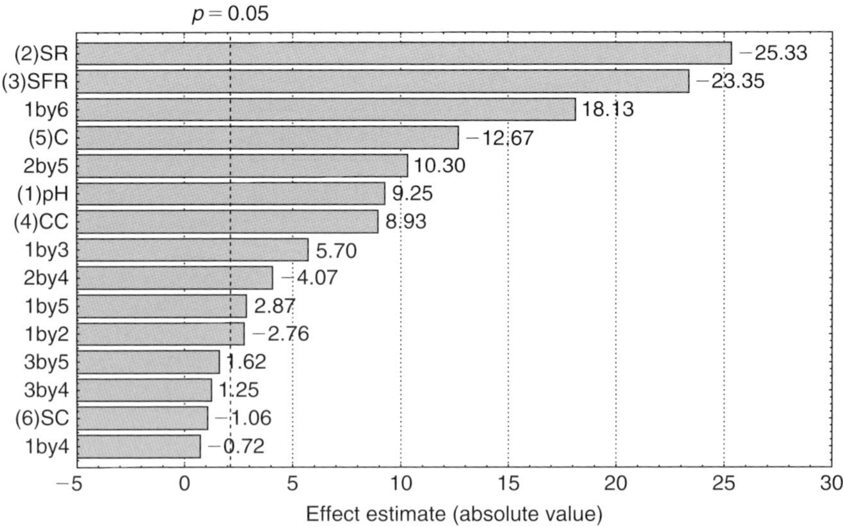

Figure 9.3 Pareto Chart obtained from 2^{6-2} fractional factorial design.

The most significant factor, SFR, which presented a negative effect (−25.33), indicates a decrease in the analytical response when working with high flow rate (3.0 mL min^{-1}). Bearing this in mind, we decided to continue the optimization using Doehlert design (Table 9.3) to obtain reagent consumption reduction and to attain better analytical signal. The SR also showed a negative effect (−23.35), thus revealing a slowly sorption process of copper ions onto MWCNT. Therefore, a better analytical response would be obtained at lower levels than 5.1 mL min^{-1}, but experiments in this direction were not tested to guarantee a reasonable sample throughput. So, SR at 5.1 mL min^{-1} was maintained for subsequent experiments. As regard, the influence of C, the negative effect (−12.67), reveals that the inclusion of coil (250 μL) in the flow system leads to dispersion of sample zone and, as a consequence, a decrease in the analytical response. Hence, the flow system was operated in the absence of the coil.

The influence of the sample pH on the analytical signal studied within the experimental domain from 3.8 to 5.5 shows, according to positive effect (9.25), that copper sorption onto MWCNT increases with increasing pH. CC presented a similar behavior to sample pH, whereas when working with its highest level (0.1% w/v), the formation of the metal chelates is favored. Under some conditions previously fixed, SC at 1.0% w/v, SR at 5.1 mL min^{-1}, and absence of coil in the flow system, Doehlert design for three factors was used as a great tool for the final simultaneous optimization

Table 9.3   Doehlert matrix for three factors used for final optimization

Runs	CC (% w/v)	pH	SFR (mL min⁻¹)	Height peak
1	0 (0.15)	0 (5.51)	0 (0.6)	0.7176
2	0 (0.15)	0 (5.51)	0 (0.6)	0.6863
3	0 (0.15)	0 (5.51)	0 (0.6)	0.7026
4	1 (0.25)	0 (5.51)	0 (0.6)	0.5416
5	0.5 (0.2)	0.866 (7.22)	0 (0.6)	0.4565
6	0.5 (0.2)	0.289 (6.08)	0.817 (0.7)	0.5805
7	−1 (0.05)	0 (5.51)	0 (0.6)	0.5863
8	−0.5 (0.1)	−0.866 (3.8)	0 (0.6)	0.5722
9	−0.5 (0.1)	−0.289 (4.94)	−0.817 (0.5)	0.5228
10	0.5 (0.2)	−0.866 (3.8)	0 (0.6)	0.5873
11	0.5 (0.2)	−0.289 (4.94)	−0.817 (0.5)	0.4944
12	−0.5 (0.1)	0.866 (7.22)	0 (0.6)	0.5139
13	0 (0.15)	0.577 (6.65)	−0.817 (0.5)	0.5323
14	−0.5 (0.1)	0.289 (6.08)	0.817 (0.7)	0.5738
15	0 (0.15)	−0.577 (4.37)	0.817 (0.7)	0.6218

Note: CC = complexant concentration; SFR = surfactant flow rate.
The values in brackets represent the real value of factor, while the first values represent the coded value of Doehlert design.

of CC, sample pH, and SFR. Their levels used in the experimental design (Table 9.3) were chosen in accordance with those results based on 2^{6-2} fractional factorial design. According to real data applied to Doehlert design, the following equation was achieved, which shows the relationship between the three factors and the analytical response:

$$\mathrm{Abs} = -5.21 + 3.29\mathrm{CC} - 13.65\mathrm{CC}^2 + 0.67\mathrm{pH} - 0.045\mathrm{pH}^2 + 12.61\mathrm{SFR}$$
$$- 9.37\mathrm{SFR}^2 - 0.21\mathrm{CC} \cdot \mathrm{pH} + 2.96\mathrm{CC} \cdot \mathrm{SFR} - 0.26\mathrm{pH} \cdot \mathrm{SFR}$$

From this equation, three response surfaces were obtained and are illustrated in Fig. 9.4. This quadratic model significance was checked by using

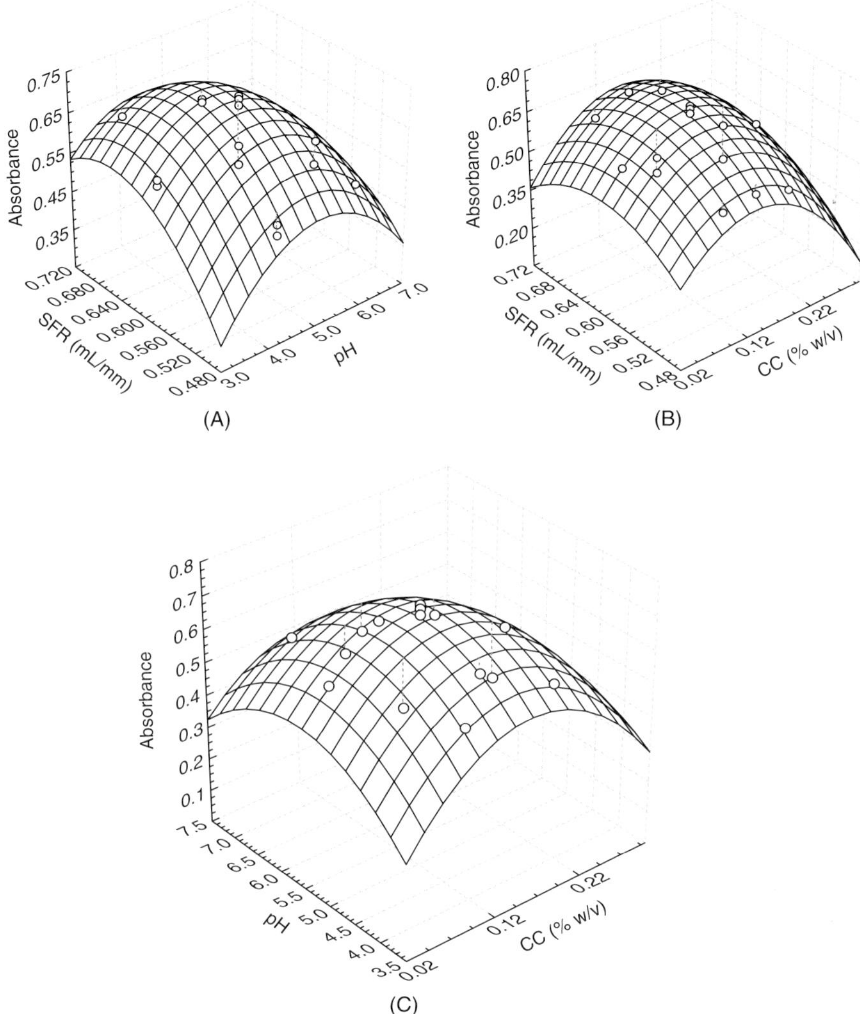

Figure 9.4 Surface responses obtained from Doehlert design for three factors.

ANOVA (Table 9.4). From MS_{lof}/MS_{pe} ratio (2.7381), in which the result is smaller than that critical $F_{3,2}$ tabled value of 19.16, it is possible to conclude that quadratic model does not present lack of fit, thus being a suitable model for adjusting with experimental results. In addition, this equation has a high determination coefficient (R^2) of 0.97745 revealing the residues of the experiment are very low. In order to check the presence of maximum points, i.e., whether in fact the critical point is the relative maximum,

Table 9.4   ANOVA from the Doehlert design shown in Table 9.3

Effect	Sum of squares (SS)	Degrees of freedom	Mean of squares	*F*-test	Probability level (*p*)
CC	0.0014711	1	0.0014711	8.3455	0.101847
CC^2	0.022375	1	0.022375	126.9652	0.007784
pH	0.007732	1	0.007732	43.8731	0.022042
pH^2	0.038242	1	0.038242	216.9996	0.004577
SFR	0.008558	1	0.008558	48.5611	0.019978
SFR^2	0.026352	1	0.026352	149.5294	0.006621
CC × pH	0.001314	1	0.001314	7.4565	0.112022
CC × SFR	0.000790	1	0.000790	4.4846	0.168389
pH × SFR	0.002394	1	0.002394	13.5853	0.066365
Lack of fit	0.001448	3	0.000483	2.7381	0.278821
Pure error	0.000352	2	0.000176		
Total SS	0.079808	14			

$R^2 = 0.97745$ and R^2 adjusted $= 0.93685$.

relative minimum, or saddle point, the Lagrange's criterion was used, which is based on calculation of Hessian determinant [4]. There is no information about the quadratic model when $\Delta_2 = 0$; there is a maximum point when $\Delta_1 < 0, \Delta_2 > 0$, and $\Delta_3 < 0$; there is a minimum point when $\Delta_1 > 0, \Delta_2 > 0$, and $\Delta_3 > 0$, and there is a saddle point when none of the above situations are observed. The Hessian determinants of a function $H(CC, pH, SFR)$ were calculated by using the following expressions:

$$\Delta_1 = \frac{\delta^2 \text{Abs}}{\delta CC^2}$$

$$\Delta_2 = \begin{vmatrix} \dfrac{\delta^2 \text{Abs}}{\delta CC^2} & \dfrac{\delta^2 \text{Abs}}{\delta pH \delta CC} \\ \dfrac{\delta^2 \text{Abs}}{\delta CC \delta pH} & \dfrac{\delta^2 \text{Abs}}{\delta pH^2} \end{vmatrix}$$

$$\Delta_3 = \begin{vmatrix} \dfrac{\delta^2 \text{Abs}}{\delta \text{CC}^2} & \dfrac{\delta^2 \text{Abs}}{\delta \text{pH} \delta \text{CC}} & \dfrac{\delta^2 \text{Abs}}{\delta \text{SFR} \delta \text{CC}} \\[2ex] \dfrac{\delta^2 \text{Abs}}{\delta \text{CC} \delta \text{pH}} & \dfrac{\delta^2 \text{Abs}}{\delta \text{pH}^2} & \dfrac{\delta^2 \text{Abs}}{\delta \text{SFR} \delta \text{pH}} \\[2ex] \dfrac{\delta^2 \text{Abs}}{\delta \text{CC} \delta \text{SFR}} & \dfrac{\delta^2 \text{Abs}}{\delta \text{pH} \delta \text{SFR}} & \dfrac{\delta^2 \text{Abs}}{\delta^2 \text{SFR}} \end{vmatrix}$$

The values found for $\Delta_1, \Delta_2,$ and Δ_3 were $-27.3, 2.4,$ and -42.2, respectively, confirming the presence of maximum points. These maximum points were obtained by solving the quadratic model as follows:

$$\frac{\partial \text{Abs}}{\partial \text{CC}} = 3.29 - 27.3\text{CC} - 0.21\text{pH} + 2.96\text{SFR} = 0$$

$$\frac{\partial \text{Abs}}{\partial \text{pH}} = 0.67 - 0.09\text{pH} - 0.21\text{CC} - 0.26\text{SFR} = 0$$

$$\frac{\partial \text{Abs}}{\partial \text{SFR}} = 12.61 - 18.74\text{SFR} + 2.96\text{CC} - 0.26\text{pH} = 0$$

This way, the maximum values for CC, pH, and SFR were found to be 0.15% (w/v), 5.20, and 0.60 mL min^{-1}, respectively, being these values fixed for the method.

3.3. Effect of Foreign Ions as Potential Interferences

Although there are some studies demonstrating interference and studies concerning selectivity of preconcentration procedures based on copper sorption onto MWCNT [21, 22], they usually use atomic absorption spectrometric measurements an inherent selective technique. Hence, the interference may occur only by competition between foreign ions and analyte for the active sites of sorbent. In the proposed method, the presence of other metallic cations could cause loss or enhancement of analytical signal due to the competition for the active sites of sorbent as well as by reaction with the DDTC complexant. So, the selectivity of the method was assessed by copper preconcentration under optimized conditions in the presence of each metallic ion Mn(II), Zn(II), Fe(II), Co(II), Ni(II), Ba(II), Pb(II), Cd(II), Al(III), and Cr(III) in concentrations up to 10-fold higher than that of the analyte (100 µg L^{-1}), while Ca(II) and Mg(II) were investigated at a concentration up to 100-fold higher. The interference was here established as a change higher or lower than 10% in the recovery of copper analytical signal

Table 9.5   Effect of foreign ion on the recovery of copper analytical signal

Foreign ion	Concentration (μg L^{-1})	Recovery (%)
Zn	1000	129
Cu	100	167
Co	1000	437
Ni	1000	138
Pb	1000	141
Cd	1000	155

Copper concentration used was 100 μg L^{-1}.

with respect to the copper signal alone. As can be seen in Table 9.5, the interference was noted only for those metals that exhibit strong sorption on MWCNT [20–22] and that react likewise with DDTC [23]. Despite the interference, the ratio analyte/foreign ion tested was higher than that observed in real samples analyzed in this work. Therefore, it suggests that the proposed method can be successfully applied for copper determination in different kinds of samples (water samples and biological material) as will be demonstrated later.

3.4. Analytical Features of the Proposed Method

An external calibration curve was built ranging from 8.0 to 500.0 μg L^{-1}, as shown in Fig. 9.5 (inset figure), resulting in the following linear equation: $\text{Abs} = 0.0013 + 0.00217[\text{Cu(II)} \, (\mu\text{g L}^{-1})]$ ($r > 0.999$). The calibration curve built without using preconcentration procedure was $\text{Abs} = -0.00768 + 0.00005709[\text{Cu(II)} \, (\mu\text{g L}^{-1})]$ with linear correlation coefficient $r = 0.9999$. PF was calculated as described by literature [24], using the ratio of the slopes of the analytical curve derived by the proposed method with that one built without preconcentration procedure. The PF achieved was found to be 38.0 taking into account a preconcentration time of 2 min and 47 s. The detection and quantification limits, calculated as three times the standard deviation of the black according to IUPAC recommendation [25], were 1.56 and 5.19 μg L^{-1}, respectively. Precision values, in terms of repeatability expressed as relative standard deviation ($n = 6$), were 6.8 and 0.21% for copper solutions at 20 and 400 μg L^{-1} concentrations, respectively.

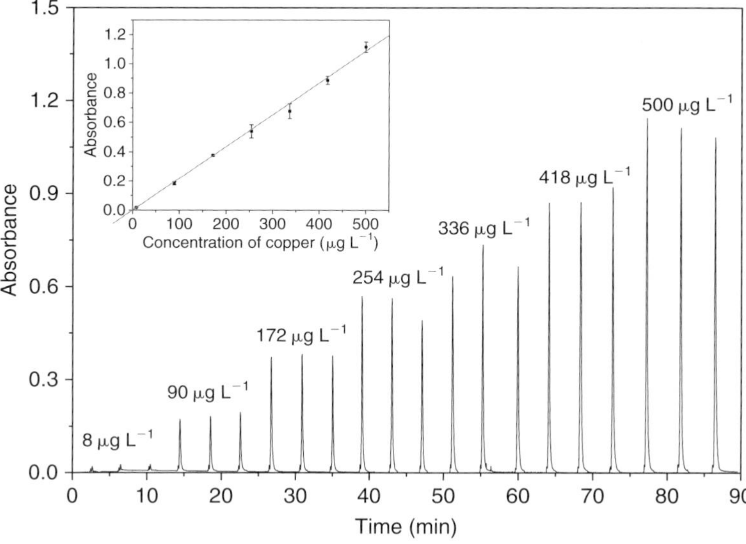

Figure 9.5 Recorder of analytical signal (fiagram) for building external calibration (inset figure).

The concentration efficiency [21], CI [22], and sample throughput were found to be $13.6\,\text{min}^{-1}$, $0.37\,\text{mL}$, and $18\,\text{h}^{-1}$, respectively. A brief comparison of the proposed preconcentration method with others previously published (Table 9.6) allowed us to develop an advantageous method, mainly owing to the lower limit of detection and low sample consumption, which leads to a high sample throughput.

3.5. Validation of the Proposed Method

The usefulness of the sorbent flow preconcentration system was evaluated by analyzing water samples and reference material. Essays based on addition and recovery tests of known copper concentration were carried out, and the results are shown in Table 9.7. Values ranging from 90.9 up to 102.3%, obtained from external calibration, prove that it is possible to achieve excellent accuracy even for those complex samples, such as synthetic seawater. Moreover, average amounts of copper found in beech leaves CRM No 100 ($12.4 \pm 1.4\,\mu\text{g g}^{-1}$) obtained by the current method agreed closely with those values derived from certified material ($12.0\,\mu\text{g g}^{-1}$) with 95% confidence level (test t-student), thereby also attesting its accuracy. Also, it is important to comment that the methodology was successfully applied for copper monitoring in river water without interference. According

Table 9.6 Comparison of different sorbent flow preconcentration systems for copper determination using UV–vis molecular spectrophotometry with the proposed method

Sorbent	Complexant agent	Preconcentration modality	Eluent	PV (mL)	LD (μg L^{-1})	References
Octadecylsilica	Neocuproine	Off-line	Isopentyl alcohol	500	0.12	[26]
DAPPS	DDTC	Online	HNO_3	13	8.4	[27]
Lewatit TP807'84 resin	Sulfarsazene	Online	HCl	100	2.0	[28]
Microcrystalline benzophenone	Cupferron	Off-line	Ethanol	100	5.5	[29]
MWCNTs	DDTC	Online	HNO_3	14.2	1.59	This work

Note: PV = preconcentration volume; LD = limit of detection; MCCAPT = 9-methyl-3-carbazolecarboxaldehyde-phenyl thiosemicarbazone; DAPPS = 1, 3-diaminepropane-3-propyl-anchored silica gel; Lewatit TP807'84 resin = resin that contain di(2,4,4-trimethylpentyl)phosphinic acid.

Table 9.7 Validation of the method for copper determination in spiked water samples

Samples	Concentration of copper added (μg L^{-1})	Concentration of copper found[a] (μg L^{-1})	Recovery (%)
Mineral water (brand 1)	0	<LD	−
	20.0	20.4 ± 1.5	102.0
	100.0	100.7 ± 1.7	100.7
Mineral water (brand 2)	0	<LD	−
	20.0	20.4 ± 1.6	102.3
	100.0	100.2 ± 1.9	100.2
Tap water	0	<LD	−
	20.0	18.5 ± 0.3	92.5
Synthetic seawater	0	<LD	−
	100.0	90.9 ± 4.9	90.9
River water	0	65.9 ± 1.4	−
	65.0	131.5 ± 1.5	100.2

Note: LD = limit of detection.
[a] The results are expressed as mean values ± SD based on three replicates ($n = 3$) determinations. Confidence interval, 95%.

to National Council of Environment (CONAMA) [31] regularized by Brazilian legislation, the allowed maximum level of copper in sweet water is 13.0 μg L^{-1}, a lower concentration than that determined. In addition, copper ions also can be determined in saline water without interference once the allowed maximum level in accordance with CONAMA is 7.8 μg L^{-1} for this kind of sample.

4. CONCLUSIONS

An attempt to coupling between flow preconcentration system and solid-phase extraction using MWCNT with spectrophotometric determination allowed a reliable and useful method for copper determination in water and biological samples. This system is the first one that demonstrates the powerful application of MWCNT in field of solid-phase preconcentration using spectrophotometric determination, a technique naturally less selective than atomic absorption spectrometry. The DDTC reagent, a nonselective complexant for copper determination, was used purposely so that the selectivity was attributed only in the solid-phase extraction, i.e., mainly owing to the high affinity of copper ions for the active sites of MWCNT. Besides the good analytical performance of such flow system including high PF, simplicity, low detection limit, and low cost of implementation, the MWCNT minicolumn has also shown excellent stability once the same packed minicolumn has been used in the development of two more methodologies [21, 22] without losses of sorption capacity. Overall, the results achieved in this work, based on accuracy and precision, also clearly showed the wide application of carbon nanotubes in the field of solid-phase preconcentration procedures for monitoring of copper ions in water samples.

ACKNOWLEDGMENTS

The authors would like to thank the Conselho Nacional de Desenvolvimento Científico e Tecnológico (CNPq), Fundação de Amparo à Pesquisa do Estado de São Paulo (FAPESP), Fundação de Amparo à Pesquisa do Estado de Minas Gerais (FAPEMIG), and Coordenação de Aperfeiçoamento de Pessoal de Nível Superior (CAPES) for the financial support and fellowships. They also would like to thank Prof. Maria Antonieta Avarenga for language assistance.

References

[1] Agency for Toxic Substances and Disease Registry. Toxicological profile for copper draft for public comment. Atlanta (GA): U.S. Department of Health and Human Services, Public Health Service; 2002.

[2] Merian E, Anke M, Ihnat M, Stoeppler M. Elements and their compounds in the environment. 2nd ed. Weinheim: John Wiley; 2004.

[3] Fleming CA, Trevors JT. Copper toxicity and chemistry in the environment: a review. Water Air Soil Pollut 1989;44:143–58.

[4] Tarley CRT, Dos Santos WNL, Dos Santos CM, Ferreira SLC, Arruda MAZ. Factorial design and Doehlert matrix in optimization of flow system for preconcentration of copper on polyurethane foam loaded with 4-(2-pyridylazo)-resorcinol. Anal. Letters 2004;37:1437–55.

[5] de Peña YP, López W, Burguera JL, Burguera M, Gallignani M, Brunetto R, et al. Synthetic zeolites as sorbent material for on-line preconcentration of copper traces and its determination using flame atomic absorption spectrometry. Anal Chim Acta 2000;403:249–58.

[6] Sweileh JA. Sorption of trace metals on human hair and application for cadmium and lead pre-concentration with flame atomic absorption determination. Anal Bioanal Chem 2003;375:450–5.

[7] Anthemidis AN, Zachariadis GA, Farastelis CG, Stratis JA. On-line liquid-liquid extraction system using a new phase separator for flame atomic absorption spectrometric determination of ultra-trace cadmium in natural waters. Talanta 2004;62:437–43.

[8] Lemos VA, Santos ES, Santos MS, Yamaki RT. Thiazolylazo dyes and their application in analytical methods. Microchim Acta 2007;158:189–204.

[9] Akl MA. An improved colorimetric determination of lead(II) in the presence of non-ionic surfactant. Anal Sci 2006;22:1227–31.

[10] Bonifácio VG, Figueiredo-Filho LC, Marcolino Jr LH, Fatibello-Filho O. An improved flow system for chloride determination in natural waters exploiting solid-phase reactor and long pathlength spectrophotometry. Talanta 2007;72:663–7.

[11] Bashammakh AS, Bahaffi SO, Al-Sibaai AA, Al-Wael HO, El-Shahawi MS. Extractive liquid-liquid spectrophotometric procedure for the determination of thiocyanate ions employing the ion pair reagent amiloride monohydrochloride. Anal Chim Acta 2007;592:16–23.

[12] Figueiredo EC, Tarley CRT, Kubota LT, Rath S, Arruda MAZ. On-line molecularly imprinted solid phase extraction for the selective spectrophotometric determination of catechol. Microchem J 2007;85:290–6.

[13] Filik H, Tayman A. A new cloud-point preconcentration approach for the spectrophotometric determination of p-aminophenol in the presence of paracetamol with 2-(2-hydroxyphenyl)-1H-benzimidazole as a coupling reagent. J Anal Chem 2007;62:530–5.

[14] Gama EM, da Silva Lima A, Lemos VA. Preconcentration system for cadmium and lead determination in environmental samples using polyurethane foam/Me-BTANC. J Hazard Mater 2006;136:757–62.

[15] Bilba D, Bejan D, Tofan L. Chelating sorbents in inorganic chemical analysis. Croat Chem Acta 1998;71:155–78.

[16] Liu G, Wang J, Zhu Y, Zhang X. Application of multiwalled carbon nanotubes as a solid-phase extraction sorbent for chlorobenzenes. Anal Lett 2004;14:3085–104.

[17] Liang P, Shi T, Jing LI. Nanometer-size titanium dioxide separation/preconcentration and FAAS determination of trace Zn and Cd in water sample. Int J Environ Anal Chem 2004;84:315–21.

[18] Hang C, Hu B, Jiang Z, Zhang N. Simultaneous on-line preconcentration and determination of trace metals in environmental samples using a modified nanometer-sized alumina packed micro-column by flow injection combined with ICP-OES. Talanta 2007;71:1239–45.

[19] Chiang IW, Brinson BE, Smalley RE, Margrave JL, Hauge RH. Purification and characterization of single-wall carbon nanotubes. J Phys Chem B 2001;105:1157–61.

[20] Liang P, Liu Y, Guo L, Zeng J, Lu H. Multiwalled carbon nanotubes as solid-phase extraction adsorbent for the preconcentration of trace metal ions and their determination by inductively coupled plasma atomic emission spectrometry. J Anal At Spectrom 2004;19:1489–92.

[21] Tarley CRT, Barbosa AF, Segatelli MG, Figueiredo EC, Luccas PO. Highly improved sensitivity of TS-FF-AAS for Cd(II) determination at ng L^{-1} levels using a simple flow injection minicolumn preconcentration system with multiwall carbon nanotubes. J Anal At Spectrom 2006;11:1305–13.

[22] Barbosa AF, Segatelli MG, Pereira AC, Santos AS, Kubota LT, Luccas PO, et al. Solid-phase extraction system of Pb (II) ions enrichment based on multiwall carbon nanotubes couple on-line to flame atomic absorption spectrometry. Talanta 2007;71:1512–9.

[23] Duran C, Gundogdu A, Bulut VN, Soylak M, Elci L, Sentürk HB, et al. Solid-phase extraction of Mn(II), Co(II), Ni(II), Cu(II), Cd(II) and Pb(II) ions from environmental samples by flame atomic absorption spectrometry (FAAS). J Hazard Mater 2007;146:347–55.

[24] dos Santos WNL, Santos CM, Ferreira SLC. Application of three-variables Doehlert matrix for optimisation of an on-line pre-concentration system for zinc determination in natural water samples by flame atomic absorption spectrometry. Microchem J 2003;75:211–21.

[25] Long GL, Winefordner JD. Limit of detection. Anal Chem 1983;55:712–5.

[26] Yamini Y, Tamaddon A. Solid-phase extraction and spectrophotometric determination of trace amounts of copper in water samples. Talanta 1999;49:119–24.

[27] de Moraes SVM, Brasil JL, Milcharek CD, Martins LC, Laranjo MT, Gallas MR, et al. Use of 1,3-diaminepropane-3-propyl grafted onto a silica gel as a sorbent for flow-injection spectrophotometric determination of copper (II) in digests of biological materials and natural waters. Spectro. Chim. Acta A 2005;62:398–406.

[28] Castilho E, Cortina JL, Beltrán J-L, Prat M-D, Granados M. Simultaneous determination of Cd(II), Cu(II) and Pb(II) in surface waters by solid phase extraction and flow injection analysis with spectrophotometric detection. Analyst 2001;126:1149–53.

[29] Lee T, Choi H. Determination and preconcentration of Cu(II) after adsorption of its cupferron onto benzophenone. Bull Korean Chem Soc 2002;23:861–4

[30] CONAMA. National Council of Environment. Resolution number 357, may 9, 2005.

CHAPTER 10

Application of Carbon Nanotubes as a Solid-Phase Extraction Material for Environmental Samples

Krystyna Pyrzynska

Department of Chemistry, Warsaw University, Warsaw, Poland

Contents

In spite of improvements in sensitivity due to modern analytical detection systems, conventional separation techniques are frequently used to overcome interferences from matrix elements and to improve detection limits through concentration of the analyte [1, 2]. Among different separation and enrichment techniques, batch and column techniques in which analytes are sorbed on different water–insoluble materials and eluted with acids or other reagents have been widely used [3]. For the materials to be beneficial in solid–phase extraction (SPE), given analytes should be easily, quantitatively, and reproducibly collected and eluted with minimum efforts exerted in experimental procedures. The SPE provides a number of important benefits in comparison with laborious classical liquid–liquid extraction, such as reduced solvent usage and exposure, low disposal costs, and short extraction times for sample preparation. Moreover, sorbent extraction can be used for selective retention of some particular chemical forms of a metal, thereby enabling speciation [4, 5].

Several materials have been proposed for SPE, such as C_{18}-bonded silica [6, 7], polymeric sorbents [8], carbon-based materials [9], zeolites [10], or polyurethane foam [11]. These have been successfully applied for various organic compounds and metal ions in different kinds of samples. The mechanism of sorption depends on the nature of the sorbent and may

DOI: 10.1016/B978-0-08-054820-3.00010-1

include simple adsorption, ion exchange, chelation, or ion-pair formation. Adsorption occurs through van der Waals forces or hydrophobic interactions, which occur when the solid sorbent is highly nonpolar. The most common sorbents of this type include octadecyl-bonded silica and styrene–divinylbenzene copolymers that provide additional π–π interactions, when π electrons are present in the analytes. The presence of appropriate functional groups in solid matrices is needed for chelating trace elements. Binding of metal ions to these groups is dependent on nature, charge, and size of the given metal ion, the nature of the donor atoms present in the group, the nature of the solid support, and the buffering conditions. The sorbent may be packaged into different formats: filled microcolumns, cartridges, syringe barrels, or discs. The choice of specific sorbent should be based on analyte, sample matrix, and technique for final detection, whereas higher enrichment factors can be obtained using adequate experimental conditions (time of loading sample, sorbent mass, etc.).

Among carbon-based sorbents, activated carbon was certainly one of the first materials applied in SPE. It has been widely used in water and wastewater treatment, primarily as an adsorbent for the removal of organic and inorganic contaminants [12]. The acceptance of separation or concentration media based on the new generations of carbon sorbents, such as graphitized carbon black (GCB) and porous graphitized carbon, has been growing during the last decade as they were shown to be appropriate for trapping polar analytes [13]. More recently, fullerenes and carbon nanotubes (CNTs) were discovered, and this opened a new perspective beyond that of carbon materials based on flat graphite-like hexagonal layers [14, 15]. CNTs possess many unique electronic, mechanical and chemical properties, and larger surface area and excellent strength. Along with the improvement of production of CNTs and characterization, progress is being made in their application. Over the past 20 years, more and more potential applications such as catalyst supports, electronic and optical devices, reinforced materials, hydrogen storage, field-emission materials, or biomedical use have been found. In many applications, including the analytical ones (such as gas sensors, electrochemical detectors, and biosensors with immobilized biomolecules), modification of their surface and interfacial engineering are essential [16, 17].

The purpose of this chapter is to summarize the analytical applications of CNTs for the enrichment and separation of organic compounds, as well as metallic species using SPE technique. The study of adsorption properties of CNTs is important in both fundamental and practical point of view.

1. SOLID-PHASE EXTRACTION OF ORGANIC COMPOUNDS

The characteristic structures and electronic properties of CNTs allow them to interact strongly with organic molecules via noncovalent forces, such as hydrogen bonding, $\pi-\pi$ stacking, electrostatic forces, van der Waals forces, and hydrophobic interactions. These interactions, as well as hollow and layered nanosized structures, make them a good candidate for use as adsorbents. The surface, made up of hexagonal arrays of carbon atoms in graphene sheets, interacts strongly with the benzene rings of aromatic compounds. In 2001, Long and Yang [18] observed that dioxins, which have two benzene rings, were strongly adsorbed on CNTs. The amounts of dioxin adsorbed were 10^4 and 10^{17} times greater than that on activated carbon and γ-Al_2O_3, respectively. Dioxins after enrichment step could be removed from the adsorbent by temperature-programmed desorption. In several other works, sorption on CNTs has been examined for different aromatic compounds, such as chlorobenzenes [19–21], three endocrine disruptors (bisphenol A, 4-*n*-nonylphenol, and 4-*tert*-octylphenol) [22], several phthalate esters [23], triazine compounds [24], sulfonylurea herbicides [25, 26], phenoxyalkanoic acids [27, 28], and organochlorine pesticides [29], as well as polychlorinated dibenzo-*p*-dioxins and related compounds [30]. These works demonstrate the application of CNTs used as the packing of SPE cartridge in trace analysis of environmental pollutants in natural water samples in combination with high-performance liquid chromatography. Fang et al. [31] applied CNTs as a sorbent for online SPE of 10 different kinds of sulfonamides before their simultaneous determination in eggs and pork.

The CNTs, C_{18}-bonded silica, and activated carbon were compared for their application for the enrichment of four chlorobenzenes under the same experimental conditions (columns packed with 15 mg of each sorbent materials, sample volume 100 mL, analytes at 5 ng/mL concentration level) [19]. The highest recoveries were found for CNTs (Fig. 10.1). Moreover, CNTs exhibit the highest adsorption ability toward these analytes, which was concluded from the obtained values of the apparent constant K in Freundlich model.

SPE of dicamba (3,6-dichloro-2-methoxybenzoic acid) using different polymeric sorbents (Oasis, Strata-X, Lichrolut SAX) and C_{18} Bond Elut or phenyl-silica gave unsatisfactory results [32]. This herbicide due to large scale of application, persistence, polar nature, and water solubility is present in several environmental compartments. CNTs have much greater herbicide adsorption capability than that of GCB and C_{18} adsorbents as it was

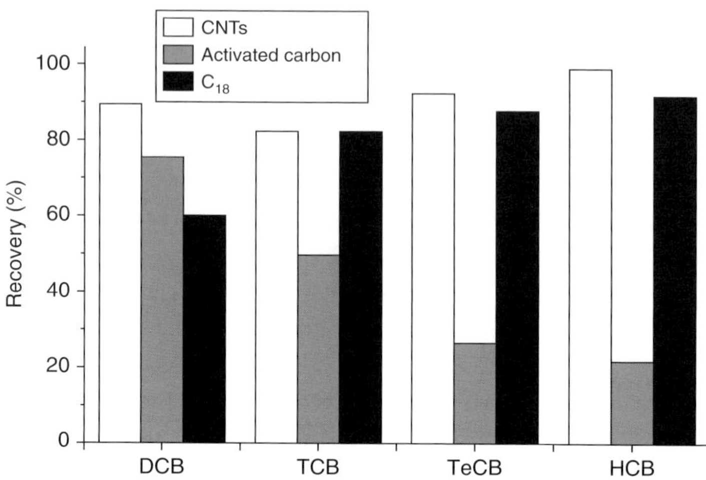

Figure 10.1   Recoveries of chlorobenzenes using CNTs, activated carbon, and C_{18} sorbents. Sample volume, 100 mL; sample concentration 5 ng/mL; eluent, 2 mL of methanol. DCB, dichlorobenzene; TCB, 1,2,4-trichlorobenzene; TeCB, 1,2,4,5-tetrachlorobenzene; HCB, hexachlorobenzene [19].

shown in Fig. 10.2A [28]. The absorption capacity increases remarkably at lower pH of sample solution as dicamba is highly an acidic compound ($pK_a = 1.9$). The decrease in pH leads to decrease in the ionization of the analyte and simultaneously to neutralization of the carbon surface charge. The HPLC chromatograms of river water samples spiked with dicamba and 2,4,5-T (2,4,5-trichlorophenoxyacetic acid) after preconcentration on C_{18} and CNT microcolumns are presented in Fig. 10.2B. The obtained results showed that CNTs were more effective than C_{18} (dicamba recovery was only 13%) for SPE of tested herbicides.

The thermodynamic characteristics of adsorption of several organic compounds on CNTs were determined by gas chromatography and compared with those for GBC and molecular crystals of fullerene C_{60} [33]. The comparison of adsorption data for compounds with different functional groups showed similarity of adsorption properties of CNTs and GCB, but there were significant differences in molecular interaction of adsorbed molecules with crystal C_{60}.

Adsorption of naphthalene, phenanthrene, and pyrene from the class of polycyclic aromatic hydrocarbons (PAHs) on fullerenes, single-walled and multiwalled CNTs was investigated to study the relationship between sorbate and sorbent properties [34]. For different PAHs, adsorption was

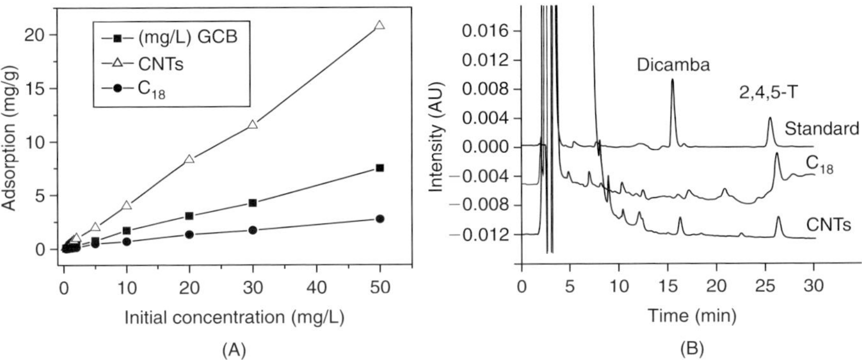

Figure 10.2 Adsorption isotherms of dicamba and 2,4,5-T herbicides onto GCB, CNTs, and C_{18} adsorbents at pH 3, and SPE/HPLC chromatograms of river water samples (250 mL) spiked with 0.04 mg/mL of dicamba and 2,4,5-T and enriched on C_{18} and CNT microcolumns [28].

in correlation with their molecular size, while for different carbon nano-materials, adsorption was in correlation with their surface area, micropore volume, and the volume ratios of mesopore to micropore.

A novel, CNT-supported micro–SPE (μ–SPE) procedure has been developed [35]. A 6 mg sample of CNTs was packed inside a sheet (2 cm $\times$ 1.5 cm) of porous polypropylene membrane whose edges were heat-sealed to secure the contents. The μ–SPE, which was wetted with dichloromethane, was then placed in a stirred sewage sludge sample solution to extract organo-phosphorus pesticides, used as model compounds. After extraction, analytes were desorbed in hexane and analyzed using gas chromatography with mass spectrometry detection. Each μ–SPE device could be used for up to 30 extractions. The comparison of CNT-supported μ–SPE) procedure with hollow fiber membrane solid–phase microextraction (HFM–SPME) and headspace solid–phase microextraction (HS–SPME) showed that this procedure is accurate, fast, and could be used as cost-effective pretreatment method for sewage sludge samples.

CNTs can also remove and preconcentrate volatile organic compounds. Saridara et al. [36] described a microtrap operating as a nanoconcentrator and injector for gas chromatography. A thin layer of CNTs was deposited by catalytic chemical vapor deposition on the inside wall of a steel capillary to fabricate the microtrap. The obtained film provides an active surface for fast adsorption and desorption of small organic molecules such as hexane and toluene. Stronger sorption of toluene in comparison with hexane could be attributed to the $\pi-\pi$ interaction between the CNTs' side-wall and the aromatic ring.

The purge-and-trap system was used for the evaluation of CNTs as an adsorbent for trapping 16 volatile organic compounds from gaseous mixtures and indirectly from water samples [37]. Due to their porous structure, CNTs were found to have much higher breakthrough volumes than that of GCB (Carbopack B) with the same surface area.

2. ENRICHMENT OF METALLIC SPECIES

CNTs have also great analytical potential as an SPE adsorbent for enrichment and recovery of metal ions [38–50]. The chemical and thermal treatment processing could have great impact on the adsorption capability of CNTs for the removal of metal ions, because the performance of carbon materials is mainly determined by the nature and concentration of the surface functional groups. Gas-phase oxidation of activated carbon increases mainly the concentration of hydroxyl and carbonyl surface groups, while the liquid-phase oxidation increases particularly the content of carboxylic acids [51]. Oxidation of CNTs with nitric acid is an effective method to remove the amorphous carbon, carbon black, and carbon particles introduced during their preparation process [38]. It is known that the oxidation of carbon surface can offer not only a more hydrophilic surface structure but also a larger number of oxygen-containing functional groups, which increase the ion-exchange capability of carbon materials. The amount of carboxyl and lactone groups on the CNTs treated with nitric acid was higher than the amount on CNTs oxidized with H_2O_2 and $KMnO_4$ [38]. At the same pH value, the zeta potential for these oxidized CNTs increased in the following order: H_2O_2 < $KMnO_4$ < HNO_3. This order indirectly suggests that the amount of acid functional groups increases in the same way. However, Cd(II) adsorption capacity for CNTs treated with $KMnO_4$ was greater (11.0 mg/g) than that for treated with H_2O_2 (2.6 mg/g) and HNO_3 (5.1 mg/g). This may be due to the adsorption role of the residual MnO_2 particles. However, the process is still unclear and needs more explanations.

Figure 10.3 shows the effect of pH on the adsorption of Co(II) on untreated CNTs and CNTs treated with HNO_3 [44]. The functional groups introduced by oxidation improve the ion-exchange capabilities of CNTs and make cobalt(II) adsorption capacity increase correspondingly. The adsorption of cobalt species increased with the increase of pH from

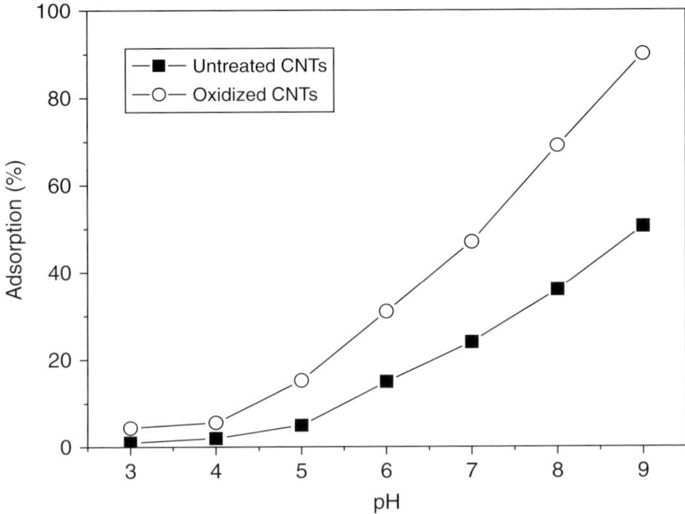

Figure 10.3   Effect of pH on the adsorption of Co(II) with untreated CNTs and CNTs oxidized with HNO$_3$. Initial metal concentration is 10 mg/L Adapted from Li YH [43].

3 to 9, but more sharp increase was observed for oxidized CNTs. The low adsorption that took place in acidic region can be attributed, in part, to competition between H$^+$ and Co^{2+} ions on the same sites. Furthermore, the charge of CNT surface becomes more negative with the increase of pH, which causes electrostatic interactions and, thus, results in higher adsorption of metal species.

The pH of the solution influences the surface charge of CNTs. When the pH is higher than the isoelectric point of the oxidized CNTs, the negative charge on the surface provides electrostatic attractions that are favorable for adsorbing cations. The affinity order of the metal ions toward CNTs at pH $\geq$ 7 is Cu(II) > Pb(II) > Zn(II) > Co(II) > Ni(II) > Cd(II) > Mn(II) (Fig. 10.4) [44]. This order agrees very well with the electronegativity of metal ion and the first stability constant of the associated metal hydroxide [51]. The general trend is that higher electronegativity and higher first constant correspond to a higher sorption level of a given metal ion.

Quantitative sorption of metal ions was usually obtained at pH $\geq$ 6 [41–45] although quantitative adsorption (>95%) of rare earth elements was obtained when the pH exceeds 3.0 [39]. As metal ions are poorly adsorbed at low pH, acid solutions such as HNO$_3$ or HCl have been used for their desorption and recovery.

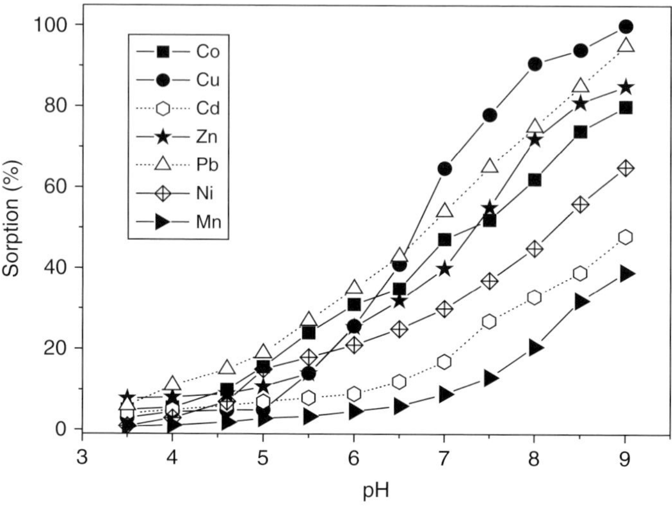

Figure 10.4   Sorption of metal ions on CNTs in the function of pH [44].

Single-metal-ion equilibrium studies usually have been conducted to investigate the maximum metal adsorption capacities of CNTs. This factor determines how much sorbent is required to quantitatively concentrate the analytes from a given solution. CNTs exhibit high affinity toward Zn(II) and Pb(II) as the adsorption capacities for these metal ions are equal to 34.4 mg/g [41] and 31.8 mg/g [49], respectively. For other divalent metals, much lower values have been reported: Cu(II), 7.5 mg/g [45]; Ni(II), 6.9 mg/g [38]; Cd(II), 5.1 mg/g [38]; Mn(II), 4.9 mg/g [42]. The adsorption capacity for each rare earth element was found to be in the range from 7.2 (for gadolinium) to 9.9 mg/g (for europium) [39]. However, the reported values depend also on the surface area of the used CNTs, as well as on experimental conditions. By applying the Langmuir equation to single-ion adsorption isotherms, Li et al. [43] calculated the maximum sorption capacities for lead, copper, and cadmium ions as 97.1, 24.5, and 10.9 mg/g, respectively. Similar experiments conducted for competitive adsorption (from the metal mixtures) showed the same affinity order of tested metal ions, e.g., Pb(II) > Cu(II) > Cd(II). However, the calculated maximum sorption capacity, corresponding to complete monolayer coverage, for Pb(II) decreased from 97.1 to 34.0 mg/g. The adsorption isotherms for single cases and for the mixture of these three ions onto CNTs are shown in Fig. 10.5. On the contrary, adsorption of Cu(II) in the presence of other divalent metal ions decreased only slightly, being the most pronounced

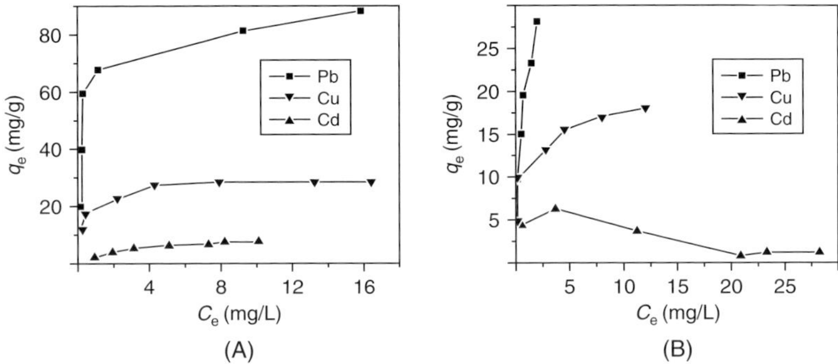

Figure 10.5 Adsorption isotherms of Pb(II), Cu(II), and Cd(II) on CNTs for single ions (A) and competitive adsorption (B); 100 mL of sample at pH 5.0 was shaken with 50 mg of CNTs for 4 h [42].

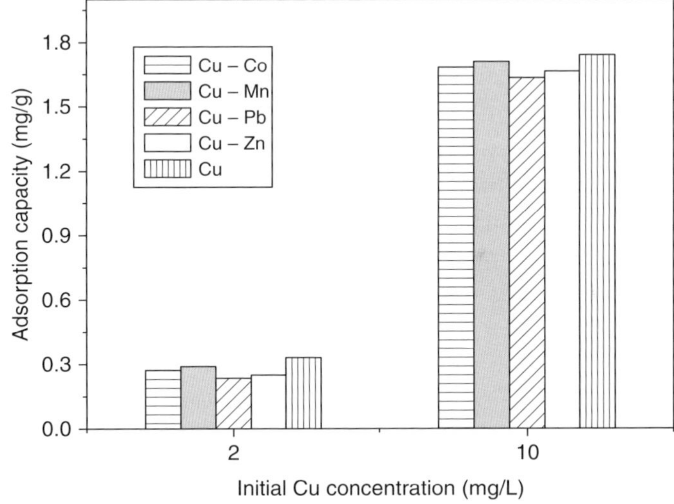

Figure 10.6 Adsorption capacity of CNTs for Cu(II) in the presence of other divalent metal ions [43].

by Pb(II) (Fig. 10.6). The decrease in absorption capacity for Cu(II) in the presence of Mn(II) was only approximately 12 and 2% at their initial concentration of 2 and 10 mg/L, respectively [44]. Thus, CNTs have good adsorption properties and high capacity for Cu(II) and can be applied for the preconcentration and purification of this element in aqueous solutions containing other metals.

For pure analytical application, CNTs packed into microcolumns have been applied for preconcentration of rare earth elements [39], Pb(II) [47],

Cd(II) [48, 50], Cu(II) [50], as well as Mn(II) and Ni(II) [52], prior to their determination by flame atomic absorption spectrometry or inductively coupled plasma atomic emission spectrometry in several environmental samples. The proposed methods were significant with respect to the good performance and preconcentration system including satisfactory preconcentration factor and enrichment efficiency, thus resulting in a low limit of detection. Moreover, addition of chelating agents and organic solvents for elution, which are most commonly used in SPE preconcentration systems with modified octadecyl silica (C_{18}) or polymeric sorbents, is not necessary.

The potential uses of fullerenes and CNTs as sorbents for preconcentration of lead, mercury, and tin organometallic compounds have been investigated [53]. These metals were complexed with diethyldithiocarbamate in a flow system and, after simultaneous enrichment on a microcolumn, were determined by gas chromatography. Comparative studies showed that CNTs, as well as C_{60} and C_{70} fullerenes, were superior to GCB and C_{18}-silica for the extraction of the 11 compounds that are studied. The sensitivity for the determination of these compounds (slope of calibration graph) was higher for carbon sorbents because they allow higher sample volume (25 mL) to be preconcentrated. The sensitivity achieved with C_{18}-silica was the lowest due to the smaller sample volume (15 mL) and lower efficiency of sorption (lower column capacity). The molecular surface area and volume for CNTs are larger than those of fullerenes, and thus, the highest sensitivities for the determination of organometallic compounds were observed for CNTs. The selectivity was lowest for GCB and C_{18}-silica; the metal ions posing the most severe interference were Cu(II) and Co(II) for all organotin species, as well as Fe(III) and Zn(II) for trimethyl lead. The proposed method was successfully applied for the determination of lead, mercury, and tin compounds in water and coastal sediment samples.

3. CONCLUSIONS

The need for efficient methods for sample concentration and cleanup in environmental application is constantly growing. In recent years, the SPE method has become the most often used technique for trace analysis of metals and organic environmental pollutants. SPE technique is fast, reproducible, cleaner extracts are obtained, and solvent consumption is reduced. Moreover, SPE can be easily incorporated into automated analytical procedures. Thus, efforts dedicated to exploring new effective adsorbents continue to grow.

CNTs have been proved to be effective adsorbents for the removal of organic compounds, particularly with those containing benzene rings. Their large sorption capacity is linked to well-developed internal pore structure and surface area. In recent years, the analytical potential of CNTs as the solid-phase adsorbents for metal ions and their chelates has been explored. The presence of the active sites on the surface, inner cavities, and inter-nanotube space contribute to the high metal removal capability of CNTs. Moreover, their strong sorption properties could be utilized in voltammetric stripping determination of transition metal cations with electrodes modified with CNTs [54]. CNTs express also high fluoride removal efficiency in a broad pH value range [55]. The comparison with other commercially available adsorbents suggests that CNTs are promising sorbents for environmental protection applications. The sorption properties of CNTs, including the possibility of their chemical functionalization [56], can be utilized in developing new microseparation methods and techniques.

References

[1] Brown RJC, Milton MJT. Analytical techniques for trace element analysis; an overview. Trends Analyt Chem 2005;24:266–74.

[2] Becker JS. Trace and ultratrace analysis in liquids by atomic spectrometry. Trends Analyt Chem 2005;24:243–54.

[3] Camel V. Solid phase extraction of elements. Spectrochim Acta Part B 2003;58:1177–233.

[4] Hirose K. Chemical speciation of trace metals in seawater: a review. Anal Sci 2006; 22:1055–63.

[5] Pyrzynska K, Trojanowicz M. Functionalized cellulose sorbents for preconcentration of trace metals in environmental analysis. Crit Rev Anal Chem 1999;29:313–21.

[6] Jak PK, Patel S, Mishra BK. Chemical modification of silica surface by immobilization of functional groups for extractive concentration of metal ions. Talanta 2004; 62:1005–28.

[7] Spivakov BY, Malofeeva GL, Petruskhin OM. Solid-phase extraction on alkyl-bonded silica gels in inorganic analysis. Anal Sci 2006;22:503–19.

[8] Rao TP, Praven RS, Daniel S. Sorption of divalent metal ions from aqueous solution by carbon nanotubes: A review. Crit Rev Anal Chem 2004;34:177–93.

[9] Pyrzynska K. Application of carbon sorbents for the concentration and separation of metal ions. Anal Sci 2007;23:631–7.

[10] Valdes MG, Perez-Cordoves AI, Diaz-Garcia ME. Zeolites and zeolite-based materials in analytical chemistry. Trends Analyt Chem 2006;25:24–30.

[11] Lemos VA, Santos MS, Santos ES, Santos MJS, dos Santos WNL, Souza AS, et al. Application of polyurethane foam as a sorbent for trace metal preconcentration – a review. Spectrochim Acta part B 2007;62:4–12.

[12] Strelko Jr V, Malik DJ, Streat M. Interpretation of transition metal sorption behaviour by oxidized active carbons and other adsorbents. Sep Sci Technol 2004;39(9):1885–905.

[13] Forgács E. Retention characteristics and practical applications of carbon sorbents. J Chromatogr A 2002;975:229–43.

[14] Trojanowicz M. Analytical applications of carbon nanotubes: a review. Trends Analyt Chem 2006;25:480–9.

[15] Valcárcel M, Simonet BM, Cárdenas S, Suárez B. Present and future applications of carbon nanotubes to analytical science. Anal Bioanal Chem 2005;382:1783–90.

[16] Lin T, Bajpai V, Ji T, Dai L. Chemistry of carbon nanotubes. Aust J Chem 2003;56: 635–51.

[17] Sun YP, Fu K, Huang W. Functionalized carbon nanotubes: properties and applications. Acc Chem Res 2002;35:1096–104.

[18] Long RQ, Yang RT. Carbon nanotubes a superior sorbent for dioxin removal. J Am Chem Soc 2001;123:2058–9.

[19] Liu G, Wang J, Zhu Y, Zhang X. Application of multiwalled carbon nanotubes as a solid-phase extraction sorbent for chlorobenzenes. Anal Lett 2004;37:3085–104.

[20] Cai Y, Cai Y, Mou S, Lu Y. Multi-walled carbon nanotubes as a solid-phase extraction adsorbent for the determination of chlorophenols in environmental water samples. J Chromatogr A 2005;1081:245–7.

[21] Peng X, Li Y, Luan Z, Di Z, Wang H, Tian B, et al. Adsorption of 1,2-dichlorobenzene from water to carbon nanotubes. Chem Phys Lett 2003;376:154–8.

[22] Cai Y, Jiang G, Zhou Q. Multiwalled carbon nanotubes as a solid-phase extraction adsorbent for the determination of bisphenol A, 4-*n*-nonylphenol and 4-*tert*-octylphenol. Anal Chem 2003;75:2517–21.

[23] Cai YQ, Jiang GB, Liu JF, Zhou QX. Multi-walled carbon nanotubes packed cartridge for the solid-phase extraction of several phthalate esters from water samples and their determination by high performance liquid chromatography. Anal Chim Acta 2003;494:149–56.

[24] Zhou Q, Xiao J, Wang W, Liu G, Shi Q, Wang J. Determination of atrazine and simazine in environmental water samples using multiwalled carbon nanotubes as the adsorbents for preconcentration prior to high performance liquid chromatography with diode array detector. Talanta 2006;68:1309–15.

[25] Zhou Q, Wang W, Xiao J. Preconcentration and determination of nicosulfuron, thifensulfuron-methyl and metsulfuron-methyl in water samples using carbon nanotubes packed cartridge in combination with high performance liquid chromatography. Anal Chim Acta 2006;559:200–6.

[26] Zhou Q, Xiao J, Wang W. Trace nalysis of triasulfuuron and bensulfuron-methyl in water samples using carbon nanotubes packed cartridge in combination with high-performance liquid chromatography. Microchim Acta 2007;157:93–8.

[27] Biesaga M, Pyrzynska K. The evaluation of carbon nanotubes as a sorbent for dicamba herbicide. J Sep Sci 2006;29:2241–4.

[28] Pyrzynska K, Stafiej A, Biesaga M. Sorption behaviour of acidic herbicides on carbon nanotubes. Microchim Acta 2007;159:293–8.

[29] Zhou Q, Xiao J, Wang W. Using multi-walled carbon nanotubes as solid phase extraction adsorbents to determine dichlorodiphenyltrichloroethane and its metabolites at trace level in water samples by high performance liquid chromatography with UV detection. J Chromatogr A 2006;1125:152–8.

[30] Fugetsu B, Satoh S, Iles A, Tanaka K, Nishi N, Watari F. Encapsulation of multi-walled carbon nanotubes (MWCNTs) in Ba^{2+}-alginate to form coated micro-beads and their application to the pre-concentration/elimination of dibenzo-*p*-dioxin, dibenzofuran, and biphenyl from contaminated water. Analyst 2004;129:565–6.

[31] Fang GZ, He JX, Wang S. Multiwalled carbon nanotubes as sorbent for on-line coupling of solid-phase extraction to high-performance liquid chromatography for simultaneous determination of 10 sulfonamides in eggs and pork. J Chromatogr A 2006;1127:12–7.

[32] Biesaga M, Jankowska A, Pyrzynska K. Comparison of different sorbents for solid-phase extraction of phenoxyalkanoic acid herbicides. Microchim Acta 2005;150:317–22.

[33] Ya Davydov V, Kalashnikova EV, Karnatsevich VI, Kirillov I. Adsorption properties of multi-wall carbon nanotubes. Full Nano Carb Nanostruct 2004;12:513–8.

[34] Yang K, Zhu L, Xing B. Adsorption of polycyclic aromatic hydrocarbons by carbon nanomaterials. Environ Sci Technol 2006;40:1855–61.

[35] Basheer C, Alnedhary AA, Rao BSM, Valliyaveettli S, Lee HK. Development and application of porous membrane-protected carbon nanotube micro-solid-phase extraction combined with gas chromatography/mass spectrometry. Anal Chem 2006;78: 2853–6.

[36] Saridara C, Brukh R, Iqbal Z, Mitra S. Preconcentration of volatile organics on self-assembled carbon nanotubes in a microtrap. Anal Chem 2005;77:1183–7.

[37] Li QL, Yuan DX, Lin QM. Evaluation of multi-walled carbon nanotubes as an adsorbent for trapping volatile organic compounds from environmental samples. J Chromatogr A 2004;1026:283–8.

[38] Li YH, Wang S, Luan Z, Ding J, Xu C, Wu D. Adsorption of cadmium(II) from aqueous solution by surface oxidized carbon nanotubes. Carbon 2003;41:1057–62.

[39] Liang P, Liu Y, Guo L. Determination of trace rare earth elements by inductively coupled plasma atomic emission spectrometry after preconcentration with multiwalled carbon nanotubes. Spectrochim Acta Part B 2005;60:125–9.

[40] Lu C, Chu H, Liu C. Removal of zinc(II) from aqueous solution by purified carbon nanotubes: kinetics and equilibrium studies. Ind Eng Chem Res 2006;45: 2850–5.

[41] Lu C, Chiu H. Adsorption of zinc(II) from water with purified carbon nanotubes. Chem Eng Sci 2006;61:1138–45.

[42] Liang P, Liu Y, Guo L, Zeng J, Lu H. Multiwalled carbon nanotubes as solid-phase extraction adsorbent fort the preconcentration of trace metal ions and their determination by inductively coupled plasma atomic emisssion spectrometry. J Anal At Spectrom 2004;19:1489–92.

[43] Li YH, Ding J, Luan Z, Di Z, Zhu Y, Xu C, et al. Competitive adsorption of Pb^{2+}, Cu^{2+} and Cd^{2+} ions from aqueous solutions by multiwalled carbon nanotubes. Carbon 2003;41:2787–92.

[44] Stafiej A, Pyrzynska K. Adsorption of heavy metal ions with carbon nanotubes. Sep Sci Technol 2007;58:49–52.

[45] Liang P, Ding Q, Song F. Application of multiwalled carbon nanotubes as solid phase extraction sorbent for preconcentration of trace copper in water samples. J Sep Sci 2005;28:2339–43.

[46] Li YH, Wang S, Wei J, Zhang X, Xu C, Luan Z, et al. Lead adsorption on carbon nanotubes. Chem Phys Lett 2002;357:263–6.

[47] Barbosa AF, Segatelli MG, Pereira AC, de Santana Santos A, Kubota LT, Luccas PO, et al. Solid-phase extraction system for Pb(II) ions enrichment based on multiwall carbon nanotubes coupled on-line to flame atomic absorption spectrometry. Talanta 2007;71:1512–9.

[48] Tarley CRT, Barbosa AF, Segatelli MG, Figueiro EC, Luccas PO. Highly improved sensitivity of TS-FF-AAS for Cd(II) determination at ng L^{-1} levels using a simple flow injection microcolumn preconcentration system with multiwall carbon nanotubes. J Anal At Spectrom 2006;21:1305–13.

[49] Li YH, Di Z, Ding J, Wu D, Luan Z, Zhu Y. Adsorption thermodynamic, kinetic and desorption studies of Pb^{2+} on carbon nanotubes. Water Res 2005;39:605–9.

[50] Liang HD, Han DM. Multi-walled carbon nanotubes as sorbent for flow injection on-line microcolumn preconcentration coupled with flame atomic absorption spectrometry for determination of cadmium and copper. Anal Lett 2006;39:2285–95.

[51] Dastgheib SA, Rockstraw DA. A model for the adsorption of single metal ion solutes in aqueous solution onto activated carbon produced from pecan shells. Carbon 2002;40:1843.

[52] Figueiredo JL, Pereira MFR, Freitas MMA, Órfão JJM. Modification of the surface chemistry of activated carbons. Carbon 1999;37:1379–89.

[53] Muñoz J, Gallego M, Valcárcel M. Speciation of organometallic compounds in environmental samples by gas chromatography after flow preconcentration on fullerenes and nanotubes. Anal Chem 2006;77:5389–95.

[54] Wu K, Hu S, Fei J, Bai W. Mercury-free simultaneous determination of cadmium and lead at a glassy carbon electrode modified with multi-wall carbon nanotubes. Anal Chim Acta 2003;489:215–21.

[55] Li YH, Wang S, Zhang X, Wie J, Xu C, Luan Z, et al. Adsorption of fluoride from water by aligned carbon nanotubes. Mater Res Bull 2003;38:469–76.

[56] An X, Zeng H. Functionalization of carbon nanobeads and their use as metal ion adsorbents. Carbon 2003;41:2889–96.

CHAPTER 11

Fire-Retarded Environmentally Friendly Flexible Foam Materials Using Nanotechnology

Richard A. Pethrick
WestCHEM, Department of Pure and Applied Chemistry,
University of Strathclyde, Glasgow, Scotland

Contents

1. INTRODUCTION

Fire kills many thousands of people a year. Death is usually a consequence of asphyxiation rather than a consequence of burning. Dense smoke and fumes emitted during the early stages of a fire can cause disorientation of person caught in the fire. High levels of carbon monoxide, carbon dioxide, and to a lesser extent sulphur dioxide, hydrogen chloride, hydrogen cyanide, and various gases, which reflect the chemical nature of the combustible fuels, are the principle cause of asphyxiation. Forensic analysis of a fire situation will usually indicate that some source of heat initiated the event. The source may be electrical wires or heaters, which have overheated, a lighted match, which is left to smolder, a cigarette butt, which was not fully extinguished, etc. In most cases, the fire is created by a flammable object fueling the fire and causing it to spread. It is therefore critical that the susceptibility of materials to undergo combustion is reduced when they are used in a domestic environment.

2. ANALYSIS OF THE VARIOUS STAGES IN WHICH A FIRE IS CREATED

To design a material that will be effective in suppressing combustion, it is useful to understand the various stages involved in the growth of a fire. We can separate the growth of the fire into three stages.

2.1. Precombustion

In this stage, the object that will form the fuel for the fire is heated by the source. Usually, the effect of heating will be to cause the material to soften and even melt. In the melt phase, the material may experience temperatures, which are above the ceiling temperature, and degradation of the material may occur. In the case of a polymer, the degradation may yield the monomer from which the polymer was formed or some other organic material of molar mass of the order of 60–200 daltons. These volatile fragments may now be at a temperature which is above the required for spontaneous combustion and a fire can start. In the initial stages, there will be sufficient oxygen present to sustain combustion and a cold flame will be generated (Fig. 11.1).

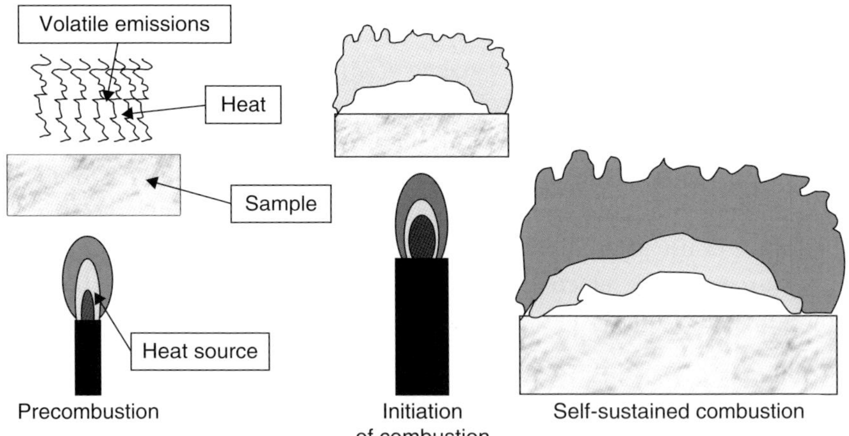

Figure 11.1   Schematic of the stages of combustion.

2.2. Initiation of Combustion

The cold flame that is created depends on a constant supply of the low molar mass fragments and plenty of oxygen. This flame can be extinguished by trapping the various radicals that are formed and by suppressing the combustion chemistry. Many fire retardants work by interrupting the combustion chemistry and damping the flame. If the combustion chemistry is allowed to establish itself, the nature of the reactions may change and become less reliant in the presence of oxygen. In a major fire, the process is able to draw air into the fire, but the process is less dominated by the oxygen chemistry. A very important factor in this stage of the growth of the fire is the amount of heat, which is radiated back onto the sample. The heat from the combustion can further melt the material and hence add fuel to the fire (Fig. 11.1). One way of suppressing the growth of a fire is to create an insulating layer on the top of the material, which is fueling the fire. Char, which is naturally formed during the degradation process, is in fact a good insulator and can effectively produce a thermal barrier and also suppress the evolution of volatiles from the material. The combination of both these effects can lead to a suppression of the growth of the fire.

2.3. Self-sustained Combustion

Once the sustained combustion process has been achieved, the growth of the fire will depend on a supply of fuel to the flame. If the volume of combustible material is small, the fire may burn to a point where it ceases to be sustainable and will go out. In practice, the flames are usually propagated by either traveling across the surface of a material or, alternatively, dropping material into the flames and increasing the volume of the material available for the fire. Some chemical additives, because they suppress the initiation of combustion, are good at controlling flame spread. Forensic analysis of a fire situation often indicates that a molten material is dripped into the source of the flame. It is therefore desirable to stop dripping, and this, in principle, can be achieved by making the polymer melt viscoelastic.

3. HOW CAN NANOTECHNOLOGY BE USED TO HELP CONTROL FIRES?

The critical stages in the development of a fire are precombustion and initiation of combustion. Although the self-sustained combustion stage is very important, at this stage the fire is usually growing at a rate that makes simple chemical control very difficult. So, what can we do with nanotechnology

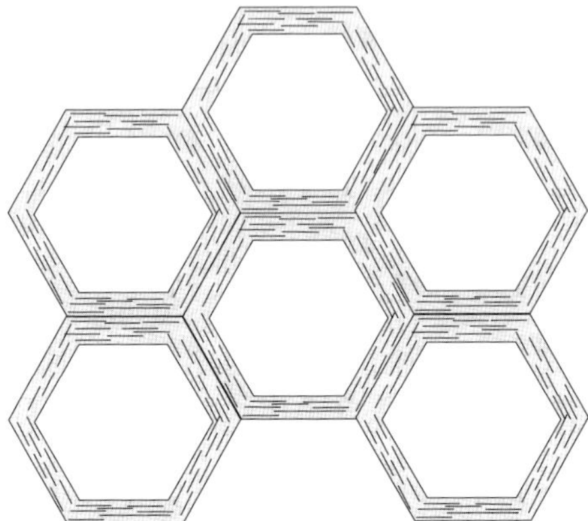

Figure 11.2   Schematic showing the nanoplatelets organized in the walls of the foam structure.

that might aid to control the growth of the precombustion and early stage combustion process? In the case of foam, we are dealing with large surface areas that are ideal for the fast release of volatile components. If we wish to control the rate of release of these low molar mass fragments, it would be desirable to introduce into the material a barrier layer. Nanotechnology provides us with that possibility. A further aid to suppression of fire is the promotion of char formation. Certain nanomaterials have a potential of aiding char formation through their ability to assist recombination reactions of low molar mass radical fragments in the solid phase. A further advantage of nanomaterials is their ability to enhance the viscosity of the melt, and this adds the desirable viscoelastic characteristics that will suppress dripping. The desirable nanomaterial would have a platelet structure and form a barrier structure in the solid, which would suppress low–molar mass emissions from escaping from the solid as it melts (Fig. 11.2).

4. CAN SUCH A STRUCTURE BE CREATED IN PRACTICE?

Over the last 10 years, a lot of interest has been directed toward the study of nanoorganically modified clay composites [1]. Clay is a naturally occurring nanomaterial that has been used for many years as a cheap additive to reduce the cost of articles made from plastics. Clay is constituted of stacks of aluminosilicate layered structures, the thickness of the layer being ~1 nm (Fig. 11.3). The chemistry of clay reflects the geological conditions that existed during its creation. One particular form montmorillonite contains low levels of impurity ions, e.g., Fe, Mg, and Li. Incorporation of these ions into the

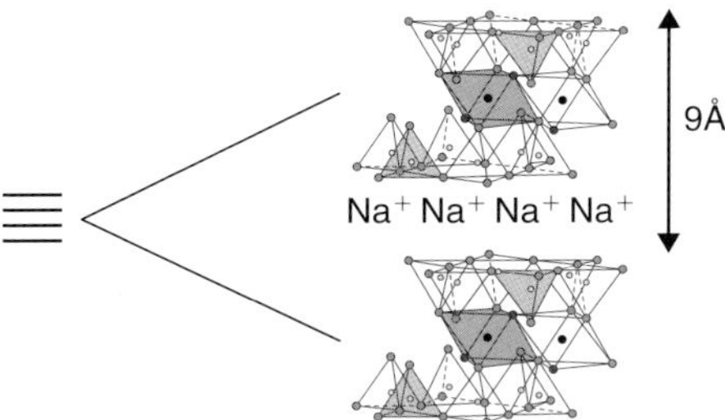

Figure 11.3 Schematic of a clay stack, showing expanded view of a typical layer formed from aluminosilicates held together by a layer of sodium ions.

layer leaves a charge deficiency that creates the binding site for the sodium ions. The characteristics slip of clays reflects the ability for platelets in a stack to be moved relative to one another when a shear force is applied. However, the relatively strong electrostatic forces keep the stacks together and clay is usually observed as nodules of approximately 1–5μ. The platelets are of length 400–100 nm and the balance of forces favors a stacked structure.

The major breakthrough with these materials came with the recognition that it was possible to exfoliate and disperse these individual platelets if the sodium ions were replaced by quaternary ammonium ions [2].

There is a very large literature on the physical property enhancement that can be achieved when the organically modified clay platelets are dispersed in the matrix [1, 2]. Flexible foams are usually obtained from polyurethanes (PUs), which is produced by the reaction of a polyether with a diisocyanate. Depending on the nature and proportions of the polyether and the diisocyanate, flexible or rigid foams will be created. Methylene diphenyl isocyanate is typically used for more rigid materials, whereas toluene diisocyanate is used predominantly for flexible foams. The nanoclay can be readily incorporated into the PU by predispersion in the polyether prior to reaction with the isocyanate [3]. To aid dispersion, it is often useful to use ultrasound. The ultrasound can interact with the clay stack, and effective exfoliation of the platelets can be achieved. A problem with the efficient exfoliation of the clay is that the platelets can form a transient structure in the liquid and significantly enhance the viscosity (Fig. 11.4).

The platelets have, as a consequence of the charge deficiency associated with the impurity ions, an "effective" negative charge on the surface. It is this negative charged region on the platelets surface which in nature clay attracts the sodium ions and in the organically modified clay is the location for the quaternary ammonium ions. What is not always recognized is that the edges of the clay platelets will also have a charge associated with the

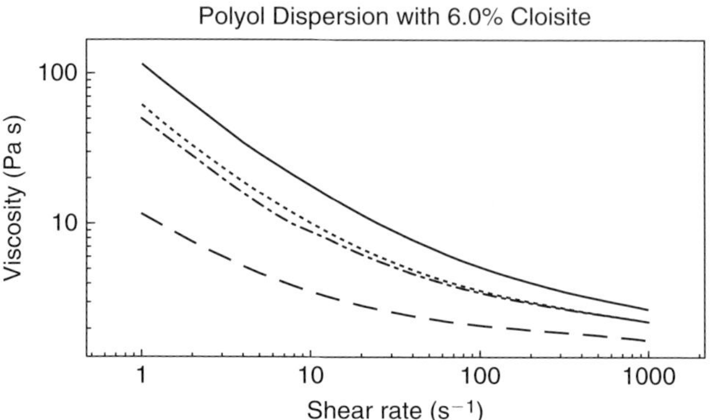

Figure 11.4 The initial viscosity and reduction of the viscosity on the addition of titanium-based coupling agents. Key: no coupling agent (–), 0.8% coupling agent (---), 1.5% coupling agent (– - –), 2.0 % coupling agent (—).

growth sites for the inorganic matrix. These edge sites will tend to have a preponderance of positive charges. The favorable interaction of the positively charged edges with the negatively charged sites on the surface of the clay platelets allows the creation of a network structure. This phenomenon has been recognized for many years and is the basis of the thickening action of clay and related system to produce pastes. Although this effect is desirable in a tooth paste, it is not at all desirable in the polyether that has to be effectively mixed with the isocyanate to form the flexible foam. To recover the desirable low viscosity, the electrostatic interactions need to be blocked. Addition of organically modified titanium compounds have been found to be effective in achieving the reduction of the viscosity (Fig. 11.4).

Using these nanomodified polyether, it is possible to create structures that mimic (Fig. 11.2). Electron microscopic examination of the walls of the flexible foams indicates that the blowing process has aligned the platelets to form the pseudo nematic order, which is desirable to create a barrier for the emission of the low–molecular mass fragments created when the polymer is heated (Fig. 11.5).

5. DO WE NEED ANYTHING ELSE TO MAKE THE FOAM FIRE RESISTANT?

The nanoclay by itself is able to reduce the emission rate of low–molecular mass fragments, but this is usually not sufficient to make the foam adequately fire resistant to pass a Crib 5 test [4]. The test involves placing two pieces of foam in the form of a chair and covering these with a loose-textured

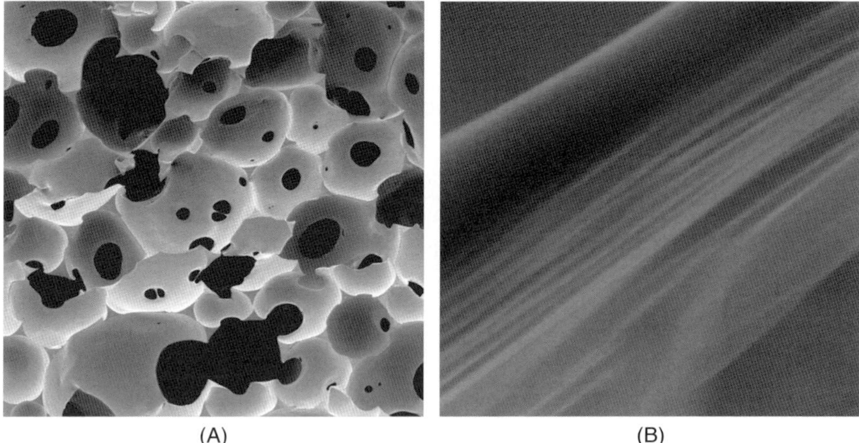

(A)   (B)

Figure 11.5 Electron micrograph of a nanomodified flexible foam (A) and a close up of the wall structure (B). The dark lines are the clay platelets aligned in the wall structure to form a barrier for low–molar mass emission.

cloth. Onto the base of the chair is placed a small, square crib formed from 20 pieces of wood so as to form a structure that is five rungs high. In the center of the crib is placed a piece of cotton wool soaked in isopentane. The cotton wool is ignited, and for the foam to pass the test, the fire must be constrained to the locality of the crib and there must not be more than 15% of the foam destroyed in the test. This is a very stringent test of a material as the crib will sustain the source of heat for several minutes and mimics the sort of fire that would be created by a cigarette butt.

To get a nano-modified flexible material to pass the Crib 5 test, it is usually necessary to add components that will aid char formation and also suppress flame spread. Char formation can be promoted by a range of inorganic nonvolatile agents that are able to interact with radicals formed during the degradation. The large surface area of the clay can be synergistic in this action and increase the efficiency of the char formation process. Once a char is formed, it will protect the polymer beneath and reduce the rate of low–molar mass emissions. This process is more efficient in rigid foams and less char promoter is required to achieve the required fire resistance. Control of the spread of the fire requires influencing the vapor-phase chemistry. Once more, this can be achieved using solid species that are only volatilized once the foam is burning. Combining a number of elements in a synergistic manner, it is possible to achieve fire-retardant characteristics without the use of the usual volatile organic compounds. This technology is disclosed by patents held by the author [5, 6].

6. SUMMARY

By the careful design of formulations and the use of an understanding of nanotechnology, it is possible to create flexible foams that have the desired fire-retardant characteristics for them to pass the Crib 5 test. The nanotechnology reduces the rate of emission of the degradation products and is synergistic in the formation of the char and reduces dripping of the molten polymer, hence reducing the spread of the fire. By itself, it is unable to achieve fire retardancy, but when combined with other agents, it achieves the desired effect. Nanotechnology has a significant role to play in the development of fire-retardant technology for the future.

ACKNOWLEDGMENTS

The author thanks EPSRC and Scottish Enterprise through its provision of a Proof of Concept grant for their support. This study was carried out in collaboration with Dr Liggat JJ, Dr Daly JH and Dr Rhoney I, and support by Mr McCulloch L.

References

[1] Krishnamoorti R, Silva AS. Rheology of polymer layered silicate nanocomposites. In: Pinnavaia TS, Beall GW, editors. Chapter 15. Chichester: Wiley Polymer Science; 2001. p. 316–43.

[2] Giannelis EP. Polymer layered silicate nanocomposites. Adv Mater 1996;8:29–356.

[3] Rhoney I, Brown S, Hudson NE. Pethrick RA. Influence of processing method on the exfoliation process for organically modified clay systems. I. Polyurethanes. J Appl Polymer Sci 2004;91(2):1335–43.

[4] British Standard 5852 (Section 4) Part II, Methods of test for assessment of the ignitability of upholstered seating by smouldering and flaming ignition sources. 1990.

[5] Liggat JJ, Daly JH, McCulloch L, Pethrick RA. British Patent 0713394.5, 2007.

[6] Liggat JJ, Rhoney I, Pethrick RA. World Patent 2006/003421, 2005.

CHAPTER 12

Simulation of Hydrogen Purification by Pressure-Swing Adsorption for Application in Fuel Cells

Qinglin Huang *and* **Mladen Eić**
Department of Chemical Engineering, University of New Brunswick, Fredericton, New Brunswick, Canada

Contents

1. INTRODUCTION

Increasing public pressure to reduce greenhouse gas emissions and limited resources in supplying fossil fuels have led to demands for diversifying energy resources as a priority. Hydrogen is a clean and sustainable energy resource that only emits water and heat resulting from the combustion. One of the important hydrogen applications is its use in proton–exchange membrane fuel cells (PEMFC), which have been considered an attractive technology for transportation, commercial building, home, and small–scale power generation. Fuel cell converts the electrochemical energy stored in hydrogen into electricity and requires high hydrogen purity, i.e., greater than 99.99%. Impurities were reported to have negative impact on fuel-cell performance and durability, and the fuel composition must therefore be carefully controlled [1, 2]. Among these impurities, CO is one of the most critical ones affecting fuel-cell operation. CO poisons the catalyst by occupying the active sites and substantially reducing the number of sites for hydrogen adsorption and oxidation. The concentration of CO in fuel-cell feed is recommended to be below 0.2, 0.5, or 10 ppm, depending on standards used [2–4]. Other impurities, such as CO_2 and CH_4, have less negative impact on fuel-cell performance compared to CO. Concentrations of CO_2 and CH_4 are proposed to be lower than 2 and 100 ppm, respectively [1, 2].

High-purity hydrogen (up to 99.999%) can be obtained by the non-cryogenic pressure-swing adsorption (PSA) process from the steam methane reforming (SMR) process, which produces approximately 95% of hydrogen in the entire hydrogen industry. The typical feed gas compositions from SMR to PSA process contain 70–80% H_2, 15–25% CO_2, 3–6% CH_4, 1–3% CO, and trace N_2 [5]. A number of experimental and theoretical studies have been carried out to investigate hydrogen separation from SMR off-gas, oven coke gas, or refinery fuel gas by PSA process using single/layered-bed configuration [6–9]. However, only few studies have been focused on controlling hydrogen impurity for fuel-cell application. Moreover, the optimization work is necessary for PSA system due to the complicated nature of dynamic cyclic process, and the fact that there are a large number of design parameters, such as step times, pressure, temperature, gas velocity, and bed dimensions, which can affect PSA separation performance. For that reason, it is not always feasible to carry out a large number of PSA experiments to determine the optimum conditions. So far, there have been several studies investigating the optimization of single-bed PSA process using gPROMS optimizer [10–13]. However, none of them studied optimization of PSA process with layered adsorbents in bed and using a feed that has more than two constituent components.

The objective of this work was to theoretically investigate the feasibility of hydrogen production from SMR off-gas by layered-bed PSA to meet the purity requirements for fuel-cell application. Activated carbon and zeolite 5A were used as adsorbents, whereas concentrations of CO, CO_2, and CH_4 were targeted to be below 10, 100 and 100 ppm, respectively. Optimization of single-bed PSA process with layered adsorbents to separate hydrogen from quaternary mixture of CO_2, CO, CH_4, and H_2 using gPROMS optimizer tool was also carried out for comparison.

2. PSA MODEL AND SOLUTION

2.1. Model Assumptions

Mathematical model was applied to describe two-layered-bed PSA system using six-step cycle (Fig. 12.1). The following assumptions were used in the model development:

1. The ideal gas law applies.
2. The pressure drop through the bed during adsorption and purge steps is described by Ergun's Equation. For other steps, the total pressure in the bed is assumed to vary linearly with time.
3. The flow pattern is described by an axially dispersed plug flow model.
4. The multicomponent adsorption equilibrium is represented by the extended Langmuir isotherm.
5. There are no radial variations in temperature, pressure, concentration, or velocity.
6. Transport and physical properties are temperature independent.
7. The adsorption rate is approximated by a linear driving force (LDF).

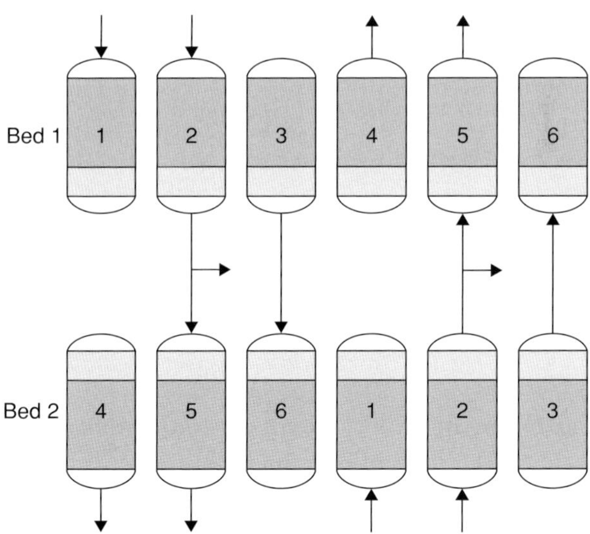

Figure 12.1 Six-step, two-layered-bed PSA cycle: 1, feed pressurization from intermediate to high pressure; 2, adsorption at high pressure; 3, pressure equalization between two beds from high to intermediate pressure; 4, depressurization from intermediate to low pressure; 5, purge with product from the other bed at low pressure; 6, pressure equalization between two beds from low to intermediate pressure.

2.2. Model Equations

The component and overall mass balances for the system are represented by Eqs. (1) and (2), respectively:

$$-D_{\mathrm{L}}\frac{\partial^2 c_i}{\partial z^2} + \frac{\partial c_i}{\partial t} + u\frac{\partial c_i}{\partial z} + c_i\frac{\partial u}{\partial z} + \rho_{\mathrm{p}}\frac{1-\varepsilon_{\mathrm{bed}}}{\varepsilon_{\mathrm{bed}}}\frac{\partial q_i}{\partial t} = 0 \tag{1}$$

$$C\frac{\partial u}{\partial z} + \frac{\partial C}{\partial t} + \rho_{\mathrm{p}}\frac{(1-\varepsilon_{\mathrm{bed}})}{\varepsilon_{\mathrm{bed}}}\sum_{i=1}^{n}\frac{\partial q_i}{\partial t} = 0 \tag{2}$$

Equation (2) determines the change of gas interstitial velocity over the length of the bed. In Eqs. (1) and (2), u is gas interstitial velocity; c_i is gas phase concentration of component i; C is total concentration in gas phase $(= P/\mathrm{R}_{\mathrm{g}}T)$; R_{g} is ideal gas constant, T is temperature; z is axial co-ordinate; q_i is adsorbed phase concentration of component i; ρ_{p} is particle density; $\varepsilon_{\mathrm{bed}}$ is bed voidage; D_{L} is the axial dispersion coefficient, which can be estimated by the following equation [14]:

$$D_{\mathrm{L}} = 0.7D_{\mathrm{m}} + \frac{1}{2}d_{\mathrm{p}}u \tag{3}$$

where d_{p} is the diameter of adsorbent particle; D_{m} is the molecular diffusivity that can be calculated from the well-known Chapman–Enskog or similar equation.

Heat effects are described by the energy balance equation:

$$\left(\varepsilon_{\mathrm{t}}\rho_{\mathrm{gas}}C_{\mathrm{pg}} + \rho_{\mathrm{bed}}C_{\mathrm{ps}}\right)\frac{\partial T}{\partial t} + \rho_{\mathrm{gas}}C_{\mathrm{pg}}\varepsilon_{\mathrm{bed}}u\frac{\partial T}{\partial z} - K_{\mathrm{L}}\frac{\partial^2 T}{\partial z^2}$$

$$-\rho_{\mathrm{bed}}\sum_{i=1}^{n}\Delta H_i\frac{\partial q_i}{\partial t} + \frac{2h}{R_{\mathrm{bed}}}\left(T - T_{\mathrm{wall}}\right) = 0 \tag{4}$$

where C_{pg} and C_{ps} are heat capacities of gas and solid particles, respectively; ρ_{bed} and ρ_{gas} are gas densities of bed and bulk, respectively; h is heat transfer coefficient; K_{L} is thermal dispersion coefficient; R_{bed} is column radius; T_{wall} is column wall temperature; ΔH_i is isosteric heat of adsorption, which is obtained from the Clausius–Clapeyron equation at the constant loading:

$$\left[\frac{d\ln P_i}{d(1/T)}\right]_q = -\frac{\Delta H_i}{\mathrm{R}_{\mathrm{g}}} \tag{5}$$

where P_i is partial pressure of component i. The adsorption rate is represented by the LDF model:

$$\frac{\partial q_i}{\partial t} = k_i\left(q_i^\star - q_i\right) \tag{6}$$

where k_i is the mass-transfer coefficient; $q_i^\star$ is the equilibrium loading of component i at time t, and can be determined by the extended Langmuir isotherm:

$$q_i^\star = \frac{q_s b_i c_i}{1 + \sum_{i=1}^{n} b_i c_i} \tag{7}$$

$$b_i = b_{1,i}\,e^{\left(b_{2,i}T\right)} \tag{8}$$

where b_i is single-component equilibrium constant; q_s is saturation capacity; $b_{1,i}$ and $b_{2,i}$ are temperature-independent parameters. The pressure drop across the bed during adsorption and purge steps is described by Ergun's equation:

$$-\frac{\partial P}{\partial Z} = 150\frac{\mu u\left(1 - \varepsilon_{\text{bed}}\right)^2}{d_p^2 \varepsilon_{\text{bed}}^2} + 1.75\frac{\left(1 - \varepsilon_{\text{bed}}\right)}{d_p \varepsilon_{\text{bed}}}u\,|\,u\,|\,\rho_{\text{gas}} \tag{9}$$

It should be noted that the layered-bed PSA system, investigated in this study, contains an initial section of activated carbon, followed by zeolite 5A in the remaining of the bed. Hence, in the PSA simulation model, the adsorbent physical properties, as well as the equilibrium parameters, mass-transfer and heat-transfer coefficients of the sorbates in the bed change at the transition location from one adsorbent to another.

All boundary conditions for the cycle steps with respect to the differential variables within the model, such as gas concentration, velocity, and temperature, are summarized in Table 12.1 [13]. The performance variables including purity, impurity, and recovery are defined as

$$\text{Purity}_{\text{H}_2} = \frac{\int_0^{t_A} \left(u\, c_{\text{H}_2}\right)\Big|_{z=L}\, dt}{\int_0^{t_A} \left(\sum_{i=1}^{n} u\, c_i\right)\Big|_{z=L}\, dt} \times 100\% \tag{10}$$

Table 12.1 Boundary conditions of six-step PSA process

Feed pressurization	Adsorption	First pressure equalization	Depressurization	Purge	Second pressure equalization
$c_i\big|_{z=0} = c_{f,i}$	$c_i\big|_{z=0} = c_{f,i}$	$\dfrac{\partial c_i}{\partial z}\Big|_{z=0} = 0$	$\dfrac{\partial c_i}{\partial z}\Big|_{z=0} = 0$	$\dfrac{\partial c_i}{\partial z}\Big|_{z=0} = 0$	$\dfrac{\partial c_i}{\partial z}\Big|_{z=0} = 0$
$\dfrac{\partial c_i}{\partial z}\Big|_{z=L} = 0$	$\dfrac{\partial c_i}{\partial z}\Big|_{z=L} = 0$	$\dfrac{\partial c_i}{\partial z}\Big|_{z=L} = 0$	$\dfrac{\partial c_i}{\partial z}\Big|_{z=L} = 0$	$c_i\big|_{z=L} = \bar{c}_{\text{product},i}$	$c_i\big|_{z=L} = \bar{c}_{\text{equ},i}$
$u\big|_{z=L} = 0$	$u\big|_{z=0} = u_{\text{feed}}$	$u\big|_{z=0} = 0$	$\dfrac{\partial u}{\partial z}\Big|_{z=0} = 0$	$\dfrac{\partial u}{\partial z}\Big|_{z=0} = 0$	$u\big|_{z=0} = 0$
$\dfrac{\partial u}{\partial z}\Big|_{z=L} = 0$	$\dfrac{\partial u}{\partial z}\Big|_{z=L} = 0$	$\dfrac{\partial u}{\partial z}\Big|_{z=L} = 0$	$u\big|_{z=L} = 0$	$u\big|_{z=L} = u_{\text{purge}}$	$\dfrac{\partial u}{\partial z}\Big|_{z=L} = 0$
$\dfrac{\partial T}{\partial z}\Big|_{z=L} = 0$	$T\big|_{z=0} = T_{\text{feed}}$	$\dfrac{\partial T}{\partial z}\Big|_{z=0} = 0$	$\dfrac{\partial T}{\partial z}\Big|_{z=0} = 0$	$\dfrac{\partial T}{\partial z}\Big|_{z=0} = 0$	$\dfrac{\partial T}{\partial z}\Big|_{z=0} = 0$
$T\big|_{z=0} = T_{\text{feed}}$	$\dfrac{\partial T}{\partial z}\Big|_{z=L} = 0$	$\dfrac{\partial T}{\partial z}\Big|_{z=L} = 0$	$\dfrac{\partial T}{\partial z}\Big|_{z=L} = 0$	$T\big|_{z=L} = \overline{T}_{\text{product}}$	$T\big|_{z=L} = \overline{T}_{\text{equ}}$

where $\bar{c}_{\text{product},i}/\overline{T}_{\text{product}}$, and $\bar{c}_{\text{equ},i}/\overline{T}_{\text{equ}}$ are average concentration/temperature of the effluent stream during the adsorption step, and the depressurizing (first) pressure-equalization step, respectively.

$$\text{Impurity}_{CO, CO_2, CH_4} = \frac{\displaystyle\int_0^{t_A} \left(u\, c_{CO, CO_2, CH_4} \right)\Big|_{z=L} \, dt}{\displaystyle\int_0^{t_A} \left(\sum_{i=1}^{n} u\, c_i \right)\Big|_{z=L} \, dt} \tag{11}$$

$$\text{Recovery}_{H_2} = \frac{\displaystyle\int_0^{t_A} \left(u\, c_{H_2} \right)\Big|_{z=L} \, dt - \int_0^{t_{Pu}} \left(u\, c_{H_2} \right)\Big|_{z=L} \, dt}{\displaystyle\int_0^{t_P} \left(u\, c_{H_2} \right)\Big|_{z=0} \, dt + \int_0^{t_A} \left(u\, c_{H_2} \right)\Big|_{z=0} \, dt} \times 100\% \tag{12}$$

where t_A, t_P, and t_{Pu} are adsorption (second step in Fig. 12.1), pressurization (first step in Fig. 12.1), and purging (regeneration; fifth step in Fig. 12.1) times, respectively.

2.3. Solution Methodology

The partial differential equations are discretized into n discretization points using the centered finite difference method. It may be noted that the number of discretization points is important for the accuracy of the numerical solution. The optimum numbers were determined by increasing discretization points until further change did not affect the solutions any more. The coupled set of ordinary differential and algebraic equations are solved by DASOLV that is based on variable time step or variable-order backward differentiation formulae. The dynamic PSA simulation is carried out using gPROMS software licensed by Process Systems Enterprise Limited. The SRQPD solver that employs a sequential quadratic programming in gPROMS was applied to solve the nonlinear programming problem for PSA optimization.

3. EXPERIMENTAL

Single-component adsorption isotherms of CO_2, CO, CH_4, and H_2 in activated carbon and zeolite 5A, which were provided by Barnebey Sutcliffe Co. and Union Carbide, respectively, were measured at different temperatures using a custom-made constant volumetric apparatus. The experimental procedure was described elsewhere [15].

4. RESULTS AND DISCUSSION

4.1. Adsorption Equilibrium, Mass-Transfer Coefficients, and Multicomponent Column Dynamics in a Layered Bed

The experimental adsorption isotherm results together with Langmuir fittings are shown in Fig. 12.2. The model fittings show reasonable agreement with experimental data. The experimental results are comparable to those reported in the literature [16–18]. The optimized equilibrium parameter values obtained from nonlinear regression method of Langmuir model fittings are summarized in Table 12.2 (columns 3–5). Henry constant values at 303.15 K were obtained from the linear portion of experimental isotherm data. Evidently, activated carbon exhibited higher saturation adsorption capacities for CO_2 and CH_4 and lower saturation adsorption capacities for CO compared to zeolite 5A. Furthermore, the adsorption affinity is stronger for CO than CH_4 in zeolite 5A. As can be seen from Fig. 12.2, zeolite 5A shows nearly rectangular isotherm for CO_2, e.g., very high Henry

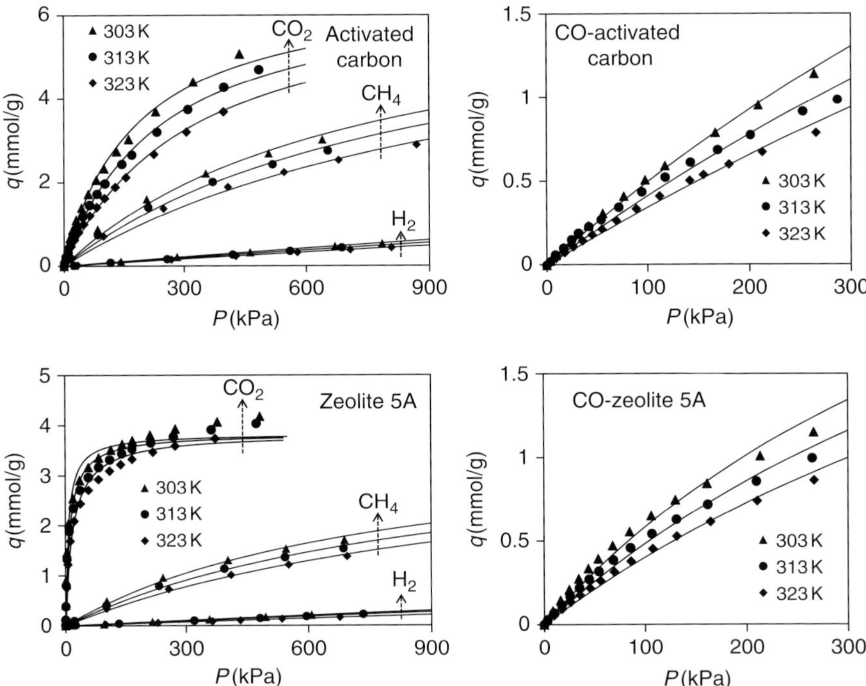

Figure 12.2   Adsorption isotherms of CO_2, CO, CH_4, and H_2 on activated carbon and zeolite 5A at three different temperatures (symbols: experiments; lines: Langmuir fit).

Table 12.2 Equilibrium and mass-transfer parameters for CO_2, CO, H_2, and CH_4 in activated carbon and zeolite 5A samples

Adsorbent	Sorbate	q_s (mmol/g)	$b_{1,i}$ (cm³/mmol)	$b_{2,i}$ (1/K)	Henry's constant at 303 K	Mass-transfer coefficient (k), measured at 298 K (1/s)
Activated carbon	CO_2	6.77	1.59×10^{-2}	-0.0295	93.4	0.125
	CO		1.14×10^{-4}	-0.0195	11.0	0.334
	H_2		2.27×10^{-6}	-0.0131	1.3	1.0
	CH_4		2.62×10^{-4}	-0.0205	38.5	0.415
Zeolite 5A	CO_2	3.82	52.60	-0.0457	1248.1	0.053
	CO		4.51×10^{-4}	-0.0214	24.1	0.145
	H_2		9.60×10^{-6}	-0.0183	0.9	1.0
	CH_4		1.80×10^{-4}	-0.0196	17.3	0.212

constant value, indicating a difficulty to perform the desorption step at low pressures. Hence, CO_2 had to be removed by activated carbon first to prevent it from breaking through into the subsequent layer of zeolite 5A.

The overall mass-transfer coefficients (k) were obtained by fitting the single-component column dynamic breakthrough data measured at low sorbate concentration with helium as carrier gas [15]. The k values for CO_2, CO, CH_4, and H_2 in activated carbon and zeolite 5A are included in Table 12.2 (last column).

Figure 12.3 shows the simulation of quaternary CO_2, CO, CH_4, and H_2 breakthrough curves at different locations on the layered column.

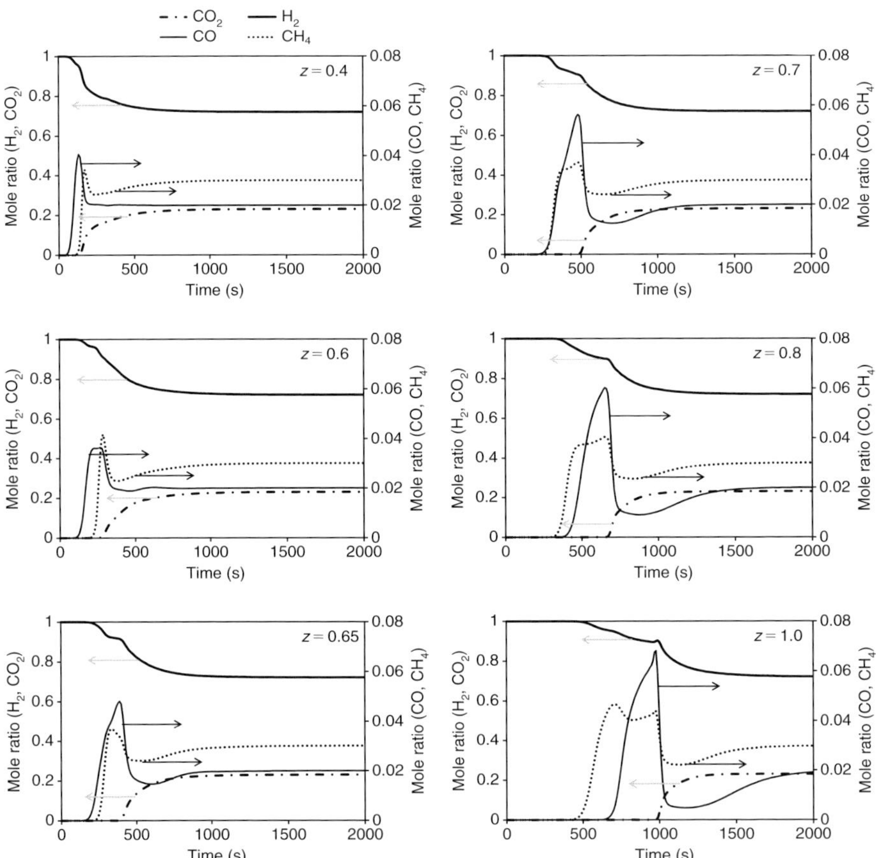

Figure 12.3 Simulated concentration breakthrough curves for CO_2, CO, CH_4, and H_2 at 298.15 K, and $z = 0.4$, 0.6, 0.65, 0.7, 0.8, and 1.0 of a layered column: column was initially saturated with H_2 at 12 atm and then fed with 23%CO_2/2%CO/3%CH_4/72%H_2 gas mixture.

The adsorbent properties and other relevant information of layered column are listed in Table 12.3. The equilibrium and mass-transfer parameters in Table 12.2 were used in the simulation. It is clear from Fig. 12.3 that the gas concentrations at different bed positions asymptotically approach the feed mixture at $23\%CO_2/2\%CO/3\%CH_4/72\%H_2$, after sufficiently long time. The breakthrough times at the column exit ($z = 1.0$) are $t_{CH_4} < t_{CO} < t_{CO_2}$, which was also reported by Park et al. [17]. As can also be observed from the same figure, CO mass transfer wave moves faster than CH_4 in the activated carbon layer ($z \leq 0.6$). After breaking through into the zeolite layer ($0.6\ 1\ z \leq 1.0$), CH_4 wave velocity increases relative to CO. CH_4 wave caught up the CO front at $z = 0.7$ and surpassed it as $z\ 2\ 0.7$, e.g., $z = 0.8$ or 1.0. This phenomenon could be explained from the definition of concentration wave velocity given by [19]:

$$\omega_c = \frac{u}{1 + \dfrac{1 - \varepsilon_{bed}}{\varepsilon_{bed}} \dfrac{dq}{dc}} \qquad (13)$$

Table 12.3   Properties of adsorbents and layered column information

	Activated carbon layer	Zeolite 5A layer
Bed length (m)	1.0	
Internal diameter (m)	0.1	
Activated carbon length/bed length	0.6	
Packing density (g/cm³)	0.427	0.761
Bed voidage	0.40	0.40
Particle density (g/cm³)	0.716	1.15
Particle radius (m)	2.0×10^{-3}	1.7×10^{-3}
Particle porosity	0.65	0.61
Particle heat capacity, C_{ps}, (J/g K)	1.05	1.16
Heat-transfer coefficient of wall (J/m²s K)	60	

where dq/dc is Henry's constant in the linear region of isotherm. Henry's constants in Table 12.2 are in the sequence CH_4 2 CO and CO 2 CH_4 for activated carbon and zeolite 5A, respectively. This demonstrated that velocity of CO wave was greater than CH_4 in the activated carbon layer, whereas the opposite was true in the zeolite layer.

From Fig. 12.3, concentration roll–ups for both CO and CH_4 can also be observed. The concentration wave of CH_4 exhibits double roll–ups at $z = 1.0$ due to the breakthrough of CO followed by CO_2. Minor transitions can also be observed from the hydrogen desorption profiles, relating to the breakthroughs of other components. For multicomponent adsorption, the relatively weakly adsorbed species are always displaced by the more strongly adsorbed components. The roll–ups of one component generally occur when the other components break through the adsorbent layer with a steep wave front [17].

Figure 12.3 shows that CO and CH_4 concentration profiles become concave after the roll–ups due to the adsorption heat effects. The temperature profiles (Fig. 12.4) in the layered bed can account for the observed phenomena if the concave section of concentration curves corresponds to the temperature peak. The adsorbed amount in the adsorbents was reduced due to the substantial increase of temperature, rendering the reduction of gas-phase concentration in the bed. Figure 12.4 shows that there are two separate temperature peaks in the zeolite layer. The first one is due to the

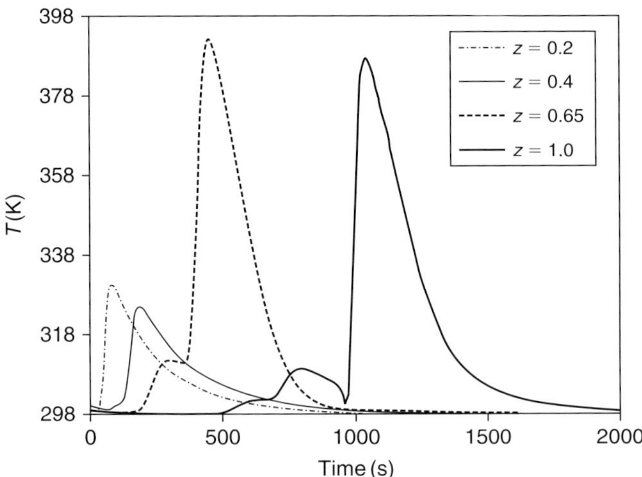

Figure 12.4   Temperature profiles at $z = 0.2$, 0.4, 0.65, and 1.0 of a layered column. For simulation conditions, see Fig. 12.3.

adsorption of CO and CH_4, whereas the second one is due to adsorption of CO_2. Because CO_2 has higher heat of adsorption in zeolite 5A than in activated carbon, the related temperature peak in zeolite layer is correspondingly larger.

4.2. Parametric Studies of PSA System

The parametric effects on PSA separation performance were theoretically investigated in this section. The results were collected at the 150th cycle of PSA simulation, which approximately reached a cyclic steady state (CSS).

4.2.1. Effects of Interstitial Velocity on the Adsorption Step, Purge Time, and Feed Temperature

In the commercial PSA system, the interstitial velocity during adsorption step is one of the most important operating parameters and, therefore, has to be carefully selected. Figure 12.5A shows the simulation results related to the effects of the interstitial velocity on the separation performance of the PSA system. The product impurity was dominated by CO and then CH_4 as the velocity increased. To obtain fuel-cell-grade hydrogen, the velocity should be maintained below 5 cm/s, under the operating conditions studied here. Hydrogen purity decreased sharply with velocity greater than 5 cm/s, whereas the recovery gradually increased with the increase of velocity. When the velocity was increased, the breakthrough times decreased; consequently, within the constant cycle time, this caused a decrease in the product purity. The decrease in purity is dependent on the extent of impurity breakthrough. For the lower velocity runs, e.g., u 1 5 cm/s, the impurity breakthrough was insignificant; therefore, the purity dropped only slightly. On the other hand, at higher velocities, the impurity breakthrough became more significant, causing a rapid drop in purity.

The effect of the purging time on the performance of the PSA process is illustrated in Fig. 12.5B. CO was dominant among impurities and could be controlled to be less than 10 ppm (mole ratio of 1.0×10^{-5} marked in the figure), when the purging time was greater than 177 s. The increase in purging time corresponds to more thorough cleaning of the adsorption bed, rendering higher purity, and lower product recovery.

The effects of the temperature on the performance of the PSA process are shown in Fig. 12.5C. Apparently, high temperatures are generally unfavorable for impurity removal. The product recovery, as well as CO and CO_2 impurity, increased with the increase of adsorption temperature (with CO dominating as the product impurity). The adsorption temperature was required to be lower than 299 K for controlling CO impurity to be

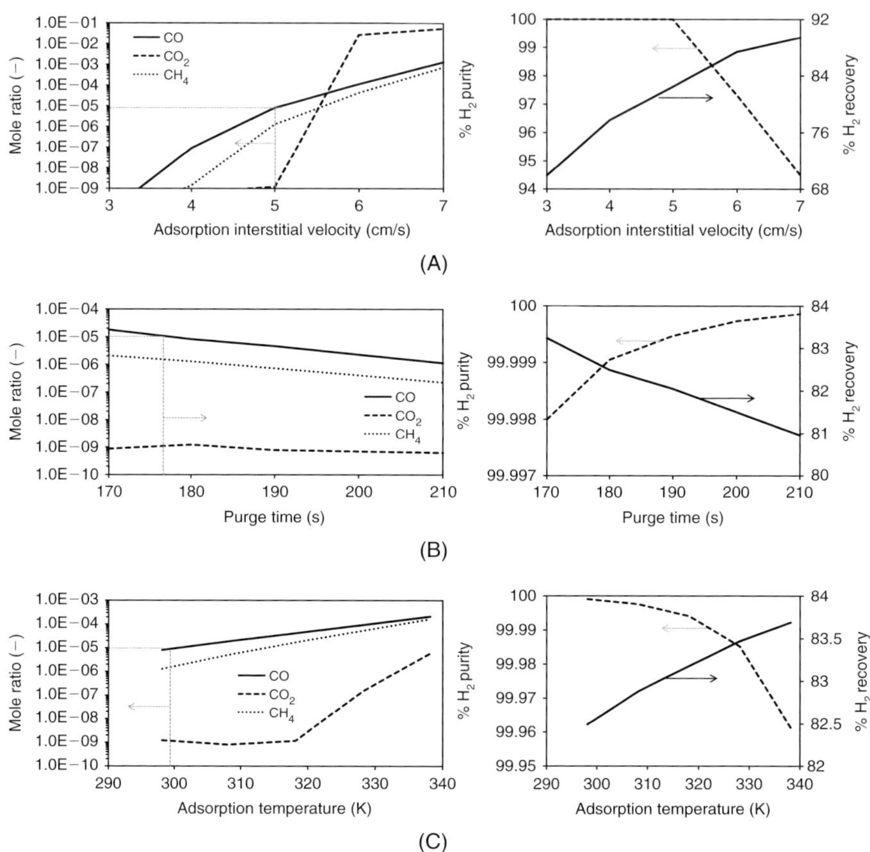

Figure 12.5 Effects of (A) adsorption velocity, (B) regeneration time, and (C) adsorption temperature on PSA separation performance. Basic PSA operating conditions, i.e., adsorbent ratio, 0.6; feed composition, $23\%CO_2/2\%CO/3\%CH_4/72\%H_2$; feed temperature, 298 K; feed pressure, 12 atm; purge pressure, 1 atm; feed pressurization time, 20 s; adsorption time, 180 s; first pressure-equalization time, 20 s; depressurization time, 20 s; purge time, 180 s; second pressure-equalization time: 20 s; adsorption and purge velocity, 5 cm/s, were used in the simulation. In the parametric studies, the investigated parameter was varied, while all others were kept constant.

lower than 10 ppm. CH_4 impurity was very low (mole ratio approximately 1.0×10^{-9}) when the adsorption temperature was below 318 K; thus, CH_4 could be considered completely removed at these temperatures. However, it increased at the adsorption temperatures above 318 K. With an increase in the temperature, the adsorption capacity of the adsorbents decreased, leading to shorter breakthrough time and hence lower product purity at the constant cycle time.

4.2.2. Effects of Adsorption Pressure and Adsorbent Ratio

The effects of adsorbent ratio and pressure on the PSA separation performance were investigated for constant feed and regeneration flow rates, as well as same step times. The results are shown in Fig. 12.6A–E. Here, the adsorbent ratio was defined as a ratio of activated carbon layer length to the packed bed length.

Figure 12.6A–C shows that high adsorbent ratio favors CO_2 and CH_4 removal, whereas low ratio is favorable for CO removal at all pressures used in this study. This is expected because the adsorption capacities of CH_4 and CO_2 are higher, whereas the capacity of CO is lower in the activated carbon compared to the zeolite 5A. The high interstitial velocity in the bed, which corresponds to the low operating pressure, e.g., $P = 8\,atm$, results in a complete or partial breakthrough of impurity, thus reducing the product purity that becomes too low to be used for fuel cell. However, at higher pressure ($P = 12, 16\,atm$), CH_4 impurity is always less than 15 ppm (mole ratio, 1.5×10^{-5}), and also, there is a range of adsorbent ratios (0.35–0.92 and 0.44–0.61 at $P = 16$ and 12 atm, respectively) that allows control of CO_2 and CO to the contents of less than 100 and 10 ppm, respectively. Such range is even broader at the higher adsorption pressures. There is also an optimum range of adsorbent ratios (0.39–0.90 and 0.50–0.80 at $P = 16$ and 12 atm, respectively, shown in Fig. 12.6B) for which minimum CO_2 impurity can be obtained, i.e., CO_2 can be effectively prevented from breaking through into the zeolite 5A layer.

As shown in Fig. 12.6D and E, hydrogen purity is increased by increasing pressure, while the product recovery is decreased. Such effect is similar to that of interstitial velocity as shown in Fig. 12.5A. With constant feed flow rate, the interstitial velocity is lower at the elevated adsorption pressure, leading to higher hydrogen purity and lower recovery. From Fig. 12.6D (and the inset), it can also be observed that for each adsorption pressure, there is one corresponding peak related to the hydrogen purity. Nearly, 100% H_2 could be obtained from layered-bed PSA within adsorbent ratio 0.39–0.7 at $P = 16\,atm$, whereas at $P = 12$ or 8 atm, there is a unique adsorbent ratio (0.48 and 0.8 at $P = 12$ and 8 atm, respectively) that generates the highest product purity. In the adsorption step of PSA operation, there is a critical impurity concentration wave velocity (see Eq. (13)), $\omega_{critical}$, where complete impurity removal is attained as $\omega_c \leq \omega_{critical}$, while impurity is present in the product as $\omega_c \, 2 \, \omega_{critical}$. At the higher adsorption pressures, e.g., $P = 16\,atm$, the interstitial velocity

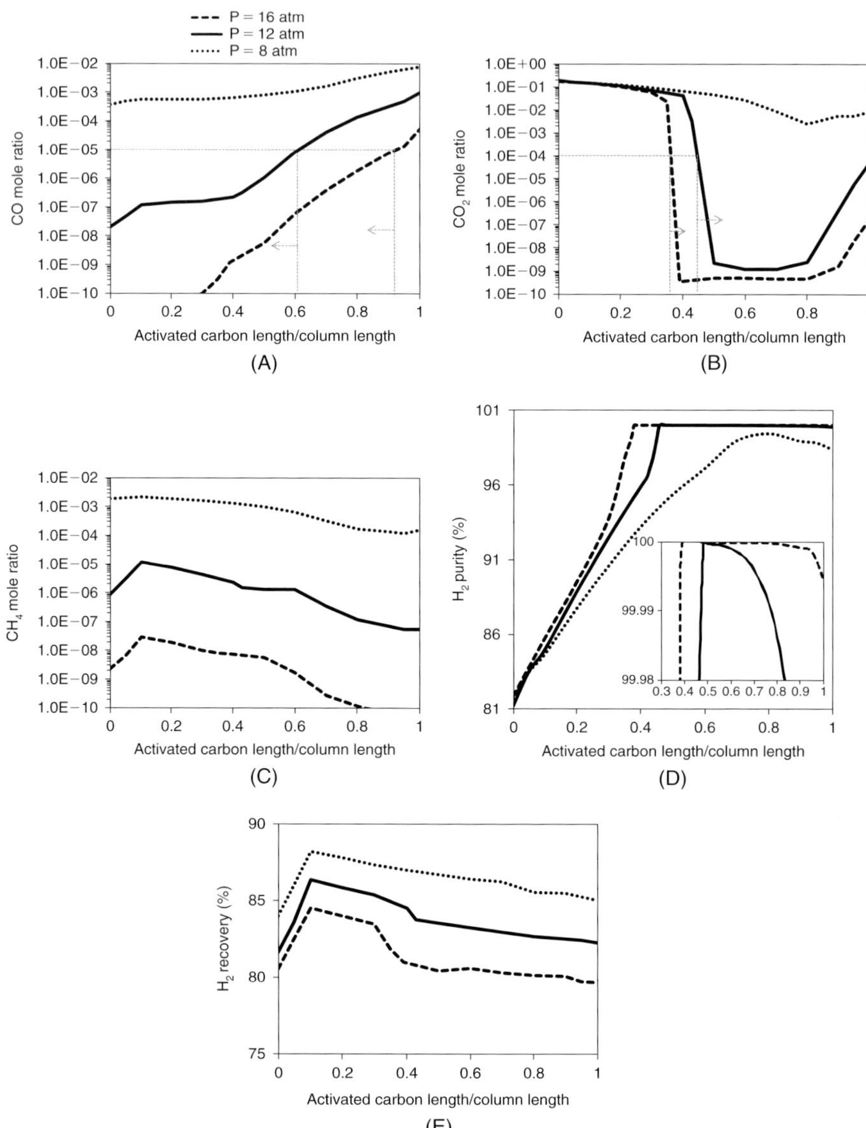

Figure 12.6 Effects of adsorption pressure and adsorbent ratio on PSA separation performance: impurities of (A) CO, (B) CO_2, and (C) CH_4, and (D) H_2 purity in the product, and (E) H_2 recovery. For PSA simulation conditions, see Fig. 12.5.

is lower and the resulting ω_c, for a range of adsorbent ratios, may be even lower than $\omega_{critical}$, indicating the complete impurity removal. On the other hand, it is impossible to obtain pure product at the lower adsorption pressures (e.g., $P = 8$ and $12\,\text{atm}$ corresponding to the higher interstitial velocities) because ω_c for all adsorbent ratios is higher than ω_{critical}. However, there may exist a unique adsorbent ratio that can generate the maximum product purity less than 100%. As can be seen from Fig. 12.6D, due to the breakthrough of impurity, the product purity dropped sharply when adsorbent ratios were lower than 0.38, 0.43, and 0.7 at $P = 16$, 12, and $8\,\text{atm}$, respectively. Furthermore, as can be seen from Fig. 12.6E, adsorbent ratio of 0.1 produced the highest hydrogen recovery for each adsorption pressure. These results show that a layered-bed PSA performs better than either single-layered activated carbon or zeolite 5A bed, with respect to purity or recovery.

4.2.3. Effects of Feed Composition and Adsorbent Ratio

Because both activated carbon and zeolite 5A exhibited the highest adsorption capacities for CO_2 among the four sorbates, three gas feeds with different compositions of CO_2, i.e., $16\%CO_2/2\%CO/3\%CH_4/79\%H_2$, $23\%CO_2/2\%CO/3\%CH_4/72\%H_2$, and $30\%CO_2/2\%CO/3\%CH_4/65\%H_2$ designated as feed A, B, and C, respectively, were used to study their effects on PSA separation performance. PSA operating conditions were kept the same in the simulation. The simulation results are illustrated in Fig. 12.7A–E.

As shown in Fig. 12.7A–C, the entire adsorbent ratio range could generate CH_4 impurity less than $15\,\text{ppm}$ for all CO_2 feed concentrations, whereas the adsorbent ratio range that could produce less than $100\,\text{ppm}$ CO_2 and $10\,\text{ppm}$ CO is more restricted at the higher CO_2 feed concentrations. The ratio ranges for production of fuel-cell-grade hydrogen are 0.23–0.50, 0.44–0.61, and 0.63–0.68 for CO_2 feed concentration of 16, 23, and 30%, respectively. The increase in CO_2 feed concentration requires an increase of the activated carbon layer, i.e., increase the adsorbent ratio for CO_2 removal. However, at the same time, this would increase CO impurity in the product. Therefore, for each feed concentration, the adsorbent ratio range of layered PSA should be carefully set for producing fuel-cell-grade hydrogen.

Similarly, as stated in the above section, an optimum adsorbent ratio exists corresponding to the highest hydrogen purity for each CO_2 concentration in the feed (see the inset in Fig. 12.7D). Hydrogen purity was found to be strongly dependent on CO_2 feed concentration and adsorbent ratio. It decreased substantially with an increase in CO_2 feed concentration, e.g., when the adsorbent ratio was below 0.26 for all CO_2 feed concentrations, and 0.26–0.4 for feeds B and C. This is attributed to the increased

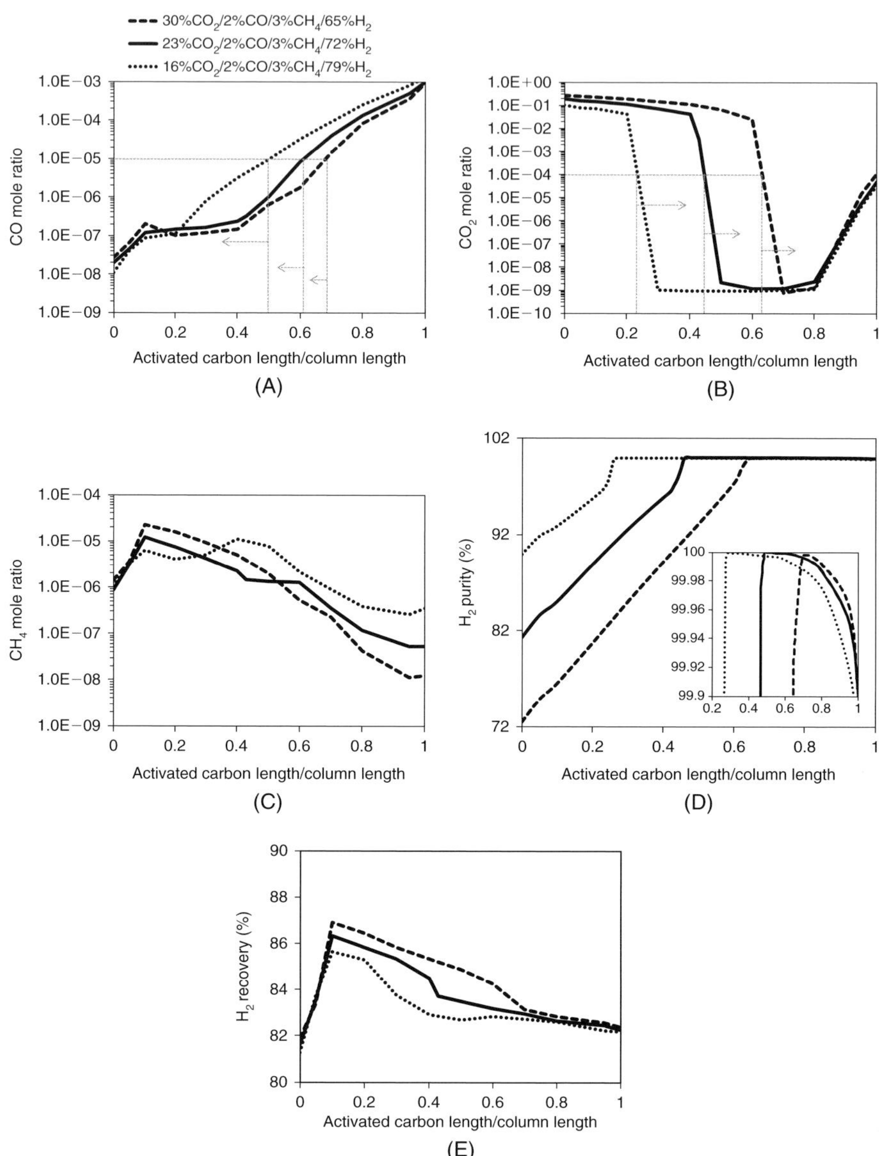

Figure 12.7 Effects of feed composition and adsorbent ratio on PSA separation performance: impurities of (A) CO, (B) CO_2 and (C) CH_4, and (D) H_2 purity in the product, and (E) H_2 recovery. For PSA simulation conditions, see Fig. 12.5.

extent of CO_2 and CH_4 breakthrough into the zeolite layer, which occurs at the higher CO_2 feed concentration and lower adsorbent ratio. However, hydrogen purities, in the case where the main impurity was CO, were in the following sequence feed C 1 feed A 1 feed B and feed A 1 feed B 1 feed C as the adsorbent ratio was set between 0.5 and 0.7 and higher than 0.7, respectively. As adsorbent ratio increased, activated carbon layer was more and more efficient in removing most CO_2 and CH_4; thus, the remaining gases entering into the zeolite layer comprises mainly CO and H_2. The lower CO_2 concentration in the feed exhibited higher interstitial velocity, due to the higher H_2 concentration. At the same time, the CO concentration into the zeolite layer was smaller compared with the case of the higher CO_2 in the feed. However, the high interstitial velocity dominated the column dynamics in the zeolite layer, leading to the earlier breakthrough of CO. Therefore, feed A case even showed slightly higher CO impurity than feed B and C cases at high adsorbent ratios.

The highest hydrogen recovery was observed, regardless of CO_2 feed concentration, if the adsorbent ratio was kept at 0.1 (see Fig. 12.7E). If the adsorbent ratio was set to a higher value than 0.1, then hydrogen recovery decreased with that ratio. Furthermore, hydrogen recovery increased at higher CO_2 feed concentration if the adsorbent ratio was kept at 0.1–0.7.

4.3. PSA Optimization

PSA processes are intrinsically dynamic, operating in a periodic fashion. The PSA separation performance depends on a number of design and operating parameters. gPROMS has been implemented in a number of PSA optimization studies [10–13]. Due to the limitation of gOPT – dynamic optimizer in gPROMS, e.g., limited optimization strategy and numerical solvers, only single-bed PSA process optimization was investigated in this study. For more complicated PSA system containing multiple beds and interactions among beds, e.g., pressure-equalization step, more specialized optimization strategy was proposed [20], which was not considered here. We attempted to apply gOPT to optimize a single-bed PSA system involving quaternary CO_2, CO, CH_4, and H_2 mixture operated in four-step Skarstrom cycle: (1) countercurrent pressurization with hydrogen product to high pressure, (2) concurrent adsorption at high pressure, (3) countercurrent depressurization to atmospheric pressure, and (4) countercurrent regeneration at atmospheric pressure. Similar to the two-layered-bed PSA system studied above, the single bed was packed with activated carbon followed by zeolite 5A. The mathematical model for the single-bed PSA process, which is somewhat different in comparison to the two-layered-bed PSA described in the theory section, was detailed by Huang et al. [13].

For single-bed PSA with product pressurization process, the equation of product recovery is different:

$$\text{Recovery}_{H_2} = \frac{\int_0^{t_A} \left(u\, c_{H_2} \right)\Big|_{z=L} dt - \int_0^{t_P} \left(u\, c_{H_2} \right)\Big|_{z=L} dt - \int_0^{t_{Pu}} \left(u\, c_{H_2} \right)\Big|_{z=L} dt}{\int_0^{t_A} \left(u\, c_{H_2} \right)\Big|_{z=0} dt} \times 100\% \quad (14)$$

In addition, the hydrogen product throughput is defined as

$$\text{Product} = \frac{\int_0^{t_A} \left(u\, c_{H_2} \right)\Big|_{z=L} dt - \int_0^{t_P} \left(u\, c_{H_2} \right)\Big|_{z=L} dt - \int_0^{t_{Pu}} \left(u\, c_{H_2} \right)\Big|_{z=L} dt}{t_{cycle}/3600} \pi R_{bed}^2 \quad (15)$$

where t_A, t_P, t_{Pu}, and t_{cycle} are adsorption, pressurization, purging, and cycle time of the four-step PSA, respectively.

The optimization of PSA model can be performed with respect to any number of performances such as product purity, product recovery, power, product throughput, etc. For performing PSA optimization, the PSA model was first simulated to CSS, which means that the mole fraction and solid concentration bed profiles were identical at the beginning and at the end of each cycle. The resulting bed profiles were fed to gOPT as an initial guess for the optimal CSS condition and were also discretized in the PSA model by method of lines. The chosen decision variables, such as cycle time, adsorption and regeneration velocities, adsorption pressure, etc. together with their constraints were added in gOPT. The optimization was then carried out by solving the objective function, as well as constraints on the variables. More details on optimization strategy were described in Ko et al. [11] and Huang et al. [13].

In this work, we investigated two optimization cases for separation of $23\%CO_2/2\%CO/3\%CH_4/72\%H_2$ based on maximum product purity and product throughput, and maximum product recovery. The constraints and objective functions of both cases are summarized in Table 12.4. Most constraints for both cases are the same except that case B includes impurity limits. It should also be noted that to find the desired optimal conditions, some constraints, such as adsorbent ratio, pressurization time, depressurization time, and regeneration pressure, were kept constant for the computation conversion. As discussed above, adsorbent ratio was an important factor affecting PSA separation performance. The value of adsorbent ratio

Table 12.4   Optimization constraints and objective functions for two cases

Case A: maximum hydrogen purity and throughput	Case B: maximum hydrogen recovery
	$0 \leq CO$ (ppm) ≤ 10
	$0 \leq CO_2$ (ppm) ≤ 100
	$0 \leq CH_4$ (ppm) ≤ 100
Column size (m): 1.0 (L) $\times$ 0.25 (ID)	
Adsorbent ratio: 0.6	
Temperature: 298 K	
Feed concentration: 23%CO_2/2%CO/3%CH_4/72%H_2	
Regeneration pressure (atm): 1.0	
$10 \leq$ adsorption pressure (atm) ≤ 14	
$0.005 \leq$ adsorption velocity (m/s) ≤ 0.15	
$-0.15 \leq$ purge velocity (m/s) ≤ -0.005	
$150 \leq$ adsorption time (s) ≤ 330	
$150 \leq$ purge time (s) ≤ 330	
Pressurization time: 20 s	
Depressurization time: 20 s	
$0\% \leq$ hydrogen recovery $\leq 100\%$	
$95\% \leq$ hydrogen purity $\leq 100\%$	
$0 \leq$ product throughput (mol/h) $\leq 1.0 \times 10^8$	

was determined beforehand in the CSS simulation so that CO_2 would not break into zeolite 5A layer, whereas CH_4 and CO would not break for the whole bed.

The optimization results are listed in Table 12.5. Compared with case B, case A required much higher computational time due to the relatively more complicated objective function. Moreover, the adsorption time and regeneration time/velocity were increased for obtaining higher product

Table 12.5   Dynamic PSA optimization results for two cases

	Case A	Case B
Adsorption pressure (atm)	12	12
Adsorption time (s)	258	183
Purge time (s)	324	172
Adsorption velocity (m/s)	0.048	0.045
Purge velocity (m/s)	-0.094	-0.046
CO impurity (ppm)	—	2.53
CO_2 impurity (ppm)	—	0.43
CH_4 impurity (ppm)	—	0.10
Hydrogen purity (%)	100	99.9997
Hydrogen recovery (%)	59.0	66.8
Product throughput (mol/h)	1135	854
Computation time on Intel Core2 Duo CPU with 2.40 GHz and 2.39 GHZ (s)	2224	873

throughput and product purity, respectively. However, the recovery for case A was lower than for case B. For case B, it was possible to obtain higher product recovery, as well as fuel-cell-grade hydrogen.

5. CONCLUSION

Single-component adsorption isotherms of CO_2, CO, CH_4, and H_2 in activated carbon and zeolite 5A were experimentally measured and described by the Langmuir model. The two adsorbents exhibited various affinities and adsorption capacities for the selected probe species. Therefore, their respective effects on impurity removal from PSA process had to be explained and clarified.

A theoretical two-layered-bed PSA model has been developed for studying impurity removal from SMR off-gas for producing fuel-cell-grade hydrogen, where CO_2, CO, and CH_4 concentrations were required to be less than 100, 10, and 100 ppm, respectively. The column investigated in this study was packed with activated carbon followed by zeolite 5A. The effects of velocity during adsorption step, regeneration time, feed temperature, pressure and composition, and adsorbent ratio on hydrogen purity, impurity, and recovery were investigated methodologically. The results showed that the layered-bed PSA process was capable of providing higher purity than activated carbon or zeolite 5A single-layered PSA bed under the same operating conditions. A range of adsorbent ratios was also determined, which was shown to provide means for controlling the impurity concentration to satisfy requirements for fuel-cell application. Such range was dependent on operating conditions, such as feed pressure and composition.

The optimization routines were also carried out to examine the optimal conditions for the separation of $23\%CO_2/2\%CO/3\%CH_4/72\%H_2$ mixture using single-bed PSA process with layered adsorbents operated in Skarstrom cycle. Using the dynamic optimizer in gPROMS, the optimal operation conditions for PSA process based on different objective functions could also be obtained. gPROMS proved to be an effective tool for PSA simulation and simple optimization. However, optimization of more complicated PSA processes, including multiple beds and operating steps, requires further study.

ACKNOWLEDGMENTS

The authors gratefully acknowledge Dr Daeho Ko from GS Engineering and Construction Corp., Korea for stimulating discussions on gPROMS optimization for his contribution. The authors also acknowledge NSERC for the financial support.

References

[1] Cheng X, Shi Z, Glass N, Zhang L, Zhang J, Song D, Liu Z-S, Wang H, Shen J. J Power Sources 2007;165: 739–56.

[2] Besancon BM, Hasanov V, Imbault-Lastapis R, Benesch R, Barrio M, Molnvik MJ. Int J Hydrogen Energy 2009;34:2350–60.

[3] ISO. Hydrogen fuel – product specification – Part 2: proton exchange membrane fuel cell applications for road vehicles, ISO/TC 197 N 352, 2006.

[4] SAE. Information report on the development of a hydrogen quality guideline for fuel cell vehicles (SAE-J2719), 2005.

[5] Sircar S, Golden TC. Sep Sci Technol 2000;35:667–87.

[6] Malek A, Farooq S. AIChE J 1998;44:1985–92.

[7] Lee C-H, Yang J, Ahn H. AIChE J 1999;45:535–45.

[8] Lee J-J, Kim M-K, Lee D-G, Ahn H, Kim M-J, Lee C-H. AIChE J 2008;54:2054–64.

[9] Ribeiro AM, Grande CA, Lopes FVS, Loureiro JM, Rodrigues AE. Chem Eng Sci 2008;63:5258–73.

[10] Ko D, Siriwardane R, Biegler LT. Ind Eng Chem Res 2003;42:339–48.

[11] Ko D, Siriwardane R, Biegler LT. Ind Eng Chem Res 2005;44:8084–94.

[12] Knaebel SP, Ko D, Biegler LT. Adsorption 2005;11(Suppl):615–20.

[13] Huang Q, Malekian A, Eic M. Sep Purif Technol 2008;62:22–31.

[14] Ruthven DM. Principles of adsorption and adsorption processes. New York: John Wiley and Sons; 1984. p. 209.

[15] Malekian A. Adsorption characterization of novel nanoporous materials and optimization of bulk PSA process to produce H_2 from plasma reactor. PhD thesis, University of New Brunswick, 2006.

[16] Yang J, Lee C-H. AIChE J 1998;44:1325–34.

[17] Park J-H, Kim J-N, Cho S-H, Kim J-D, Yang RT. Chem Eng Sci 1998;53:3951–63.

[18] Grande CA, F. Lopes VS, Ribeiro AM, Loureiro JM, Rodrigues AE. Sep Purif Technol 2008;48:1338–64.

[19] Farooq S, Huang Q, Karimi IA. Ind Eng Chem Res 2002;41:1098–106.

[20] Biegler LT, Jiang L, Fox VG. Sep Purif Rev 2004;33:1–39.

CHAPTER 13

Removal of Fine Particles on Fibrous Filters: A Review

Lucija Boskovic *and* **Igor E. Agranovski**
Griffith School of Engineering, Griffith University, Brisbane, Queensland, Australia

Contents

Modern societies have become more complex, and now require industrial processes that generate finer aerosols. Fibrous filters have long been recognized as an efficient means of removing aerosol particles from the carrier gas, or air stream. There are various mechanisms by which the aerosol particles are captured on a filter material, and these include interception, inertia, diffusion, electrostatic attraction, and gravity. The aerosol particle size, as well as filter properties, determines the removal mechanisms that dominate for a particular filtration application.

Nanoparticles have attracted the attention of an increasing number of researchers from several disciplines in the past 10 years. The term *nanoparticle* came into frequent use in the early 1990s together with the related concepts of "nanoscaled" or "nanosized" particles. Until then, the more general terms *submicron* and *ultrafine* particles were used. Nanoparticles, smaller than 100 nm, have various properties that differ from the corresponding bulk material, and this makes them attractive for many new electronic, optical, or magnetic applications. Nanoparticles with various geometric shapes can be produced by well-controlled processes [1–3]. Along with scientific excitement over the prospects of nanoparticles and nanotechnology, there have been increasing concerns regarding the risks this science may pose. Nanoparticles administered to the lung produce more potent adverse effects in the form of inflammation and subsequent tumors compared to larger-sized particles of identical chemical composition [4].

DOI: 10.1016/B978-0-08-054820-3.00013-7

Many theories have been developed to describe the particle interaction with the surface of the filter and to estimate the probability of the particle adhesion onto a surface [5–10]. However, the experimental data and the collection prediction for these particles are still inadequate [11]. Apart from that, some of the theories mentioned above have been proven to be misleading. So, according to Wang and Kasper theory [10], the thermal impact velocity of a particle will exceed the critical sticking velocity of particles, if the size of the particle is smaller than 10 nm. However, many authors [12–15] have shown that no thermal rebound could be detected in the particle size range below 10 nm, and that the above theory is erroneous. Balazy et al. [16] reported that thermal bounce occurred for 20-nm particles.

The collection efficiency of a fibrous filter depends on the structure of the filter, (porosity, fiber diameter, and filter thickness), on the operational conditions (filtration velocity, temperature, and humidity), and, in particular, on the filtering aerosol characteristics (particle density, size, and shape) [17–19].

Obviously, the particle size (and correspondingly the surface area) plays a crucial role in the bouncing processes. It was found that adhesion forces are greater when acting on larger size particles. At the same time, a drag force acting on the larger particles is also larger detaching them from the filter fibers at lower air velocity compared to the smaller ones [20].

The effect of particle shape, although very important, has not been investigated in detail. The classic filtration theory only considers spherical particles because fibrous filtration is a very complex process, and adding another particle-shape-related uncertainty would make theoretical approach far too complicated. As a result, most investigations on nonspherical particles consider simplest regular particle shapes [18]. In addition, the filtration theory does not account for possible particle bounce, assuming that it perfectly adheres to the fiber after initial collision. However, for some process conditions, this assumption is not valid and the particle bouncing could significantly alter the filter efficiency [21].

As discussed above, despite its demonstrated importance in filtration research, the detailed mechanisms of particle–fiber interaction after collision and the possibility for the particle to bounce have not been fully investigated. Earlier experimental studies of the collision process have focused on a droplet-surface collision that is quite different from a particle-surface collision [22]. In experiments conducted by [23–25], a more advanced technique, such as laser Doppler velocimetry, was used to measure the bounce of polystyrene latex (PSL) particles on polished quartz and stainless-steel surfaces. Some techniques for studying the impaction of aerosols onto a flat plate to compare bounce properties have been developed [26, 27]. The particle deformation during adhesion to a flat surface was examined by Tsai et al. [28] and Rimai et al. [29]. Although these investigations were important for understanding the general principles involved, they were not very useful in understanding the real processes occurring in filtration.

In cases discussed above, the mass of the particle was assumed to be negligible compared to the mass of the particle hosting surface; so, the surface was assumed to behave as a rigid body. However, in a case of collision of particle with filter fibers, they are able to deform or move, which alters the particle-surface interaction compared to the case of collision onto rigid surface.

The possibility of the particle bouncing depends on its composition, its shape, its velocity, and the type of impaction surface. When a solid particle contacts a surface at low velocity, the particle loses its kinetic energy by deforming itself and surface. At higher velocities, part of the kinetic energy is dissipated in the deformation process (plastic deformation), and part is converted elastically to kinetic energy of rebound. If the rebound energy exceeds the adhesion energy − the energy required to overcome the adhesive forces − a particle will bounce away from the surface. When the materials comprising the particle and surface become harder, the particle becomes larger, or its velocity becomes greater, then the likelihood of the particle bouncing from the surface also becomes greater [19].

There are two classical approaches to describe particle bounce. The first approach defines a critical velocity V_c, above which bounce will occur [30, 31],

$$V_c = \frac{\beta}{d_p} = \frac{1}{d_p} \frac{\left(1 - e_{pl}^2\right)}{e_{pl}^2} \frac{A}{\pi x^2 \left(6 p_{pl} \rho\right)^{\frac{1}{2}}} \tag{1}$$

where β is a constant for a particular impaction surface, d_p is the particle diameter, e_{pl} is the coefficient of restitution (for plastic deformation only), A is the Hamaker constant, p_{pl} is the microscopic yield pressure, ρ is the particle density, and x is the distance between the particle surface and its adjacent surface. Hamaker constants are given in the literature for a limited number of elements and compounds [28].

The other method involves the kinetic energy (KE_b) required for bounce to occur when a particle (d_p) collides with a surface [23], and the magnitude of kinetic energy can be calculated as

$$KE_b = \frac{d_p A\left(1 - e^2\right)}{2xe^2} \tag{2}$$

where e is the coefficient of restitution (plastic and elastic deformation), which is equal to the ratio of the rebound velocity to the approach velocity. The value of e is reported to range from 0.73 to 0.81, although these values were derived using hard impaction surfaces [9]. It is reported that A and e ought to be determined experimentally because it is very hard to determine them theoretically.

The nature of the particle motion along the fiber surface is important. For perfectly spherical particles, the type of motion of the particle in contact with fiber is either sliding or rolling. However, for both forms of the motion of the spherical particle, the area of the particle contact with the fiber does not change during the motion. Any deformation will increase the time in contact and thus provide greater opportunity for sliding motion to convert into rolling motion. However, even for large spheres, it is difficult to visually distinguish sliding and rolling objects.

A totally different picture would occur in the case of nonspherically shaped particles (cubes, for example), which either slide on a face, or tumble on a corner or edge. During tumbling, the area of the contact changes significantly depending on the form of contact, and may affect the bounce probability, and therefore, the filtration efficiency.

Boskovic et al. [32] investigated the influence of particle shape on filtration processes. Two types of particles, including spherical PSL and iron oxide, and perfect cubes of magnesium oxide of the same electrical mobility diameter in the size range of 50–300 nm, were examined. It was found that the removal efficiency of spherical particles on fibrous filters is very similar for corresponding sizes regardless to the fact that the densities of PSL and iron oxide differ by a factor of five. On the other hand, the removal efficiency of MgO cubic particles was measured to be much lower compared to the similar spheres. This difference becomes more obvious with an increase of particle size from 50 to 300 nm. All experiments were conducted at a filtration velocity of 2 cm/s to avoid the influence of inertial effects. This disparity was ascribed to the different nature of motion of the spherical and cubic particles along the fiber surface, following the initial collision. After touching the fiber surface and before coming to rest, the spherical particles could either slide or roll compared to the cubic ones, which could slide or tumble. During tumbling, the area of contact between the particle and the fiber changes significantly, thus affecting the bounce probability, whereas for the spheres, the area of contact remains the same for any point of particle trajectory. The particle kinetic energy in the flow was chosen to indicate when the tumbling is significant for the filtration efficiency. This parameter does not play any role for the actual filtration of the particles, which could not tumble, as the inertial mechanism for particle removal is negligibly small within the entire range of particle sizes investigated by Boskovic et al. [32].

The experimental work has also been extended to include sodium chloride (NaCl) particles of intermediate shape (cubic particles with rounded edges), and all the three types of particles (PSL, MgO, and NaCl) have been tested at filtration velocities ranging from 5 to 20 cm/s [33]. Particles of NaCl are being captured with efficiencies lower than those of PSL particles but higher than the efficiencies of cubic MgO particles, at the lowest filtration velocity. The difference between the filter efficiencies for the collection of MgO and NaCl particles decreases with an increase in velocity. With velocity increase, the filtration efficiency of the cubic MgO particles exceeds the filtration efficiency for the intermediate-shaped NaCl particles because of the dominating inertial effects of the denser MgO particles of similar size.

The nature of the contact surface is also important in the bouncing processes. There has been some research on the effect of coating filter fibers with a liquid, and its influence on particle bounce. Walkenhorst [34] examined the effect of coating of model wire filters with vegetable oils, mineral oils, and Vaseline. He reported that the former two substances increased filtration efficiency while the latter did not. Although oil-coated filters are suitable for laboratory processes, using such liquids in industry would not be advantageous. Agranovski and Braddock [35] stated that liquid films coating the fiber could reduce the likelihood of particle bounce, although this has yet to be proven experimentally. Agranovski et al. [36] developed a process in which the filter is coated with a thin layer of water, allowing collection of aerosols onto the water film rather than directly by the solid fiber surface. These technologies are industrially applicable as quantities of water for filter irrigation are more readily available and recyclable in industry than oils and greases.

Mullins et al. [21] examined the effect of particle bounce in fibrous filter systems by comparing the filtration efficiencies of solid PSL particles with liquid diethylhexyl sebacate particles of the same size and shape factor under identical filtration conditions. The experiments were conducted first in dry filter systems, then in wet systems. They concluded that for the wet filtration regime, the efficiencies do not significantly differ with changing particle type, but bouncing was greatly reduced in wet filter systems. The water film acts to inhibit bounce by either aiding energy dissipation or preventing the particle being repelled from the fiber. Factors such as surface chemistry and surface roughness of the fibers could have a significant effect on wetting and wettability. However, most models assume a homogeneous fiber substrate because this is the least complex case [37].

Boskovic et al. [33] tested the influence of particle shape and filtration velocity on the filtration efficiency of oil-coated fiber filters at different filtration velocities. The results of that investigation are important for better understanding of causes of substantial difference in filtration of particles with cubic and spherical shapes. It gives a clear answer to the question: Does the difference in filtration efficiency relate to the primary filtration mechanisms or secondary effects of bouncing and re-entrainment? The mineral oil was selected as it gives a uniform layer on the fiber surface and does not readily run down the fiber, as water would do, over the period of the experiment.

A thin coating of the polypropylene filter with a mineral oil was used to absorb the energy of collision and to minimize the particle bounce. The filtration efficiency of spherical PSL, cubic MgO, and intermediate NaCl particles was measured in the same size range of 50–300 nm, for a filtration velocity of 10 and 20 cm/s. It was found that tested particles, regardless of shape, have very similar filtration efficiencies. The curves are all close together and the standard deviations have overlap, showing insignificant discrepancy in the results compared to the findings of [32, 38]. Also, the experimental results were compared with theoretical filter efficiency estimated according to the classic approach [19]. For calculations, all parameters of filter were adjusted to take into account the alterations, due to fiber coating. The theoretical predictions are in good agreement with experimental results. The emission of oil was not influencing the process as was verified by "zero" emission from the coated filter in absence of test aerosol supply. The results of this experiment have shown that the oil coating minimizes the amount of particle motion along the fiber after initial collision, making results for all particle shapes similar.

As mentioned above, collection of particles in fibrous filters is a very complex problem and due to that the particle shape is generally assumed to be spherical [17, 18]. However, aerosols consisting of irregular agglomerates are frequently formed in various processes, including coal combustion for power generation, pigment production, pharmaceutical drug delivery, diesel emissions, and others [39, 40]. The largest amount of atmospheric carbon black particles is emitted by diesel engines [41]. It has been shown that particle surface area is the most important parameter in terms of nanoparticle toxicity [42]. Thus, agglomerates are also expected to be more toxic than spherical particles with the same mass. It is well known that the agglomerated diesel exhausts are particularly hazardous as they can contribute to the development of lung cancer, the reduction of visibility, global climate change, soiling of buildings, damage to the materials exposed, and reduction of the ground water quality [43–45]. To account for nonspherical particle shape, some investigations have been recently performed, and their results are discussed here.

Fu et al. [46] studied filtration measurements of chain-aggregated aerosols passing through a screen-type diffusion battery at various face velocities. Also, they investigated the average orientation angle of the particles with respect to the flow field. The iron oxide chain aggregates had mean lengths in the range of 1500–3200 nm. The primary sphere size ranged from 17 to 41 nm. The results were compared with predictions based on the model derived by Chan and Dahneke [47] for the friction coefficient for chain aggregates. The only unknown parameter required for predicting of the single-fiber efficiency was the interception parameter. This parameter was estimated by matching the theoretical single-fiber efficiency with experimental data. They found that interception was the dominant deposition mode and that with increasing face velocity, both the filtration efficiency and average orientation angle decrease.

Using this method for determination of the interception equivalent diameter is very complicated as it requires realistic description of all deposition mechanisms for every kind of nonspherical particle. On the other hand, the results show that although the particles have a simple shape, they show complex deposition behavior due to particle alignment in the flow.

Lange et al. [48] compared the filtration efficiency of fibrous filter for carbon agglomerates and for spherical particles as a function of the mobility diameter. They derived a filtration model based on interception equivalent diameter of agglomerates, which was defined as a diameter of a spherical particle having the same penetration as that of the agglomerate. They showed that the concept of equivalent diameters can be used to predict filter penetration for nonspherical particles if both the interception and mobility diameters could be obtained. However, the authors made the assumption that the particles are not subject to any alignment. They expected that the concept of equivalent diameters will lead to satisfactory results if the particles are elongated.

Kim et al. [49] studied the filtration efficiency of silver agglomerates, with the agglomerate structure controlled by changing the temperature of a sintering furnace. Irregular agglomerates were spherically shaped by heat treatment at $600\,^{\circ}\mathrm{C}$, allowing direct comparison of the filtration of the agglomerate and a sphere with the same mobility diameter. The authors compared experimental and theoretical results in terms of impaction and interception, and similarly to Lange et al. [48] assumed identical diffusion parameters for spherical and nonspherical objects. An optimal (for filtration process) particle orientation to the flow was considered at the point of contact with the filter as L was taken as a maximum projected length of the agglomerate. However, neglecting the influence of diffusion deposition is quite questionable assumption, as an oscillating nonspherical particle is certainly different compared to the spherical one for filtration processes. As discussed above, even particles with a simple shape may show very complex deposition behavior due to the particle alignment in the flow [46, 50].

There are few different techniques that have been used lately to evaluate dynamic behavior of nonspherical particles. One of them considers the particle as porous sphere characterized by an intrinsic permeability, k, which is defined by Darcy's law and its Brinkman modification [51, 52]. However, this approach is strictly limited to isotropic structures [53].

The second approach is based on developing a Brownian dynamic algorithm, taking into account hydrodynamic coupling between the particles [54, 55]. However, this approach is restricted to relatively simple aggregate shapes. A novel simulator capable to characterize translational and rotational motion of aggregates was proposed by Moskal and Payatakes [53]. The simulator was used to determine the dynamics of aerosol aggregates, shape, and internal structure to find the corresponding diffusion coefficient values. Each aerosol aggregate was assumed to be a rigid body composed of identical primary particles. Balazy and Podgorski [45] theoretically investigated an influence of various aggregate parameters (fractal dimension, primary particle radius), as well as filter-fiber diameter and filtration velocity on the removal efficiency. Their results indicate that an aggregate morphology, filter structure, and process conditions strongly influence the aggregate deposition efficiency; however, no experimental verification of the findings was provided.

The third way of modeling of nonspherical particle motion and deposition is based on the consideration of equivalent spherical particle, which has the same translation properties as the nonspherical object. Goldenberg and Shapiro [56] tested the applicability of the equivalent diameter model to describe the deposition of fibrous aerosols from turbulent air flow in a pipe using a correlation technique developed for spherical particles. They found that particle deposition was underestimated because interception was not taken into account correctly.

Boskovic et al. [57] utilized recently developed method of nanoparticle generation by metal combustion, which allows production of loose and agglomerated aerosols with a regular morphology, enabling some theoretical characterization of both types of objects. They compared experimental filter efficiency results for loose and agglomerated MgO particles against theoretical predictions in the region where diffusion and interception filtration mechanisms are dominant. It was shown that nanoparticle agglomerates are removed much more efficiently compared to the regularly shaped nanoparticle at two filtration velocities tested. Then it was shown that the theoretical predictions, which are based on the assumption that the particles are spherical, significantly underestimate filtration efficiency values for the removal of elongated agglomerates of the same mobility. To account for such difference, the following issues were considered. The interception mechanism is crucial for nonspherical particles because such particles have larger spatial extension due to orientation effects compared with spherical particles of the same mobility diameter. The probability of capturing a rotating particle is proportional to the surface area of the particle. In the case of diffusion filtration, this is a total particle surface area, while in the case of interception, this is the particle projection onto a plane perpendicular to a streamline.

The probability of capturing of a particle revolving in the gas carrier has to be considered. In the case of spherical particle, the total single-fiber efficiency is calculated by the following equation:

$$E_\Sigma = E_\mathrm{D} + E_\mathrm{R} \tag{3}$$

where E_D and E_R are the single-fiber efficiencies due to diffusion and interception, respectively [19].

As particles used in the experiments were not spherical, Eq. (3) needs to be modified in some way to account for the particle shape. Deriving theoretical forms for agglomerated particles for E_D and E_R seems to be impossible with the current state of knowledge. However, some outcomes may be gleaned by fitting parameters to better match experimental results. It was considered that

$$E_{\Sigma}^{ns} = k_1 E_D + k_2 E_R \tag{4}$$

where E_{Σ}^{ns} is the single-fiber efficiency for nonspherical particles, k_1 and k_2 are coefficients to be obtained by fitting to the experimental data. Then, the filter efficiency for nonspherical particles, E_T^{ns}, can be calculated as

$$E_T^{ns} = 1 - \exp\left[\frac{-4E_{\Sigma}^{ns}\alpha t}{\pi d_f(1-\alpha)}\right] \tag{5}$$

and this expression can be fitted to the experimental data.

The suggested approach looks very promising for a filter design procedure for industrial applications. To use it, particles released from a particular source could be collected and characterized by using widely available techniques, including surface microscopy, NMR, and others. Then images could be digitized, and both three-dimensional surface and two-dimensional projection to any plane could be precisely determined by various types of software (for example, AutoCAD). Based on these results, k_1 and k_2 fitting coefficients could be obtained and used for theoretical predictions of the filter efficiency for particular process. Obviously, in many applications, it could lead to utilization of a less-expensive filter media, which would still be meeting all air-quality requirements applied to a particular source of aerosol emission.

It should also be stressed that suggested time-consuming procedure of particle characterization does not need to be repeated for each site. It could be done once for particles released from industrial process of interest and then used for similar processes at other locations.

References

[1] Altman IS, Jang Y-H, Agranovski IE, Yang S, Mansoo C. Stabilization of spinel structure during combustion synthesis of iron nanooxides. J Nanopart Res 2004;6:633–7.
[2] Altman IS, Agranovski IE, Choi M. On nanoparticle surface growth: MgO nanoparticle formation during a Mg particle combustion. Appl Phys Lett 2004;84:5130–2.
[3] Altman IS, Agranovski IE, Choi M. Mechanism of nanoparticle agglomeration during the combustion synthesis. Appl Phys Lett 2005;87:53–104.
[4] Stern ST, McNeil SE. Nanotechnology safety concerns revisited. Toxicol Sci 2008;101(1):4–21.
[5] Bradley RS. The cohesive force between solid surfaces and the surface energy of solids. Philos Mag 1932;13:853–62.

[6] Hamaker HC. The London–Van der Waals attraction between spherical particles. Physica 1937;4:1058–72.

[7] Johnson K, Kendall K, Roberts AD. Surface energy and the contact of elastic solids. Proc R Soc Lond A 1971;324:301–13.

[8] Dahneke B. Particle bounce or capture – search for an adequate theory: I. Conservation-of-energy model for a simple collision process. Aerosol Sci Technol 1995;23:25–39.

[9] Wall S, John W, Wang H-C, Goren S. Measurements of kinetic energy loss for particles impacting surfaces. Aerosol Sci Technol 1990;12:926–46.

[10] Wang H-C, Kasper G. Filtration efficiency on nanometer-size aerosol particles. J Aerosol Sci 1991;22:31–41.

[11] Friedlander SK, Pui DYH. Emerging issues in nanoparticle aerosol science and technology. J Nanopart Res 2004;6:313–20.

[12] Heim M, Mullins BJ, Wild M, Meyer J, Kasper G. Nano-scale aspects of aerosol filtration in fibrous filters. Aerosol Sci Technol 2005;39(8):782–9.

[13] Wang J, Chen D-R, Pui DYH. Modeling of filtration efficiency of nanoparticles in standard filter media. J Nanopart Res 2007;9(1):109–15.

[14] Japuntich DA, Franklin LM, Pui DYH, Kuehn TH, Kim SC, Viner AS. A comparison of two nano-sized particle air filtration tests in the diameter range of 10 to 400 nanometers. J Nanopart Res 2007;9:93–107.

[15] Huang S-H, Chen C-W, Chang C-P, Lai C-Y, Chen C-C. Penetration of 4.5 nm to 10 um aerosol particles through fibrous filters. J Aerosol Sci 2007;38:719–27.

[16] Balazy A, Podgorski A, Gradon L. Filtration of nanosized aerosol particles in fibrous filters. I. Experimental results. J Aerosol Sci EAC Proc 2004;II:S967–80.

[17] Davies CN. Air filtration. London: Academic Press; 1973.

[18] Brown RC. Air filtration: an integrated approach to the theory and applications of fibrous filters. Oxford: Pergamon Press; 1993.

[19] Hinds WC. Aerosol technology: properties, behaviour, and measurement of airborne particles. New York: John Wiley and Sons; 1999.

[20] Mullins ME, Michaels LP, Menon V, Locke B, Ranade MB. Effect of geometry on particle adhesion. Aerosol Sci Technol 1992;17:105–18.

[21] Mullins BJ, Agranovski IE, Braddock RD. Particle bounce during filtration of particles on wet and dry filters. Aerosol Sci Technol 2003;37:1–14.

[22] Gillespie T, Rideal E. On the adhesion of drops and particles on impact at solid surfaces. J Colloid Sci 1955;10:281–98.

[23] Dahneke B. The capture of aerosol particles by surfaces. J Colloid Interface Sci 1971;37:342–53.

[24] Dahneke B. Measurements of the bouncing of small latex spheres. J Colloid Interface Sci 1973;45(3):584–90.

[25] Dahneke B. Further measurements of the bouncing of small latex spheres. J Colloid Interface Sci 1975;51(1):58–65.

[26] Wang H-C, Walter J. Comparative bounce properties of particle materials. Aerosol Sci Technol 1987;7(3):285–99.

[27] Xu M, Willeke K. Technique development for particle bounce monitoring of unknown aerosol particles. Aerosol Sci Technol 1993;18:129–42.

[28] Tsai C-J, Pui DYH, Liu BYH. Elastic flattening and particle adhesion. Aerosol Sci Technol 1991;15:239–55.

[29] Rimai DS, Demejo LP, Bowen RC. Mechanics of particle adhesion. J Adhes Sci Technol 1994;8(11):1333–55.

[30] Cheng YS, Yeh YC. Particle bounce in cascade impactors. Environ Sci Technol 1979;13(11):1392–6.

[31] Hiller R, Loeffler F. On the degree of impact and the degree of adhesion on particle separation on filter fibres. Staub Reinhalt Luft 1980;40(9):405–11.

[32] Boskovic L, Altman IS, Agranovski IE, Braddock RD, Myojo T, Choi M. Influence of particle shape on filtration processes. Aerosol Sci Technol 2005;39:1184–90.

[33] Boskovic L, Agranovski IE, Braddock R. Filtration of nanosized particles with different shape on oil coated fibers. J Aerosol Sci 2007;38:1200–29.

[34] Walkenhorst W. Investigations on the degree of adhesion of dust particles. Staub Reinhalt Luft 1974;34:149–53.

[35] Agranovski IE, Braddock RD. Filtration of mists on wettable fibrous filters. AIChE J 1998;44(12):2775–83.

[36] Agranovski IE, Braddock RD, Myojo T. Removal of aerosols by bubbling through porous media. Aerosol Sci Technol 1999;31(4):249–57.

[37] McHale G, Newton M, Carroll B. The shape and stability of small liquid drops of fibers. Oil Gas Sci Technol 2001;26(1):47–54.

[38] Boskovic L, Agranovski IE, Altman IS, Braddock RD. Filter efficiency as a function of nanoparticle velocity and shape. J Aerosol Sci 2008;39:635–44.

[39] Brasil AM, Farias TL, Carvalho MG. Evaluation of the fractal properties of cluster–cluster aggregates. Aerosol Sci Technol 2000;33:440–54.

[40] DeCarlo PF, Slowik JG, Worsnop DR, Davidovits P, Jimenez JL. Particle morphology and density characterization by combined mobility and aerodynamic diameter measurements. Part 1: theory. Aerosol Sci Technol 2004;38:1185–205.

[41] Xiong C, Friedlander SK. Morphological properties of atmospheric aerosol aggregates. Proc Natl Acad Sci 2001;98:11851–6.

[42] Oberdorster G, Oberdorster E, Oberdorster J. Invited review: nanotoxicology: an emerging discipline evolving from studies of ultrafine particles. Environ Health Perspect 2005;113:823–39.

[43] Neeft JPA, Makkee M, Moulijn JA. Diesel particulate emission control. Fuel Process Technol 1996;47:1–69.

[44] Mauderly JL. Diesel emmision: is more health research still needed? Toxicol Sci 2001;62:6–9.

[45] Balazy A, Podgorski A. Deposition efficiency of fractal-like aggregates in fibrous filters calculated using Brownian dynamic method. J Colloid Interface Sci 2007;311:323–37.

[46] Fu T-H, Cheng M-T, Shaw DT. Filtration of chain aggregate aerosols by model screen filter. Aerosol Sci Technol 1990;13:151–61.

[47] Chan PF, Dahneke B. Free molecule drag on straight chains of uniform spheres. J Appl Phys 1981;52:3106–10.

[48] Lange R, Fissan HJ, Schmidt-Ott A. Predicting the collection efficiency of agglomerates in fibrous filters. Part Part Syst Charact 1999;16:60–5.

[49] Kim SC, Wang H-C, Emery MS, Shin WG, Mulholland GW, Pui DYH. Structural property effect of nanoparticle agglomerates on particle penetration through fibrous filter. Aerosol Sci Technol 2009;43:344–55.

[50] Kousaka Y, Endo Y, Ichitsubo H, Alonso M. Orientation specific dynamic shape factors for doublets and triplets of spheres in the transition regime. Aerosol Sci Technol 1996;24:36–44.

[51] Kim AS, Stolzenbach KD. The permeability of synthetic fractal aggregates with realistic three-dimension structure. J Colloid Interface Sci 2002;253:315–28.

[52] Vainshtain P, Shapiro M, Gutfinger C. Drag resistance of porous aggregates in slip regime. Abstr Eur Aerosol Conf 2003;1:S569–70.

[53] Moskal A, Payatakes AC. Estimation of the diffusion coefficient of aerosol particle aggregates using Brownian simulation in the continuum regime. J Aerosol Sci 2006;37:1081–101.

[54] Hijazi A, Zoaeter M. Brownian dynamics simulation of rod-like particles in the dilute flowing solution. Eur Polym J 2002;38:2207–11.

[55] Beard DA, Schlick T. Inertial stochastic dynamics. I. Long-time-step methods for Langevin dynamics. J Chem Phys 2000;112(17):7313–22.

[56] Goldenberg M, Shapiro M. Applicability of the equivalent diameter model for description of deposition of nonspherical particles from turbulent air flows. J Aerosol Sci 1991;22(Suppl. 1):S157–60.

[57] Boskovic L, Altman IS, Braddock RD, Agranovski IE. Removal of elongated particle aggregates on fibrous filters. Clean-Soil Air Water 2009;37(11):843–9.

CHAPTER 14

On the Relationship between Social Ethics and Environmental Nanotechnology

Janne Nikkinen

Postdoctoral Researcher, Faculty of Theology, University of Helsinki, Finland

Contents

1. INTRODUCTION

During the last decade, billions of dollars have been invested in nanotechnology research. With a $1 trillion impact on the world economy, and the potential to employ two million workers of whom 50% would work in the United States [1], it is understandable that expectations are high for this "revolution" to enhance productivity and provide both public and individual profits. One commentator has noted that "[a]ccording to some of its proponents, nanotechnology will cure cancer and heart disease, reverse pollution, feed the world and provide cheap – even free – consumer goods … it sometimes seems as though there is almost nothing that it [nanotechnology] cannot do [2]." In addition, nanotechnology is said to be have "tremendous potential" for helping the global poor, especially those approximately

2.5 billion people living on less than \$2 ppp per day (ppp = purchasing power parity, the long-term equilibrium exchange rate of two currencies to equalize their purchasing power) [3]. The list continues with claims that "[n]anomachines will repair damaged human cells on the molecular level, thus healing injury, curing disease, prolonging life, or perhaps annihilating death altogether [4]." With these and other grandiose promises, it is understandable that the public is anxious to see progress in this field.

There are obstacles, however, that may hinder the expected progress. So far, warnings of threats to public health have not included the hazards caused by nanotechnology. For example, the top 10 public health hazards published by the World Health Organization (WHO) in the World Health Report 2002 consist of being underweight, unsafe sex, high blood pressure, and similar issues, but not nanotechnology [5]. However, in a Global Risk Network Report published in early 2009, nanotechnology risks were included for the first time [6]. After the heated debate on genetically modified (GM) food during the 1990s, the general public views all new technologies with a certain amount of weariness. At best, the reaction to nanotechnology is somewhat mixed, perhaps, in part, because of its great promise and also because the public hopes that nanotechnologies may offer solutions to profound problems.

There are also differences in how the public understands the claims of nanotechnology researchers. Those working in the field may see their task as of one merely communicate empirical research results with education of the public a matter for the media and other interested parties. However, in a subconscious way, there are several quasi-ideological claims that are advanced within the scientific community. These include purely ideological beliefs, as well as the claims that may be true, but so far are without solid evidence. This chapter points out some problem areas in the current discussion, with remarks about furthering the discussion of how to combine ethical aspects with research. I hope thereby to advance discussion of the proper way to interact with the public and show whether there are grounds for further research in this area.

2. GENERAL OVERVIEW

2.1. The Issues

There are many unresolved issues within the scientific community concerning how nanotech research should be understood. The first major issue is the proper definition of nanotechnology. It is practically impossible to find a conference or a meeting with "nanotechnology" in its title, without also finding a discussion about defining the field. There is also the question about the proper identity of those who work with nanotechnologies. Many of these individuals identify themselves as chemists, physicists, or other types of specialists, but not as "nanotechnologists." Further complicating the matter is that the field of nanotechnology combines researchers from different disciplines, different countries, and different personal ideologies and interests. Whether nanotechnology will radically alter life on earth as we know it or merely improve our daily living is also a matter of some disputes. Furthermore, philosophical disagreement exists over whether chemistry will be totally mechanized or whether molecules are to be controlled [7]. Some researchers have also pointed out that in chemistry, matter has been manipulated on an atomic scale for hundreds of years [8]. Thus, aside from new labels, there is nothing new under the sun.

I will not deal with all these aforementioned issues here, but defining nanotechnology is relevant for an ethical evaluation of nanotechnology. According to Joachim Schummer [9], a definition (or avoiding a clear definition) shapes one's perception of issues in an ethical sense. Schummer states that the term *nanotechnology* in itself "does not denote an established research field but is rather a term used by governments to describe their research funding priorities, definitions may be tailored so as to cope with the ethical sensitivities of their publics [...]." Furthermore, because nanotechnology has been hailed as the next industrial revolution, "any new technological product associated with nanotechnology may be supposed to bear new kinds of risks and to require new regimes of evaluation" [9]. In addition, to claim that "nanotechnology will be the next industrial revolution" is already a positive socioethical evaluation [10].

The origins of nanotechnology are also important to pin down in this context because they have at least some implicit influence on the identity of those who are considered nanotechnologists or who see themselves as working in this field. In the literature about nanotechnology, there are many references to an ambiguous dinner speech made by physicist Richard Feynman entitled "There Is Plenty of Room at the Bottom" in 1959 as the starting point for the field of nanotechnology. However, as Colin Milburn, Curys Mody, and Ed Regis [4, 11, 12] have noted on different occasions, Feynman apparently obtained at least some of the ideas expressed in his speech from Robert A. Heinlein's short story *Waldo*, published in 1942. In that science fiction novella, Heinlein writes about microscopic surgery using microscopic machines. Although Feynman credited Al Hibbs with this idea, it seems that Feynman actually got his inspiration from the science fiction stories of the 1940s. It is certain that a professor from Tokyo Science University, Norio Taniguchi, used the term to describe the precision manufacture of materials with nanometer tolerances, but few would have known about Taniguchi without Eric Drexler's reference to him in his book *Engines of Creation* in 1986 [13]. Science and science fiction therefore mix in the history of nanotechnology.

The identity of those working in the field of nanotechnology also relates to other dimensions besides the question of whether nanotechnology can be traced back to Feynman, Taniguchi, or Drexler or its researchers' occupational titles. The role of nationality is one such dimension. What is in the interest of countries may not be the interest of individuals. In nanotechnology publications, it is often suggested that there is "a competition" in investments between countries or even regions, such as between the United States and the European Union. It is unlikely that many researchers will see the situation as so competitive, partly, because the role of nation-states is diminishing in the current political climate and, partly, because many Europeans do not consider themselves supervisors of the interests of other nations located on the same continent. Those involved in academic research often move from one institution to another based on factors related to the academic working environment, without paying much attention to the true or expected increase in competition that nanotechnology research gives their host countries. Individual profit and personal research interests may have significant roles in such personal decisions than those what governments regard as important.

Furthermore, in the debate about the social and ethical impact of nanotechnology, one hope consistently expressed by scientists has been "educating the public to understand nanotechnology" [10]. Although this may important, it is worth discussing of those challenges that researchers in this field face when interacting with each other in the scientific community, and only then turn the relations with the public. There are only rare occasions when those involved in basic nanotechnology research work engage with the public. There have been some experiments in the field of bionanotechnology or nanomedicine (the molecular-level manipulation of human biosystems), but perhaps less in other sectors in involving the public [14]. The assumption in this context is that with adequate information and accurate reporting, the public will comprehend the issue of nanotechnology and skeptics will eventually be convinced about the benefits of the technologies and their applications developed in this context.

The problem with this approach is that scientific reports produced from the different fields of nanotechnology are generally tedious to read and accessible only to experts working within the same area of research. The media often present research results in a way that is not scientific in nature. Newspapers, radio, and TV stations have to work according to commercial interests, and science education is not their primary concern. They are primarily interested in "breakthroughs," but real breakthroughs are quite rare in any field of science. The media also have a tendency to focus on colorful personalities and individual researchers (such as Nobel laureates including late Richard Feynman). In the field of nanotechnology, research is often done in teams that may consist of dozens of individuals. For example, microchip design teams often comprise 250–300 individuals [15]. It may not be possible to attribute the merit for a particular innovation to individuals or even small groups.

Besides, in the debate about science and technology, there are relatively few forums in a modern society where experts and laypeople may interact in a fruitful way or get together for meaningful debate. One commentator has even stated:

> **Contemporary technological discussion is shameful. Leaders who wish to recommend options and sometimes policy call upon the experts. Heavily biased by personal and professional interests, experts craft their messages so they are resistant to most counterclaims. ... by using excessively technical vocabulary, their arguments become arguments from authority. ... As a result, citizen-consumers are frozen out of depthful discussions involving science and technology, especially those related to decision-making. When an occasional miscreant speaks out, he is derided, labelled, or patronizingly dismissed. Unsurprisingly, when citizen-consumers are involved in science and technology decision-making, it is usually during the post-decision implementation phase [16].**

Although this view is rather drastic, it indicates something of the difficulties that modern scientific discussion faces today. Communication between interested parties could at least be enhanced. The discussion about social, legal, and ethical implications of nanotechnology is also quite undefined in scope, further complicating the dialogue. The participants are often even unaware of what the subject entails. For example, nanoethics has been defined as "a shorthand turn of phrase that encompasses the development, study, practice, and enforcement of a set of culturally accepted beliefs, mores, guidelines, standards, regulations, and even laws for governing rapidly advancing nanotechnologies across multiple economic sectors" [9]. This kind of broad definition may be adequate to define the purview of nanoethics, but many commentators have also cast doubt on whether this kind of specific branch of ethics even exists in the first place [17]. In their view, there are no novel questions raised that could not be answered by conventional ethics.

Even if one accepts this viewpoint, however, there are profound reasons to suggest that the implications of nanotechnology will be to raise issues that demand serious ethical attention. The question is more of how to understand the relevance of the issues to one's own research and work context. Perhaps the most important thing is to have an outsider's view of the work that is conducted and to understand why some people regard such research not only as beneficial but also potentially dangerous. It is clear that simple creation of material of an artificial kind cannot be the reason for the objections and sometimes the opposition by environmentalists, owing to the fact that synthetic chemists produce approximately 900,000 new chemical substances globally per annum [18].

Here, I follow a somewhat looser definition, in which nanoethics addresses the ethical, social, and policy issues, including legal ones, associated with nanoscience and nanotechnology [19]. Because the domain of ethical discussion in nanotechnology is somewhat fragmented, it is useful to define at least a few of the relevant problem areas. Norbert Jömann and Johann S. Ach have identified the relevant ethical concerns in this context as anthropological, biomedical (i.e., nanomedicine), socioethical, and environmental [20]. Although, as Jömann and Ach also readily admit, this division is not without problems, it indicates how the issues at hand might be classified. Because the focus of this publication is on the environmental aspects of nanotechnology, I will not deal much with the first two categories, although I will use some examples related to molecular biology and nanomedicine to illustrate certain ethical questions relevant to the last two categories.

For example, there is a vast amount of nanoethical literature discussing possible developments (at the moment highly unlikely) that may lead to certain undesirable consequences in the endeavor to enhance human nature. Usually, the concern is related to issues of privacy, such as the loss of personal privacy or the possibility that nanotechnology may lead to an ability to read minds. In neuroethics, the discussion relates to developing methods to tell when someone is lying through neuroimaging, which is a theme that nanotechnology shares with neuroscience [19]. This technological advancement would in turn, according to some doomsday scenarios, lead to authoritarian rule and, in an unforeseen way, enable a police state to be put into place.

Even if we accept that there are consequences to realizing human capacities, it is not plausible that this scenario will ever materialize, owing to the functionality of the brain and other limiting factors. There are profound reasons, therefore, why this approach has been labeled "speculative" nano-ethics, and it is not the kind of debate worth analyzing here in detail. The second issue, that is, biomedical research in relation to nanotechnology, is more difficult to combine with environmental and ethical aspects. The sectors of nanomedicine as given by NanoBio-RAISE (a subprogram of the previous 6th Framework Programme Science and Society Co-ordination Action funded by the European Commission) are drug delivery, biomaterials, in-vivo imaging and in-vitro diagnostics, active implants, and therapy. Progress in this field leads to questions of medical ethics, such as whether information stored by implanted devices developed for medical purposes may be given to a employer or to insurance companies.

It should be noted, however, that in most nanotechnologies, the research does not directly affect the environment, so environmental harm, bodily exposure, or other health hazards are not relevant concerns. The case is remarkably different with bionanotechnology or nanomedicine [10]. Since the issues raised in the context of bionanotechnology are often connected to (human) health hazards, it is possible to combine them with a more general discussion dealing with environmental and socio-ethical matters. For example, the medicalization of society is a wide phenomenon that manifests itself in many areas, and it is not caused by research in nanotechnology alone. However, research in bionanotechnology has the ability to expedite this process too, unless ethical attention is drawn to the applications that are launched for consumers and customers.

2.2. The Public and the Scientist

Curiously, scientists worry more about the health risks than does the general public, but such seems to be the case in this context. According to Dietram Scheufele and colleagues, as a research field, nanotechnology may be the first emerging technology for which scientists may have to explain to the public why people should be more concerned rather than less about the potential health and environmental risks involved [21]. The public seems more concerned about issues related to privacy; in one survey, the difference was nearly one of two members of the public (45%) shared such a concern compared with fewer than one-third (30%) of nanoscientists [22]. Studies in general show that most of the members of the general public know little or nothing about nanotechnology. Still, they are confident that its future benefits will outweigh its risks [23]. However, the public is quite concerned that those doing commercial research would not be in position to control the safety of their work. In a survey carried out in 2006, only

12% of US respondents expressed the opinion that companies should be responsible of regulatory safety of nanomaterials. Instead, universities and the government should be in charge of nanomaterial safety. However, most government funding is spent on developing new products instead of on toxicology and other health-related research [24].

Kerstin Schrader-Frechette [24] has noted that there are clear examples of a conflict of interest in this context. In 2000, the independent Science Advisory Board of the US Environmental Protection Agency (EPA) studied all available health and safety research provided by American pesticide manufacturers to see whether they are accurately researched and statistically sound. The conclusion was that all studies conducted with humans were "scientifically invalid." The basic shortcoming was the use of small sample sizes, from 7 to 50 subjects, when at least 2500 subjects would have been needed for each and every group to draw the conclusions claimed in these studies about the health impacts of the pesticides. According to Kerstin Schrader-Frechette, "[t]hey [i.e., the pesticide studies] were predetermined to generate false-negative conclusions, false conclusions that the pesticides were not harmful. … Because virtually all nanotechnology research is done by those who expect profit from it, mostly chemical companies, there are few grounds for believing that this research, done with a clear conflict of interest, is likely to produce results that are any more reliable than the pesticide studies evaluated by the EPA Science Advisory Board in 2000" [24].

The interface between the public (often taxpayers who in many countries fund the research and also consumers who use the products of private companies doing research in this area) and those who do the research is especially difficult. In an explanation of one survey conducted in the United States in 2005 about the opposition to nanotechnology, those who oppose where characterized as having "strong religious beliefs, lower educational attainment, and a general mistrust of institutions" [25]. This is perhaps not the population group to which most scientists feel a sense of belonging. However, as Ronald Sandler has pointed out in the context of the Woodrow Wilson Project on Emerging Nanotechnologies, even concerns that are not scientific should be taken into account. Sandler states:

> **Among the fundamental principles of a liberal democratic society such as the United States is that ideas not be excluded from social and political domains on the basis of the worldviews from which they emerge. … That some worldviews are not 'scientifically informed,' as commonly understood in science and technology communities, is not justification for their being marginalized in public policy and regulatory discussion, even regarding science and technology [3].**

Besides criticism of a religious nature, Schummer has pointed out that there are differences in understanding concepts that have ethical significance, such as human dignity and integrity. Thus, the perceptions of discussion participants have on nanotechnology also differ. For example, developing artificial intelligence and enhancing human brain operations through nanotechnology are considered more problematic in the United States than in Europe, where operational capacities are seen more as instrumental. Privacy concerns are also more prevalent in the United States than in the United Kingdom, where surveillance with cameras in public areas is accepted. Schummer notes, however, that there is a difference between Europeans in this sense; when different continents are compared, more examples of cultural variables may stand out, variables that also influence perceptions of nanotechnology [9]. The difference between the public and experts is that, as social studies have shown, experts in any field tend to express perceived risks in quantitative terms, whereas the public uses terms of a more qualitative nature [26].

It is also uncertain how the public will deal with information expressed in quantitative terms. For example, it is well-known that approximately 50% of tobacco smokers will eventually die as a result of some tobacco-related illness, resulting in a 50% risk of death due to smoking. By comparison, the probability in winning a common form of lottery, a draw of 6/49, may be remarkably low, requiring the purchase of 7-million lottery tickets to have a similar probability of winning a first prize. Nevertheless, most smokers think that they will belong to the 50% who will not be affected by tobacco-related illness, and those buying a lottery ticket imagine that they will win with a single ticket. Thus, the ability of an individual to evaluate personal risks and benefits in the real world is quite limited in scope (I thank Professor Jaana Hallamaa from the University of Helsinki for pointing this out for me). In nanotechnology, as Tsjalling Swierstra and Arie Rip have pointed out, a further problem is that risks and possible harms are in many cases speculative, concerning yet unidentifiable individuals and stakeholders of a collective nature. Therefore, Swierstra and Rip note that "[t]he asymmetry of benefits and harms is almost unavoidable, structures not just argumentation but also action, and has given rise to increasing recognition of the need for early warning" [26]. This need for early warning combined with lack of scientific data regarding emerging technologies and their applications leave much room for speculation and misinterpretation when weighing the possible risks. It is thus necessary to discuss the societal context of health and environmental risks and the overall impact of nanotechnology on society as a whole.

3. ANALYSIS

In terms of the relationship between nanotechnology and environment, societal aspects are by far the widest implications. The environmental impact of those materials and products developed through research in nanotechnology may be divided into positive and negative. A positive evaluation demands that the end product truly assist in making our lives easier and our ecosystems cleaner. If nanotechnology increases pollution and creates health hazards, then risk assessment is usually taken into account. This is an oversimplified description, but one that it is often used. There are consequences both from doing and from not doing something. Thus, the argument goes, developing a specific nanotechnology application is in a sense imperative because it may enhance productivity and serve the greater human good. Or at least, the amount of good it produces should overcome the negative side effects that different technologies may have. At least, the development should create new knowledge, thus advancing science. It seems, however, that bio-nanotechnology especially creates moral and legal controversies that are not easily solved. They may also serve as a point of comparison of the nature of controversies that all nanotechnologies could be accounted for. Toxicity is a common theme, but there are also certain legal and other perspectives.

3.1. Legal Perspectives

Perhaps the clearest example of the legal issues that nanotechnology raises is the one of patenting. Here, the discussion focuses mostly on the United Kingdom and the United States, with whose practices readers of this book are most familiar. There are notable court cases that deal with patents in the context of GM microorganisms, such as Diamond vs. Shakrabarty (447 US 303 1980, a US Supreme Court case, dealing with the question whether GM microorganisms can be patented at all). It was decided that a live, man-made microorganism is a patentable subject. At the heart of all patent systems (to this day), there is a basic agreement between inventors and governments. However, what was effective in the past may not be effective today. Technologies are changing, the context of technology is different, and, most important, certain technologies are viewed as so fundamental to human existence that governments view those technologies with suspicion or even with distaste (as in the case with cloning, for example).

For these reasons, governments are constantly revising the specifics of the patent system in this field of inquiry. In Europe, governments have concluded that methods of treating living bodies are so fundamental to health that patent applications covering them are often declined. In the United States, the tendency is to move in a direction where patents may be granted. The reasoning is that there is otherwise a danger that research methods might be suppressed and concealed by inventors if they are not rewarded for their disclosure. This is also the reasoning behind the practice of some countries to refuse patents for organisms created in laboratory, whereas other countries have taken the opposite stance. The case of the OncoMouse (or the Harvard mouse), the mouse that was genetically modified by researchers Philip Leder and Timothy Stewart at Harvard Medical School in the early 1980s, was a watershed in the debate over whether it is possible to patent animals or animal variations. It is worthwhile asking, as Patrick M. Boucher [27] has, whether society "is better off," after few decades of this legal permission, "by having allowed Philip Leder to patent the oncomouse." To certain extent, the answer is yes (although actually the patent belongs to Harvard College). The OncoMouse enabled researchers to acquire a better understanding of cancer without having to gather data from years or even decades of human suffering. What might be a moral issue here is the suffering caused to the transgenic animal. From a legal perspective, the basic idea of patenting is not to assure the inventor of an unfettered right to produce, market and sell the invention, in the official parlance, and the right to practice the innovation [27].

Although the general public often has the commercial view, such a view is mistaken. What the patent rights grant is the ability to prevent others from engaging in these same activities, and thus, patenting an organic matter is a highly problematic issue for a society. In addition, there is a question of who has the moral right to pursue this kind of research. The scope of nanotechnology in which matter is transformed into living creatures poses a different problem from ordinary human enhancement, namely, whether it is morally right to produce creatures that do not inherently belong to the human species but are a mixture of human, animal, or machine (or a mix of all the aforementioned three) that do not exist in the current ecosystem. Leaving this decision to the self-regulation of researchers may be questionable ethically because not all are willing to accept limitations on their work. In addition, in the competition for research grants and funding, extraordinary claims and controversial research may sometimes be advantageous. There are almost no self-imposed limits on when a researcher of nanotechnology (or any other researcher) would give up his research.

To present a concrete example from this kind of pursuit of research in morally questionable areas, in the United Kingdom, there is a Chair of

Professor of Cybernetics, located in the University of Reading in South East England. The current holder of the chair carries out research in artificial intelligence, control, robotics, and biomedical engineering. He also claims to be "the world's first human cyborg (an organism that has both natural and artificial systems)," owing to several chips and other implants that he has (and his wife also has had at his insistence). Without judging whether the professor in question is truly the cyborg he claims to be, I would like to mention one interesting experiment he has been conducting: an experiment with brain cells dissolved from the brain of a fetal rat. The free-floating brain cells (or neurons) are transferred into an electrode-ringed Petri dish. The cells quickly re-associate with each other and begin randomly sending out electrical signals. The justification is that this kind of research has implications for certain neurodegenerative diseases. It may also assist in the creation of neural interfaces, thus assisting the severely impaired in improved communication.

In the long run, however, this kind of research may also develop hybrid life forms, especially if human stem cell research acquired from human embryos is permitted. There have been applications for research funding in the United Kingdom that raise this possibility, but so far research with human embryonic stem cells has not been permitted. However, after public consultation, the Human Fertilisation and Embryology Authority permitted research on cytoplasmic hybrids (a type of research in which human DNA is inserted into animal eggs). Other chimera research has so far not been authorized [28]. Other European nations have taken a mixed approach to such research. In Switzerland, the law recognizes that an aborted fetus may still be alive at the time of tissue removal and acknowledges the difficulties that may arise when the status of the fetus is reduced to a supplier of raw material for research purposes [29]. In the Netherlands, a remarkable proportion of the population opposes the creation of embryos solely for research purposes. However, although this practice is prohibited by Section 24 of the Dutch Embryo Act, Section 33.2 states that the prohibition will be assessed every five years to determinate whether the moral attitude of population has changed [26].

With this kind of variation in legislation, the question of whether to allow human stem cells to be used in any kind of research in which hybrid life forms may be created advisedly cannot be adequately answered. One might, however, reflect a question of what could be regarded as a possible harm. If stem cell research using cells obtained from human embryos is permitted, then it is difficult to determine exactly when the possible harm takes place. Such research is grounded in advancing the human good in the sense that those suffering from devastating diseases might be helped, and sensible persons would not object to that. Creation of hybrid life forms or even nanomachines, including human nuclei, would not then be problematic at all.

Even if such an approach were to be accepted, it would still remain uncertain what exactly creates a nanomachine. Bernadette Bensaude-Vincent has noted that "[a]lthough biologists generally agree that living systems are the product of evolution rather than of design, they describe them as devices designed for specific tasks [30]." Living cell is a minifactory, "full with numerous bionanomachines in action [30]." According to Bensaude-Vincent, the machine has become an all-pervading metaphor, not only in biology but also in chemistry and materials science. In this view, it is easy to understand why those developing hybrid life forms from living matter and machines understand their work as advancing the process of evolution. The idea is that advancement will benefit all mankind and provide a greater good for everyone. There is the question, however, of what constitutes a beneficial increase in the good of a modern society. Something that increases the individual happiness of one person may not have the same effect on another. There is a broad consensus that hunger and sickness are social ills that should be addressed by responsible societies. Nanotechnology is often combined with the aspirations to reduce hunger and improve the economic situation in developing countries [26]. In a similar way, examining risk-free organic matter detached from the human body is presented as straightforward research.

3.2. The Definition of Harm

The wider issue relevant to our discussion here would be more of what constitutes harm in the context of research in nanotechnology. Is there a line across which research in this field becomes morally blameworthy? One might question whether combining the organic matter with synthetic one has always, of necessity, only positive results. Are the life forms that are created from human stem cells still machines that can be destroyed at will? If these life forms develop higher functions, should they be assigned legal rights and protection? What if they develop to a stage in which they express their own desires and wishes? There have been predictions that by 2050, it might be possible to develop a robot that could be very human-like. Robots might be able even to marry humans, or so artificial intelligence researcher David Levy at the University of Maastricht in the Netherlands stated in an interview with *LiveScience* on 12 October, 2007 (Levy has also predicted that the state of Massachusetts in the United States will be the first jurisdiction to legalize marriage with robots, given its liberal approach to other similar issues) [31].

It is understandable that researchers in molecular biology or in materials science would not readily accept limitations on their work. If the main objective of nanotechnology is to create applications and solutions that accomplish future tasks better than with conventional machines, then one would have to accept the idea that research should be as unregulated as possible. The underlying ideology is that knowledge is good for its own sake, even though there might be risk involved in using the knowledge acquired. Certain inventions thus become inevitable, most notoriously, nuclear weapons (although nuclear technology has its peacetime applications as well). It is widely accepted that nuclear weapons and other weapons of mass destruction are contrary to the good of mankind. Nevertheless, many people would be reluctant to blame those involved in research and development of such weaponry. Blame would bring into question the freedom of scientific inquiry and the role of science in conducting wars. The Institute for Soldier Nanotechnologies (ISN) at MIT has the mission of increasing the survivability of the US soldiers, with no mention of decreasing the capability of those on the opposing side to survive. In fact, the public mission statement of the ISN is to investigate "how technology can make soldiers less vulnerable to enemy and environmental threats" [32].

Naturally, many other fields of science serve the interests of the military: nanotechnology alone cannot be blamed for this. For example, Richard Feynman, represented as an initiator of nanotechnology research, was involved in the Manhattan Project in the early stages of his career. Later, for moral reasons, he moved into genetics, along with several other colleagues. There are possible environmental pollutions and health hazards that may result from military uses of nanotechnology, but the Military interest is currently focused on those who use the applications and their immediate surroundings. If, for example, nanomaterials used in future uniforms were to disseminate nanoparticles, then the channels of exposure are quite foreseeable. However, it is a different thing to evaluate whether weapons developed in the future with the assistance of nanotechnology have a different ecological impact than earlier weapons. For example, during the Vietnam War, the goal of the so-called Agent Orange campaign was the destruction of the mangrove forests and the long-term poisoning of the soil and crops.

This kind of warfare could be enhanced with nanotechnology and lead to increased environmental pollution. Although the political climate does not support the use of nanotechnology in such manner, in western democratic nations, there may be other regimes in the world that have an interest in developing this kind of weaponry. There have been calls for a moratorium on developing independently active micronanosystems for future battlefields [33, 26].

Even though there may not be any major military involvement in this area of research, it may justifiably be asked whether nanotechnology has the capability to advance the good of mankind on a global scale. There is a strong possibility that nanotechnology will, at least in the short term, benefit those who have assets and are already in a privileged position. It is also noteworthy that possibly toxic substances that are banned in western countries are recommended for use in developing countries. The use of dichlorodiphenyltrichloroethane (DDT) is regulated by an international agreement known as the Stockholm Convention on Persistent Organic Pollutants, but in 2006, the WHO decided to back away from its position of three decades and recommend the use of DDT to prevent malaria, a recommendation based on lack of other cost-effective insecticides [34]. Naturally, the WHO simultaneously urged the need to develop effective insecticide alternatives. But since the problem is mainly in Third World countries, it remains to be seen whether the response to calls by the WHO is as effective as when the lives of citizens in western countries are at risk.

These specific examples do not, however, give a clear answer on what exactly is ethically problematic in nanotechnology research in relation to environmental nanotechnology, so closer scrutiny is needed. In the context of the relationship between nanotechnology and the environment, it seems that ethical issues are not specific to case of environmental nanotechnology. The issues are more general ones, relating to all research on emerging technologies. This is visible in one of the most influential documents in this regard was a publication by the Royal Society and the Royal Academy of Engineering in the United Kingdom, which concluded among other things that "most nanotechnologies pose no new risks, but highlight uncertainties about the potential effects on human health and the environment of manufactured 'nanoparticles' and 'nanotubes' if they are released" [35, 36]. The only difference that nanotechnology makes in this context is that with the change in size, nanoparticles have novel properties in the physical or chemical sense. These changes occur in relation to such things as electrical conductivity, magnetism, or catalysis, among others. The positive or negative effects that altered properties have on humans or animals or the whole ecosystem are a matter for some discussion [37].

On the one hand, to associate ethical issues in relation to environment only with this domain is clearly too narrow a viewpoint. Ethical debate about nanotechnology cannot be something that risk analysis or safety research is capable of doing by the steps of hazard identification, exposure, or dose-response assessment and risk characterization. On the other hand, science does not occur in a vacuum, as implied when the National Nanotechnology Initiative (NNI) in the United States expression "societal and ethical implications" of nanotechnology is used. Bruce Lewenstein has pointed out that a perspective in which science and technology come first and "implications" to follow is not plausible perspective if the history of science is taken into account [38].

There are, however, certain ethically relevant issues in nanotechnology that are common to all scientific inquiry and which may culminate in this kind of high-profile branch of inquiry such as nanotechnology, with high public expectations. The political and economic consequences of possible breakthroughs create an atmosphere that adds challenges to nanoscience and to development of sustainable technologies. In particular, the interface between the general public, their values, and scientific community is important to guarantee interaction. This interface is also the place where moral and ethical values contrast and may affect decision making. In this area, special attention should be paid to research ethics so that the information that the public obtains is as accurate as possible. There have been widely publicized issues of misconduct in research that may create suspicion about any new field of technology.

3.3. Commercial and Conflicting Interests

The research results are increasingly used for commercial purposes, instead of dissemination of ideas, or they are presented in a manner that serves certain research agenda. As an example of the former, Bruce Lewenstein has pointed to the US Bayh-Doyle Act (or the University and Small Business Patent Procedures Act, as it is also called) from the year 1980, which was intended to motivate the US universities to make patent applications on the basis of their research. Studies of the effect of the Bayh-Doyle Act have noted that there have been some regrettable consequences for research. Nanotechnology has the ability to create intellectual property deemed valuable in modern society, and its free dissemination for general public is not as interesting as it used to be for many of those operating in the area of research. According to Lewenstein, the negative consequences of the Bayh-Doyle Act include "restricted dissemination of faculty research, delays in publication, deleted information, and – most ominous to those who believe academic research should be 'pure' in its motivations – a change in direction of faculty research toward projects with commercial potential [38]."

As an example of the latter, Joseph Pitt [39] has noted that the pictures taken with one of the iconic tools of nanotechnology, namely the scanning tunneling (electronic) microscope (STM), are actually not images in the sense that they are portrayed. The invention of STM is considered an important step in the emergence of nanotechnology, and the letters "IBM" formed by IBM fellow Don Eigler in 1989 by positioning 35 xenon atoms on a surface is perhaps one of the most well-known images of nanotechnology research. Still, according to Pitt, most of the STM images are more of "imaginings." In many cases, STM images are presented and interpreted in a way that supports the overall paradigm of nanotechnology to show the simple world that exists "in the bottom." It is difficult to say that something occurs at the nanolevel and those images taken with STM show that something accurately. In fact, we do not know when the occurrences exactly take place in the molecular level, and thus, in most cases, it is impossible to present the pictures for general public as representative.

Even the pictures shown at scientific conferences are constructed with the assistance of a computer, enhanced with colors, and manipulated in other ways. In addition, scanning probe microscopics as a whole are known for their vulnerability to error and are dependent on accurate interpretation of the results [40]. For this reason, there is a line on one side of which it could be claimed that STM images describe reality and, on the other side of the images, serve the purposes of those advancing their own viewpoints. Furthermore, Jochen Henning has examined changes in the design of scanning tunneling microscopic images from 1980 to 1990; he notes that "[i]n their daily experiments, tunnelling microscopists deal with numerous images … it is also part of their daily practice to design these images in such a way that they conform to conventional expectations and nanotechnological utopias. … The status of the tunnelling microscope as one of the central instruments in nanotechnology rests on its areas of application and technical possibilities, but also in the power and effect of its images and an image design that emerged out of a dynamic process during the 1980s" [41].

It is understandable that in the area in which commercial interests are high, violations of research ethics violations and other types of misdemeanors increase. Although Pitt or Hennig do not claim that the tendency to present STM images in a certain way constitute any kind of serious transgression, it should be noted that the information the public has about science is limited. Therefore, extra attention should be paid in communicating the research results in such way that they are explained as accurately as possible for lay-people. If the interpretation of scientific results is left to journalists, even those working for the science magazines, then there is a strong possibility that scientific and technical information will not be accurately understood. There is also a tendency in the scientific community to claim credit for discoveries of a positive nature, including those that are widely seen as advancing the common good, and to avoid responsibility for the negative cases. Jerome Ravetz has noted that "[s]cience takes credit for penicillin, while society takes the blame for the [atomic] Bomb [42, 8]." Thus any manipulation of STM images should be done in a way that does not leave room for misunderstanding when the research described in pictures is evaluated.

3.4. The Precautionary Principle

One of the key principles of environmental law and its application to nanotechnology is thought to be the precautionary principle. There are many references to this principle and its moral and political significances. Originally, it emerged in Germany in the 1970s. One of its well-known formulations is the Principle 15 Rio Declaration from the year 1992, which stated that "[i]n order to protect the environment, the precautionary approach shall be widely applied by States according to their capabilities. Where there are threats of serious or irreversible damage, lack of full scientific certainty shall not be used as a reason for postponing cost-effective measures to prevent environmental degradation" [43]. The use of the precautionary principle, however, has been quite limited, mostly on certain types of cost-benefit analyses [44]. There is an application of the precautionary principle in French legislation. The French Barnier law, accepted in 1995 and named for the then − Minister for Agriculture and Fisheries, Michel Barnier, lists certain axiomatic principles for environmental jurisprudence in France. It states *inter alia* that "[t]he absence of certainties, given the current state of scientific and technological knowledge, must not delay the adoption of effective and proportionate preventive measures aimed at forestalling a risk of grave and irreversible damage to the environment at an economically calculable cost" [8].

Philosophically, this application of the precautionary principle to legislation is weakly grounded, especially in relation to the conceptual footing of the notion of "precaution," as shown by Jean-Pierre Dupuy and Alexei Grinham, who write:

> **When the precautionary principle states that the 'absence of certainties, given the current state of scientific and technical knowledge, must not delay etc.,' it is clear that it places itself from the outset within the framework of epistemic uncertainty. The assumption is that we know that we are in a situation of uncertainty. It is an axiom of epistemic logic that if I do not know P, then I know that I do not know P. Yet, as soon as we depart from this framework, we must entertain the possibility that we do not know that we do not know something. In cases where uncertainty is such that it entails that uncertainty itself is uncertain, it is impossible to know whether or not the conditions for application of the precautionary principle have been met. If we apply the principle to itself, it will invalidate itself before our eyes [8].**

It could be argued that, in specific cases, the precautionary principle has a certain practical value. The principle has been studied by Thomas Faunce and his colleagues, who have explored the principle in an Australian context, evaluating how health hazards of engineered nanoparticles in sunscreens containing titanium dioxide (TiO_2) and zinc oxide (ZnO) have been addressed. In 2006, the US Food and Drug Administration (FDA) addressed the risks caused to humans and the environment with regard to different consumer products, such as sunscreens. After this, Faunce and his colleagues in Australia specifically evaluated whether the precautionary principle was utilized by the Australian Therapeutic Goods Administration (TGA) regarding the permission to approve marketing of the sunscreens. The specific problem of the precautionary principle is that it is extremely difficult to assess what level of risk is acceptable. The principle is ambiguous, allowing random application and possibly preventing research and development that might be useful in alleviating the environmental harm it causes. Furthermore, Faunce and his coauthors note that "the precautionary principle, on any rational analysis, does not express some incontrovertible, monolithic regulatory truth, but rather sets framework within which precautionary measures practicably may be taken" [45].

In their conclusion, the authors note that assessing the legal framework with the precautionary principle presents challenges. It is practically impossible to evaluate what measures exist for legislators that could be based on the precautionary principle. In the case of introducing sunscreens with engineered nanoparticles to the consumer market, the alternatives are limited. It would be possible to introduce a total ban on the use of engineered nanoparticles or to introduce a requirement that engineered nanoparticles are considered new ingredients when sunscreens are evaluated. A third option would be to introduce labels that inform consumers to notice that engineered nanoparticles were used to develop the sunscreen. Finally, it would be possible to have a sort of panel or other bureaucratic institution to follow the marketing and sales of sunscreens that contain engineered nanoparticles, such as TiO_2 and ZnO. However, none of these options could be considered very meaningful, according to Faunce and his coauthors. Thus, the precautionary principle lacks usability and is too vague for the purposes for which it was intended [45].

As to whether TiO_2 is dangerous to human health, mixed results have been reported in the literature. In *Nanotechnology: Health and Enviromental Risks*, Jo Anne Shatkin concludes that "[t]he available data do not appear to be sufficient at present to derive quantitative hazard assessment for nano-TiO_2 or for nanomaterials in general" [46]. Similarly, Deb Bennett-Woods notes in *Nanotechnology: Ethics and Society* that "… sunscreens using nanoparticles of titanium dioxide have been on the market for some time and there is evidence that particles cannot penetrate otherwise healthy skin to a depth that would allow absorption into the body" [25]. Nevertheless, both authors state the need for further research in this area. The problem with sunscreens is that they are not always used in ideal conditions, and taking this factor into account would affect the results of toxicity studies. For example, the users are not limited to healthy individuals, and sunscreens are often used on abnormal skin. In addition, Thomas Faunce and his colleagues note that when there is a lack of data, as well as evidence that is contradictory, there is a danger that the precautionary principle becomes only "a formula for doing nothing" [45]. Thus, from the perspective of environmental hazards and nanotechnology, there is strong need to develop other conceptual tools to assist in risk assessment and ethical evaluation in regulating the research and development of new consumer applications. With a different approach, say Faunce and his colleagues, TGA's decision that those sunscreens including engineered nanoparticles such as TiO_2 and ZnO do not require new testing because they are considered the equivalents of their counterparts might have had different results [45].

4. CONCLUSIONS

The discussion of social, ethical, and legal aspects of nanotechnology is a broad one, and it is not possible to deal with all the relevant issues in one article. For purposes of the present volume, however, I hope that I have shed some light on those issues that are currently among most debated. The difficulty of analyzing ethical discussions is that many nanoethics publications quote each other and are, therefore, repetitious. After proposals from the Human Genome Project (a 13-year project completed in 2003), one of was the suggestion by James Watson that 3–5% of research funds be devoted to research on socioethical aspects, it was perhaps understandable that those participating to involved in nanotechnology research would readily accept the idea that serious attention should also be paid to ethical concerns. Those in charge of steering the NNI in the United States have been especially vocal in their support of research into ethical questions with investigations of a more technical nature. In the NNI framework, close to 3% of funds have been allocated for ethical research.

With these things in mind, it may be that the demand for publications in nanoethics may have exceeded the capacity of researchers to produce new and relevant research results. Many publications that deal with ethical aspects include material that started as conference proceedings or has been published elsewhere. There is nothing inherently wrong with reprints of quality material, but it is difficult to advance scientific debate or to come up with new ideas if same publications are constantly being republished in various formats. Part of the difficulty might be the limited number of researchers working in this area but the situation also suggests that funding may be temporary or combined with some specific nanotechnology project that has ethical research attached. What are needed in this context are long-term research projects that could accumulate over the years and a widening of perspectives from an evaluation of NNI to international comparisons. This would enable clarification of what is already known and show where the attention should be aimed in future research.

There is a strong need to analyze and explicate the ideological content, which is often implicit, in nanotechnology research. The phrase "ideological content" refers to viewpoints and assumptions expressed in the debate about nanotechnology. These views may be widespread and accepted by researchers, even though the views lack support in the realm of science. From time to time, there are references to several problematic and scientifically largely unproven notions, such as visions of what nanotechnology could do in the future. For example, eliminating suffering and death might be laudable aims in any field of science, but these are not very likely scenarios in the real world.

When there is the possibility of science fiction scenarios, the best cure is to clarify what is actually known and what claims may be true, but still lack solid research evidence. It is understandable that in funding applications, a researcher has to have a positive view of the likely outcomes. It is another thing to write for publications of a scientific nature that require accurate research reports. Having many ideas for an application is not the same thing as producing safe and tested real-world solutions. Furthermore, it is known that negative research results (i.e., those that do not produce any significant or positive results) are in many cases published only after a delay or are not published at all. In addition (with a reference to pesticide studies) it seems that more attention should be paid to the statistical validity of research.

Nevertheless, the situation reflects a wider problem in current academic research, creating bias and complicating systematic reviews. Systematic analysis of the current status of nanotechnology research by evaluating all available data and information is fundamental to avoid mistaken conceptions and perhaps erroneous research priorities. Furthermore, the precautionary principle seems to continue its existence as a useful concept, at least on the basis of the frequency with which it is mentioned, but its shortcomings are now well documented. As the case study from Australia demonstrated, the precautionary principle is quite insufficient to guide environmental law and regulation of nanotechnology. There is a clear need for developing better tools and concepts to cope with the challenges that nanotechnology research creates.

ACKNOWLEDGMENT

This work has been supported by a grant from the Ella and Georg Ehrnrooth Foundation (Finland).

References

[1] Department of Health and Human Services in the US. Approaches to safe nanotechnology. Managing the health and safety concerns associated with engineered nanomaterials. <http://www.cdc.gov/niosh/docs/2009-125/>; 2009 [accessed 26.06.09].
[2] Sparrow R. Revolutionary and familiar, inevitable and precarious: rhetorical contradiction in enthusiasm for nanotechnology. Nanoethics 2007;1:57–68.
[3] Sandler R. PEN 16 – Nanotechnology: the social and ethical issues. Woodrow Wilson Center for Scholars; 27 Jan. 2009.
[4] Milburn C. Nanotechnology in the age of posthuman engineering: science fiction as science. Configurations 2002;10:261–95.
[5] WHO. World Health Report. Reducing risks, promoting healthy life. Geneva, Switzerland: World Health Organization; 2002.
[6] World Economic Forum. Global risk report. Geneva, Switzerland: WEF; 2009.
[7] Shew A. Nanotech's history. An interesting, interdisciplinary, ideological split. Bull Sci Technol Soc 2008;28:390–9.

 [8] Dupuy J-P, Grinbaum A. Living with uncertainty: toward the ongoing normative assessment of nanotechnology. Techne 2004;8(2):4–25.
 [9] Schummer J. Cultural diversity in nanotechnology ethics. Interdiscip Sci Rev 2006;31(3):217–30.
[10] Sandler R, Kay WD. The National nanotechnology initiative and the social good. J Law Med Ethics 2006;34:675–81.
[11] Mody CCM. Small but determined: technological determinism in nanoscience. In: Schummer J, Baird D, editors. Nanotechnology challenges. Implications for philosophy, ethics and society. Singapore: World Scientific Publishing; 2006. p. 95–130.
[12] Regis E. Nano: the emerging science of nanotechnology. Boston, US: Little, Brown; 1995.
[13] Drexler KE. The engines of creation. The coming era of nanotechnology. New York, US: Anchor Books; 1986.
[14] Jones R. When it pays to ask the public. Nature 2008;3:578–9.
[15] European Commission. Nanotechnology: innovation for tomorrows world. EU: CORDIS. 2004
[16] Berube DM. The Rhetoric of nanotechnology. In: Baird D, Nordmann A, Schummer J, editors. Discovering the nanoscale. Amsterdam, Netherlands: IOS Press; 2004. p. 173–92.
[17] Litton P. "Nanoethics"? What's new? Hastings Cent Rep 2007;37(1):22–5.
[18] Preston CJ. The promise and threat of nanotechnology – can environmental ethics guide us? In: Schummer J, Baird D, editors. Nanotechnology challenges. Implications for philosophy, ethics and society. Singapore: World Scientific Publishing; 2006. p. 217–48.
[19] Alpert S. Neuroethics and nanoethics: do we risk ethical myopia? Neuroethics 2008;1:55–68.
[20] Jömann N, Ach JS. Ethical implications of nanobiotechnology. In: Ach JS, Siep L, editors. Nano-bio-ethics. Ethical dimensions of nanobiotechnology. Berlin, Germany: LIT Verlag; 2006. p. 13–62.
[21] Scheufele DA. Scientists worry about some risks more than the public. Commentary. Nat Nanotechnol 2007;2:732–4.
[22] Editorial. A little knowledge. Nat Nanotechnol 2007;2:731.
[23] Pidgeon N, Harthorn BH, Bryant K, Rogers-Hayden T. Deliberating the risks of nano-technologies for energy and health applications in the United States and United Kingdom. Nat Nanotechnol 2009;4:95–8.
[24] Shrader-Frechette K. Nanotoxicology and ethical conditions for informed consent. Nanoethics 2007;1:47–56.
[25] Bennett-Woods D. Nanotechnology: ethics and society. Boca Raton, US: CRC Press (Taylor & Francis Group); 2008.
[26] Swierstra T, Rip A. Nano-ethics as NEST-ethics: patterns of moral argumentation about new and emerging science and technology. Nanoethics 2007;1:3–20.
[27] Boucher PM. Nanotechnology – legal aspects. Boca Raton, US: CRC Press (Taylor & Francis Group); 2008.
[28] Human Fertilization and Embryology Authority in the United Kingdom. HFEA statement on licensing of applications to carry out research using Human-Animal Cytoplasmic Hybrid Embyos. Minutes 2007 and 2008 <http://www.hfea.gov.uk/418.html>; [accessed 13.06.09].
[29] Buechler A. The transplantation of human fetal brain tissue. In: Brownsword R, Yeung K, editors. Regulating technologies. Legal futures, regulatory frameworks and technological fixes. Oxford, UK and Oregon, US: Hart Publishing; 2008. p. 243–62.
[30] Bensaude-Vincent B. Two cultures of nanotechnology? In: Schummer J, Baird D, editors. Nanotechnology challenges. Implications for philosophy, ethics and society. Singapore: World Scientific Publishing; 2006. p. 7–28.
[31] Choi CQ. Forecast: sex and marriage with Robots by 2050. Special to LiveScience posted on 12 October 2007. <http://www.livescience.com/technology/071012-robot-marriage.html>; 2007 [accessed 11.06.09].

[32] The website of MIT Institute for Soldier Nanotechnologies. <http://web.mit.edu/isn/> [accessed 04.06.09].

[33] Altmann J, Gubrud M. Military arms control and security aspects of nanotechnology. In: Baird D, Nordmann A, Schummer J, editors. Discovering the nanoscale. Amsterdam, Netherlands: IOS Press; 2004. p. 269–77.

[34] WHO. The Use of DDT in Malaria Vector Control. Position statement. <http://apps.who.int/malaria/docs/IRS/DDT/DDTposition.pdf>; 2007 [accessed 15.06.09].

[35] Sweeney A. Social and ethical dimensions of nanoscale science and engineering research. Sci Eng Ethics 2006;12(3):435–64.

[36] The Royal Society and the Royal Academy of Engineering. Nanoscience and nano-technologies: opportunities and uncertainties. London, UK: Royal Society and Royal Academy of Engineering; 2004. <http://www.nanotec.org.uk/finalReport.htm>; [accessed 26.06.09].

[37] Powers TM. Environmental holism and nanotechnology. In: Allhoff F, Lin P, editors. Nanotechnology and society. Current and emerging ethical issues. New York, US: Springer; 2009. p. 109–23.

[38] Lewenstein BV. What counts as a 'Social and Ethical Issue.' In: Schummer J, Baird D, editors. Nanotechnology challenges. Implications for philosophy, ethics and society. Singapore: World Scientific Publishing; 2006. p. 201–15.

[39] Pitt J. When image is not an image? In: Schummer J, Baird D, editors. Nanotechnology challenges. Implications for philosophy, ethics and society. Singapore: World Scientific Publishing; 2006. p. 131–41.

[40] Nanotechnology & Ethics Group, Report of the First Meeting. UNESCO Paris 5–6 July 2005, 10. <http://portal.unesco.org/shs/en/files/8958/11441614301Nanotech Report1.pdf/NanotechReport1.pdf>; 2005 [accessed 26.06.09].

[41] Hennig J. Changes in the design of scanning tunneling microscopic images from 1980 to 1990. In: Schummer J, Baird D, editors. Nanotechnology challenges. Implications for philosophy, ethics and society. Singapore: World Scientific Publishing; 2006. p. 143–61.

[42] Ravetz JR. … et augebitur scientia. In: Harre R, editor. Problems of scientific revolution. Progress and obstacles to progress in the sciences. Oxford, UK: Oxford University Press; 1975. p. 42–57.

[43] The United Nations Conference on Environment and Development in Rio de Janeiro from 3rd June to 14th 1992. Rio declaration on environment and development, UN development programme. <http://www.unep.org>; 1992 [accessed 09.06.09].

[44] Demissie HT. Taming matter for the welfare of humanity: regulating nanotechnology. In: Brownsword R, Yeung K, editors. Regulating technologies. Legal futures, regulatory frames and technological fixes. Oxford, UK and Oregon, US: Hart Publishing; 2008. p. 327–56.

[45] Faunce T, Nasu H, Bowman D. Sunscreen safety: the precautionary principle, the Australian therapeutic goods administration and nanoparticles in sunscreens. Nanoethics 2008;2:231–40.

[46] Shatkin JA. Nanotechnology – health and environmental risks. Boca Raton, US: CRC Press (Taylor & Francis Group); 2008.